PROCEEDINGS OF THE
FIRST INTERNATIONAL SYMPOSIUM ON
CYCLODEXTRINS

Budapest, Hungary,
30 September—2 October, 1981

ADVANCES IN INCLUSION SCIENCE

The Cyclodextrin Symposium was held under the auspices of the
Medical Chemistry Section of the Hungarian Pharmacological Society

and

The Chemical Division of the
Hungarian Academy of Sciences

organized by the

Congress Bureau of the MOTESZ
(Association of the Hungarian Medical Societies)

and sponsored by the

CHINOIN Pharmaceutical-Chemical Works, Budapest

PROCEEDINGS OF THE FIRST INTERNATIONAL SYMPOSIUM ON CYCLODEXTRINS

Budapest, Hungary,
30 September—2 October, 1981

Edited by
J. SZEJTLI
CHINOIN Pharmaceutical-Chemical Works, Budapest, Hungary

D. REIDEL PUBLISHING COMPANY
DORDRECHT, HOLLAND / BOSTON, U.S.A. / LONDON, ENGLAND

Library of Congress Cataloging in Publication Data

CIP

International Symposium on Cyclodextrins (1st: 1981: Budapest, Hungary).
Proceedings of the First International Symposium on Cyclodextrins,
Budapest, Hungary, 30 September–2 October, 1981.
(Advances in inclusion science)
"Held under the auspices of the Medical Chemistry Section of the Hungarian Pharmacological Society and the Chemical Division of the Hungarian Academy of Sciences, organized by the Congress Bureau of the MOTESZ (Association of the Hungarian Medical Societies) and sponsored by the CHINOIN Pharmaceutical-Chemical Works, Budapest."
Includes indexes.
1. Cyclodextrins—Congresses. 2. Cyclodextrins—Therapeutic use Congresses. I. Szejtli, József. II. Magyar Farmakológiai Társaság Medical Chemistry Section. III. Title. IV. Series.
QD320.I585 1981 547.7'815 82-5359
ISBN 90-277-1433-9 (Reidel) AACR2

Distributors for the U.S.A. and Canada
Kluwer Boston Inc.,
190 Old Derby Street, Hingham, MA 02043, U.S.A.

Distributors for Albania, Bulgaria, Chinese People's Republic, Cuba, Czechoslovakia, German Democratic Republic, Hungary, Korean People's Republic, Mongolia, Poland, Rumania, the U.S.S.R., Vietnam, and Yugoslavia
Kultura Hungarian Foreign Trading Company
P.O.B. 149
H-1389 Budapest 62, Hungary

Distributors for all remaining countries
Kluwer Academic Publishers Group,
P.O. Box 322, 3300 AH Dordrecht, Holland

D. Reidel Publishing Company is a member of the Kluwer Group

Printed in Hungary

CONTENTS

5. CYCLODEXTRINS IN PHARMACEUTICALS

Address of welcome by prof. J. Szejtli:

Ladies and Gentlemen,

First of all I owe you a short explanation, how we arrived at the decision to organize this Symposium. It is known for all of you, that Cyclodextrins have already been known for a long time. Great scientists: as Freudenberg, French, Cramer and many others, have dedicated years and years to the investigations of these amazing group of oligosaccharides, creating the fundaments of molecular encapsulation.

We are standing on their shoulders, their publications form the real basis of our present knowledge, our efforts and our successes. I have the honour to announce that the opening lecture will be delivered by professor Cramer. Professor Cramer thank you, for your coming and participating in this Symposium.

By the end of the sixties such an enormous amount of knowledge had accumulated, that the industrial realization of CD production, the application of CD-s on industrial scale seemed to be within reach. As professor Nagai will tell you, around that time an intensive research on CD-s was launched forth in Japan and independently also here in Hungary.

Since that time twice a year half a day's workshops are organized in Budapest to discuss recent results of our team. This team is a rather complex and mixed one; more than 12 institutions - universities, research institutes and industrial laboratories - more than 70 people are somehow - fully of partially, indirectly or directly - involved in the activity of this team. Realizing the exponentially growing number of patents and papers, the break-through of CD-s is expected for this decade. This break-

-through means production of cyclodextrins at reasonable prices on industrial scale, their legislation in food and pharmaceutical industry, and later on also in agrochemistry.

As a very preliminary illustration, you have received some samples of spice extracts stabilized with β-CD, and a French vanillin pudding, which contains vanillin also as CD-complex.

The possibilities of application of CD-s are almost inexhaustible. Hundreds of scientists and research workers in the industry will have much to do in this field during the next 10-20 years.

According to the Chemical Abstracts, during the last 30 months 239 papers and patents were published on CD-s. Reidel Publishing Company of Dordrecht, Holland and Boston, USA decided to launch a new journal, "Journal of Inclusion Phenomena" in 1983.

Ladies and Gentlemen, participants from 17 countries are present here. A cordial wellcome to all of you, on behalf of the Hungarian Pharmacological Society, and of the CHINOIN Pharmaceutical-Chemical Works. And now may I ask professor Beck, chairman of the Chemical Division of the Hungarian Academy of Sciences to open the Symposium.

Accepting the suggestion of professor Beck the following telegram was sent to

Professor K.FREUDENBERG
Heidelberg

The participants of the first Symposium on Cyclodextrins send their regards and express their highest respect to the old master of Cyclodextrin Chemistry.

Budapest 30.9.81.

J.Szejtli F.Cramer

CHEMISTRY AND PRODUCTION OF CYCLODEXTRINS

I.Int.Symp. on Cyclodextrins
Budapest, 1981

CYCLODEXTRIN - A PARADIGMATIC MODEL

F. CRAMER

Max-Planck-Institut für experimentelle Medizin
Hermann-Rein-Strasse 3
D-3400 Göttingen, FRG

SUMMARY

Since Emil Fischer's postulate of the 'lock and key' specificity of enzymes with respect to substrates one tries to understand this specificity in terms of stereochemistry, interaction of specific groups, weak interactions, hydrophobic binding or rates of reactions. Even with the knowledge of the complete three-dimensional structure of many enzymes, their extremely high specificity in most cases is hard to understand. In this situation cyclodextrins have served as extremely useful models for enzyme like interactions even in an early stage of cyclodextrin research. They form stereospecific complexes. They can be used in order to separate enantiomers. They show hydrophobic interactions. They provide an inner surface with dielectric properties different from the outside solution. They show off- and on-rates in complex formation similar to those of enzymes. They can accelerate certain chemical transfer reactions to a considerable extent. Therefore during the past 30 years they have served as models for enzyme specificity, enzymatic catalysis, weak interactions, cavities in solution. They offer an ideal set of models and therefore in themselves they are models for models: paradigmatic models.

1. What are cyclodextrins?

Cyclodextrins have been discovered by Schardinger about 80 years ago (1) when he tried to degrade starch with Bacillus macerans. About 40 years later Freudenberg determined the structure of the Schardinger dextrins (2, 3, 4, 5). I wanted to show you the portraits of these pioneers of cyclodextrin research. Unfortunately I couldn't obtain a picture of F. Schardinger. In Fig. 1 you see the portrait of Karl Freudenberg,

Fig. 1: K. Freudenberg
born 1886

who still lives in Heidelberg at the age of 95 years in good health and excellent mental condition. The cyclodextrins are formed from amylose by an enzyme cutting off about one loop of the helical molecule as shown in Fig. 2.

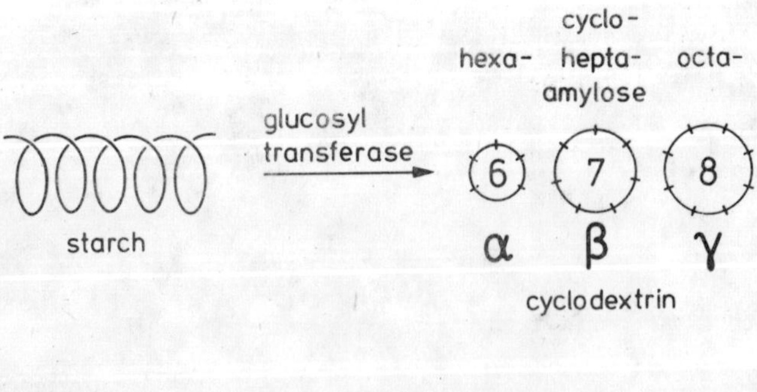

Fig. 2: Formation of α,β- and γ-cyclodextrin from amylose.

The biochemistry of this process has been studied in detail (6). With pure enzyme only very few byproducts are formed (Fig. 3).

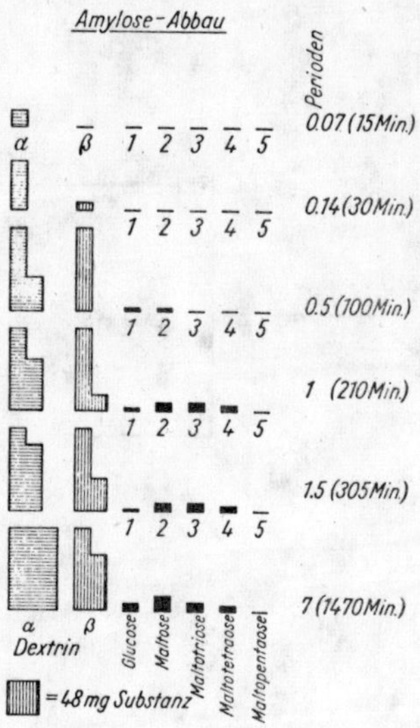

Fig. 3: Formation of cyclodextrins from amylose according to F. Cramer and H. Steinle (6).

The enzyme therefore is a specific glycosyl transferase. The remarkable thing about the cyclodextrins is that they can form inclusion compounds (7, 8).

The formation of these compounds is mainly determined by steric parameters and weak interactions such as hydrophobic interactions and hydrogen bonds, but not at all by covalent bond formation. For example bromobenzene does not fit into the small cavity of α-cyclodextrin and therefore does not form an inclusion complex. In this way, α- and β-cyclodextrin can be separated. With this property cyclodextrins served as models for a new type of noncovalent compounds.

2. What is a model?

Our scientific enterprise to describe the real world would be a hopeless task if we really would aim at a description of the total world. The vast multitude of parameters to be described and depending on each other makes it impossible to describe everything at the same time or even from the same point of view. Scientific description therefore has to reduce reality. This is called scientific reductionism (Fig. 4).

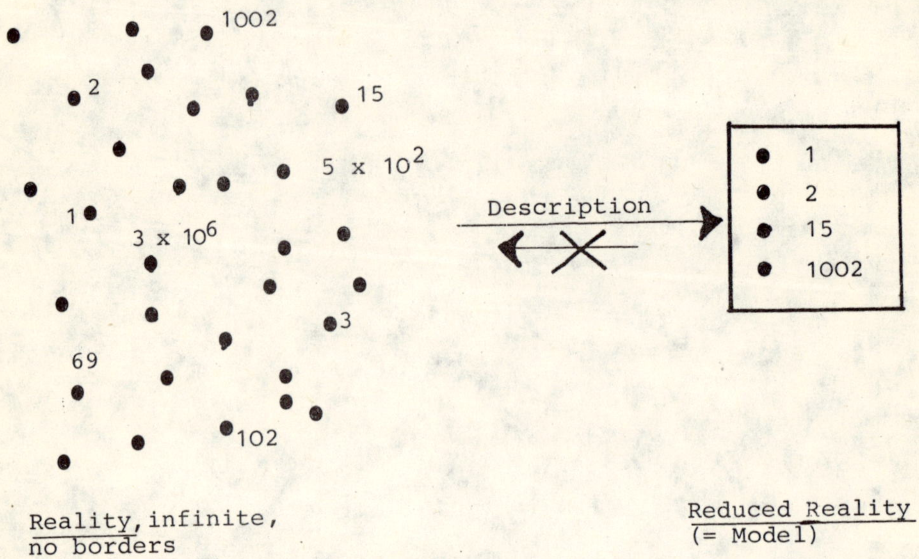

Fig. 4: Scheme of scientific reductionism. The reduced reality should describe a few isolated phenomena, however correctly.

A few describable parameters are selected from the total reality which is infinite and therefore undescribable. These few parameters are then described in great detail. If the selection of the parameters is adequate and if the description is good enough and close enough, this reduced reality represents a model of a certain aspect of nature. Whereas the total reality is infinite and multidimensional and has no sharp borders, the model is limited, describes a low number of parameters (in our case 1, 2, 15 and 1003) and is finite. The process of description is vectorial. The model cannot directly be transported back and fitted into the reality. It can, however, be used in order to understand one or more aspects of reality. In this manner cyclodextrins can be used as models to understand chemical reality. In the following I shall describe a few selected aspects of the model.

3. Model I: Cavity in solution

Since this review is partially a historical introduction I may insert a personal memory. When I was a young postdoc I tried to titrate something in presence of cyclodextrins both in an acidimetric and in iodimetric reaction. The results were very puzzling and obviously quite wrong. I checked all solutions, blanks etc. again and again - the same 'unrealistic' result. After a sleepless night I came to the conclusion that the cavity of the cyclodextrins can trap suitable molecules even in solution. This would mean a homogenous solution could have two phases. I made a few actually quite easy experiments to this extent which confirmed my hypothesis of the 'cavity in solution' (9). 30 years ago this was a rather revolutionary claim. There was, of course, the Debye-theory about the structure of solvent in the environment of an ion, but not the slightest concept of an 'inner'surface of a homogeneous solution. Correspondingly the new paradigm caused a considerable controversy. A whole maffia of physicochemists was united against the young greenhorn from Heidelberg. I believe the concept was finally successful.

Nitrophenol and a series of azo dyes were chosen to determine the equilibrium rate constants and the thermodynamic and geometric parameters for adduct formation. p-Nitrophenol as well as its anion exhibit a spectral shift of about 15 mμ in aqueous solution on adding α-cyclodextrin. Both species complex with cyclodextrins, but at different rates and with different equilibrium constants. Fig. 5 shows the changes in the p-nitrophenolate spectrum (pH=11) with increasing amounts of cyclodextrin. Isosbestic points at 398 and 446 mμ indicate a 1:1 equilibrium. Evaluation of the spectral data yields the equilibrium constant of dissociation

$K = 3.55 \times 10^{-4}$ M at 20° for the reaction

α-cyclodextrin + p-nitrophenolate $\rightleftarrows$ inclusion complex.

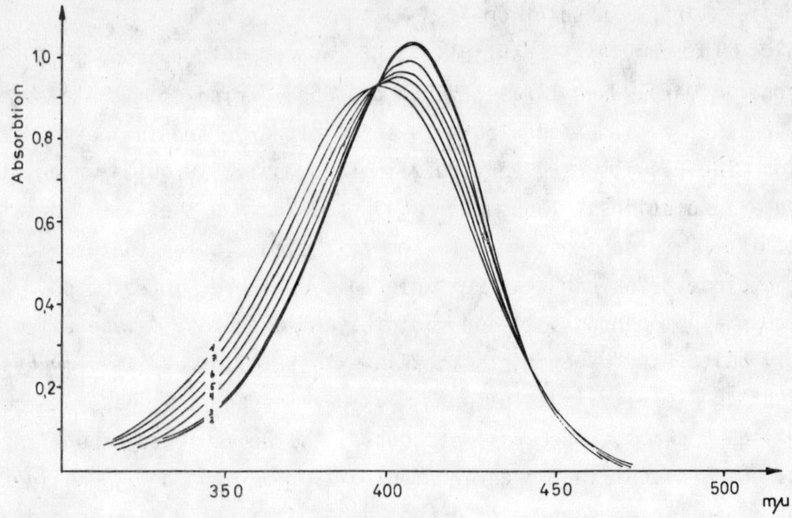

Fig. 5: p-nitrophenol spectrum at pH 11 with varying amounts of cyclodextrin (see ref. 10).

With azo dyes pronounced spectral shift are observed; in fact this was the original observation of an inclusion in solution with the indicator methylorange. With these substances also the kinetic parameters were determined by the temperature jump method (10). A mechanistic picture for adduct formation with azo dyes was put forward which takes into account the similarity of the equilibrium constants and reaction enthalpies on the one hand and the vastly different rate constants on the other. The mechanism includes the following steps:

1. The approach of the substrate to the cyclodextrin molecule.
2. Breakdown of the water structure inside the cyclodextrin ring and transport of some water molecules out of the ring.
3. Breakdown of the water structure around that part of the substrate molecule which is to be included, and release of some water molecules into the solution.
4. Interaction of the substituents of the substrate molecules with groups at the rim or at the inside of the cyclodextrin.
5. Possible formation of hydrogen bonds (known to be extremely fast).

6. Reconstitution of the water structure around the exposed parts of the substrate after the inclusion process.

Since steps 1 and 2 are common to all inclusion processes, no matter what type of substrate is involved, the substrate specificity must be attributed to steps 3 or 4 (and 6 respectively) or, with less likelihood, to 5. The rate constants of the monosubstituted azo dyes decrease in the substituent order

$- NO_2 > - OH > - N(CH_3)_2 > \cdot O^{\ominus}$.

This order reflects not so much the relative size, i.e. steric hindrance, of the substituents but rather their type of hydration. Removal of the tightly bound water around groups which carry a charge or form hydrogen bonds, such as OH and $N(CH_3)_2$, is a slower process than the melting of 'iceberg' water. In these cases step 3 should be rate determining (10).

It is after all not a surprise that inclusion compounds in solution also exhibit chiralic specificity. This has been observed in the early days of cyclodextrin chemistry (11). Most clearly this is shown when observing the ORD curves of cyclodextrin in the region of the main adsorption band of an otherwise achiralic azo dye (Fig. 6).

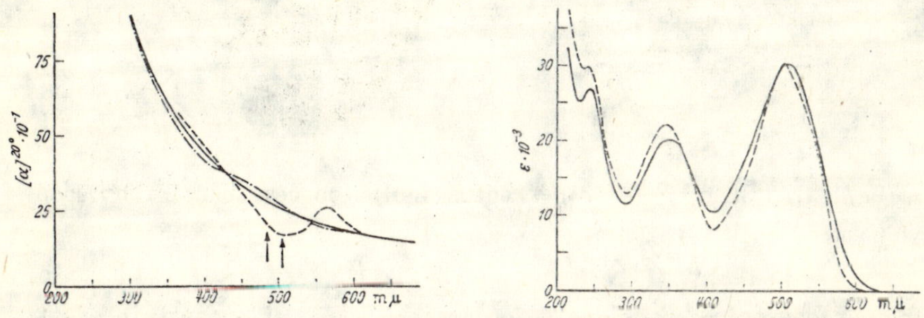

Fig. 6: Left: ORD-Curves for the system γ-cyclodextrin Congo red (---) and γ-cyclodextrin naphthol-(1)-4-azo-4-naphthaline sulfonic acid-(1) (-·-·); γ-cyclodextrin only (——). Right: Absorption spectra of Congo red with (---) and without (——) γ-cyclodextrin (14).

The chiralic cyclodextrin imposes asymmetry onto the electronic system of the azo dye. It is therefore understandable that reactions at chiralic centres in presence of cyclodextrin are directed in an asymmetric fashion.

4. Model II: Cavity in crystals

The concept of inclusion compounds has been developed with crystals of polyphenolic compounds (12) and with urea (13). Cyclodextrins provide a molecular hollow space which either can be filled with water or other solvents or can take up a suitable guest molecule. Again the incorporation into the crystal is mainly determined by stereochemical factors. A typical and rather surprising example is the formation of the krypton-α-cyclodextrin complex. When α-cyclodextrin is recrystallized under high pressure of krypton, the noble gas is trapped inside the cavity. On dissolving the crystal the gas is liberated (Fig. 7).

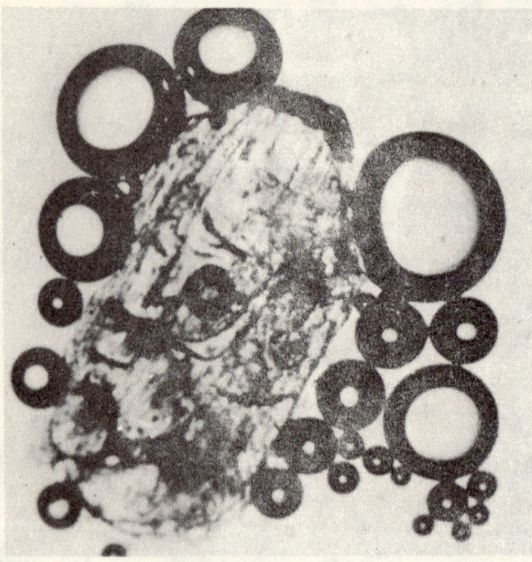

Fig. 7: Crystal of α-cyclodextrin-Krypton inclusion compound (right) upon liberating Krypton (left) (14).

The cyclodextrin-iodine inclusion compounds are of special interest. They can serve as models for the fact that molecules in a defined environment and because of a geometric arrangement imposed by the geometry of the environment can adopt different physical properties. Thus the iodine is forced into a linear arrangement which changes spectral and electronic properties throughout. In his contribution later in the meeting Dr. W. Saenger will discuss these complexes in detail.

5. Model III: Enzyme like action

Probably the most valuable model function was the observation that cyclodextrins can exert enzyme like catalysis. The first example of this was the acceleration of the oxidation of furoin (15). The microenvironment inside the cyclodextrin shifts the keto-enol equilibrium towards the enol form which is the oxidizable form. Therefore cyclodextrin catalyses the oxidation. Since the main binding contribution between cyclodextrin and its partners is the geometrical fitting, it is not surprising that catalyses brought about by cyclodextrin occur in a stereospecific fashion. Cyclodextrins, being themselves chiralic, can induce chirality during synthesis. This has been shown in the synthesis of 4-chloromandelic acid from 4-chlorobenzaldehyde and hydrocyanic acid. The reaction is accelerated 1.38 times and at the same time a partially optically active mandelic acid is formed (11).

Decarboxylation of phenylmethylcyanoacetic acids, substituted acetoacetic acids and trihalogenoacetic acids is catalysed by cyclodextrins (16). Depending on the substituent, the catalytic acceleration factors are between one and fifteen. Probably a bifunctional mechanism as shown in Fig. 8 is in operation. Decarboxylation is a monomolecular decomposition reaction and thus should follow the first order rate law. If, however, the reaction pathway goes entirely via the catalytic action of the cyclodextrin or, in terms of enzymology, when the catalytic site is completely filled (the enzyme is saturated) zero order kinetics should be observed. This is indeed the case. Therefore, the system can be treated according to the Michaelis-Menten theory and a K_m can be calculated from the kinetic data (17).

In order to calculate the change in activation energy for reactions catalysed by cyclodextrins, the rates of decarboxylation at different temperatures were measured. For the decarboxylation of the 4-chlorophenyl compound the values shown in Table 1 were found.

Table 1: Thermodynamic data for the decarboxylation of 4-chlorophenylmethylcyanoacetic acid.

	ΔH	ΔS	ΔF
non-catalysed	30	25	22.3
catalysed	21	1.5	21.0
Δ	9	23.5	1.3

In all cases the activation energy is lowered in the presence of ß-cyclodextrin. When the ΔH values are split into the values for activation entropy and activation enthalpy it is seen that a decrease in activation entropy occurs in the same direction whereas the free energy of activation remains almost constant. The entropy term is essentially responsible for the small extent of cyclodextrin catalysis which according to the Arrhenius equation should otherwise be in the order of 10^6 if based on the ΔH value alone.

Pyrophosphates are cleaved at pH 12 in the presence of cyclodextrins at high catalytic rates (18). Again a high substrate specificity is observed. At low Ca^{2+} concentrations, when no spontaneous hydrolysis of pyrophosphate is observed, cyclodextrin still catalyses the reaction, thus exerting an all-or-nothing effect.

In this catalysis a direct involvement of the catalysing cyclodextrin can be seen since a covalently phosphorylated cyclodextrin is isolated as an intermediate. This would correspond to the acylated enzyme [E·S'] encountered, for example, in chymotrypsin catalysis. The mechanism is depicted in Fig. 9.

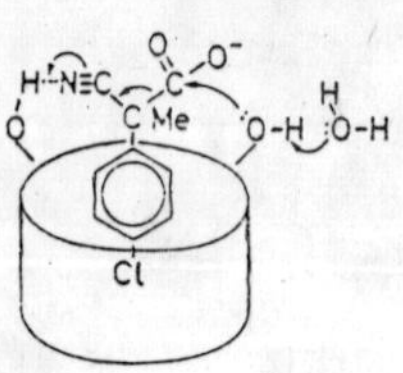

Fig. 8: Mechanism of decarboxylation (14)

Fig. 9: Mechanism of phosphorylation (14)

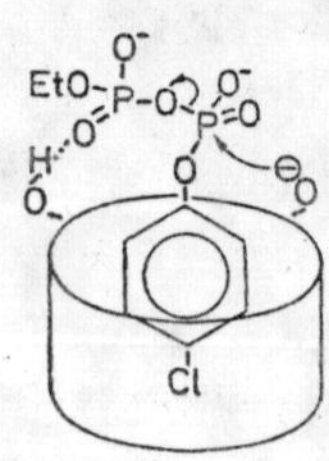

A number of highly interesting enzyme-like catalyses have been discovered with substituted cyclodextrins and various substrates in the group of Bender (8), Breslow (19,21), Tabushi (20,22) and I expect that we shall hear much of it during this symposium.

In order to achieve a chymotrypsin-like catalysis at neutral pH, cyclodextrin was furnished with imidazole groups (23). Cyclodextrin was reacted with various imidazole derivatives, e.g. with chloromethylimidazole. These compounds do in fact have a high catalytic effect upon the hydrolysis of p-nitrophenyl acetate at pH 7.5. Fig. 10 gives a pictorial comparison of the hydrolysis of p-nitrophenyl acetate with water, cyclodextrin, the imidazolyl-cyclodextrin model, and actual chymotrypsin.

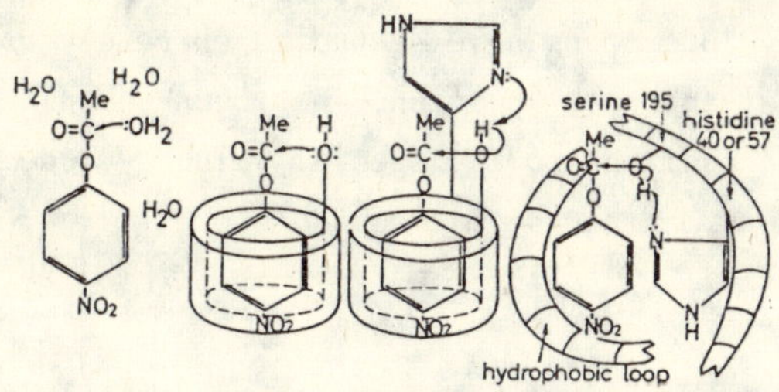

	water	α-CD-OH	β-CD-imidazole	chymotrypsin
M.W.	18	1000	1400	25000
p_K of cat. group	14.0	12.0	6.95	6.95
K_m	—	2.6×10^{-3}	$\sim 10^{-3}$	10^{-5}
k_R[a]	—	$>4 \times 10^{7}$	$\sim 10^{7}$	10^{7}
k_{cat}	0.13×10^{-4}	0.32×10^{-4}	38.4×10^{-4}	3.0
k_{cat}/k_{H_2O}	1.0	2.5	300	~ 10000

→ towards an enzyme →

[a] = rate of formation of the enzyme-substrate complex

Fig. 10: Comparison of various catalysts for the hydrolysis of p-nitrophenyl acetate (14).

In enzymes a more sophisticated fixation of the various reactants is probably necessary. This can only be achieved in a molecule of some rigidity with a certain minimum molecular weight. This it is likely that in a given system of chemical compounds - such as proteins - a certain minimum molecular weight is necessary to achieve the required refinement of the machinery. By evolution, chymotrypsin, with a molecular weight of 25 000, might even have developed a minimum molecular weight at maximal efficiency. In any cyclodextrin - since it is a symmetrical torus - there is no fixation of the substrate other than its entrenchment in the void space. The imidazolyl group, too, is not rigidly fixed and, therefore, entropy factors disfavour the necessary arrangement for the catalytic step. Thus we are still at least two orders of magnitude below the catalytic capacity of the natural enzyme.

6. The paradigm of models

As discussed above scientific models are made in order to focus on certain aspects of reality. If the aspect is relevant and if the focus is sharp, the model-like description of nature can lead to a better understanding not only of the particular phenomenon but also of larger areas of nature. In this respect cyclodextrins have served as extremely useful typical models by which the model approach in science has been demonstrated with high efficiency. Thus in addition to the special value cyclodextrins can serve as models for models, as paradigmatic models.

Literature cited

(1) F. Schardinger: Z. Untersuchung Nahrungs u. Genußmittel 6 874 (1903)
(2) K. Freudenberg, R. Jacobi: Liebigs Ann. Chem. 518 102 (1935)
(3) K. Freudenberg, M. Meyer-Delius: Ber. 71 1596 (1938)
(4) K. Freudenberg, E. Plankenhorn, H. Krauber: Liebigs Ann. Chem. 558/1 (1945)
(5) K. Freudenberg, F. Cramer: Chem.Ber. 83 296 (1950)
(6) F. Cramer, H. Steinle: Liebigs Ann. Chem. 595 81 (1955)
(7) For the first extended summary: see F. Cramer, Einschlußverbindungen, Springer Verlag, Berlin, Göttingen, Heidelberg (1954)
(8) M. Bender, M. Komiyama: Cyclodextrin Chemistry, Springer Verlag, Berlin, Heidelberg, New York (1978)
(9) F. Cramer: Chem.Ber. 84 851 (1951)
(10) F. Cramer, W. Saenger, H.-Ch. Spatz: J.Amer.Chem.Soc. 89 14 (1967)
(11) F. Cramer, W. Dietsche: Chem.Ber. 92 1739 (1959)
(12) A.M. Powell: J.Chem.Soc.London 1950 298, 300, 468
(13) W. Schlenk jr.: Fortschr.Chem.Forschung 2 92 (1951)
(14) F. Cramer, H. Hettler: Naturw. 54 625 (1967)
(15) F. Cramer: Chem.Ber. 86 1576 (1953)
(16) F. Cramer, W. Kampe: J.Amer.Chem.Soc. 87 1115 (1965)
(17) W. Kampe: Ph.D.Thesis, Heidelberg (1961)
(18) N. Hennrich, F. Cramer: J.Amer.Chem.Soc. 87 1121 (1965)
(19) R. Breslow, M. Hammond, M. Lauer: J.Amer.Chem.Soc. 102 421 (1980)
(20) I. Tabushi, Y. Kuroda, K. Shimokawa: J.Amer.Chem.Soc. 101 4759 (1979)
(21) R. Breslow, P. Bovy, C.L. Hersh: J.Amer.Chem.Soc. 102 2115 (1980)
(22) I. Tabushi, Y. Kuroda, A. Mochizuki: J.Amer.Chem.Soc. 102 1152 (1980)
(23) F. Cramer, G. Mackensen: Chem.Ber. 103 2138 (1970)

I.Int.Symp. on Cyclodextrins 1.2.
Budapest, 1981

CYCLODEXTRIN RESEARCH IN JAPAN

T. NAGAI

Hoshi Institute of Pharmaceutical Sciences, Shinagawa-ku
Tokyo, Japan

SUMMARY

A start in research of cyclodextrin (CDX) in Japan was actually made around the end of 1960's. The activity in the research field has been enhanced since several chemical companies started to produce CDX of high quality and comparatively low price in a large scale around 1971. As the application of CDX is expanding to various field, the price is expected to get cheaper in future, accompanying more active development in the research.

Concerning the method of production of CDX, a CDX-producing enzyme was found recently from an alkalophilic bacteria of Bacillus genus. The other new approaches have also been made.

As the safety of CDX has been confirmed, researches for pharmaceutical application are very active at the present as follows: solubilization, enhancement of absorption, suppression of irritation to stomach, and preparation of inclusion compounds with drugs. Solid inclusion compounds with drugs can be prepared by freeze-drying, cogrinding, and so on. The products of tablets and injections of prostaglandins included by CDX are now on market.

In addition to such fundamental studies of CDX as on the formation mechanism and X-ray analysis of inclusion compounds, noticeable ones are made recently concerning an application of CDX or its derivatives to investigations for an understanding of enzymic or chemical reactions, for example, using capped CDX.

Most recent representative researches in Japan are introduced here by the respective participants from Japan.

We don't forget the names of Villiers, Schardinger and Cramer and so on as the pioneers in researches of cyclodextrin (CDX). During the past twenty years, so many papers have been published concerning CDX. Actually, however, there was no remarkable contribution to these researches by Japanese workers at the early stage. Researches of CDX in Japan started really since the end of 1960's by Kuge, Koizumi, and so on. Around that time, Matsutani Kagaku Kogyo Co. and CPC International Inc. started a production of β-CDX in a large scale. Now, Nihon Shokuhin Kako Co. produces CDX on the commercial basis in Japan. As the application of CDX is expanding to various fields, the price is expected to get cheaper in future, accompanying more active development in the research.

It is very difficult for me to introduce here the whole works in CDX researches in Japan. Therefore, I like to talk about some examples of them.

I. Method of production of CDX

Okada improved a method of production of β-CDX using B. megaterium (1). Shiosaka produced α-CDX efficiently by using B. stearothermophilus at optimal temperature 70° (2). Recently, Horikoshi of the Institute of Physical and Chemical Research found an alkalophilic bacteria Bacillus No. 38-2, which holds an enzyme with an optimal activity at pH 9.0 to produce β-CDX in a 75-80% yield at 70° (3). This enzyme is now applied to the commercial production by Nihon Shokuhin Kako Co.

II. Safety of CDX

The research group of Teijin Ltd. reported LD_{50} (g/kg) of β-CDX as: >12.5, >0.9 and >0.9 for oral, subcutaneous and intraperitoneal administration in mice, respectively; >12, >1.5 and > 1.2 for oral, subcutaneous and intraperitoneal administration in rats, respectively; giving the proof that β-CDX is a substance of high safety (4). They studied also the chronic toxicity of β-CDX for 6 months, showing there were no adverse effects concerning the body weight, urine analysis, hematology, biochemistry, organ weight, autoscopy and microscopic examination and also no toxicological symptom during the administration (5). Recently, Hatano Research Institute of Food and Drug Safety Center made a systematic toxicological investigation on the product of β-CDX of Nihon Shokuhin Kako Co. mentioned above, concluding that it is hardly toxic (6).

III. Preparation of CDX inclusion compounds

The most important part of CDX research is occupied concerning the inclusion compounds. The existing three methods are first taken into consideration for the preparation of solid inclusion compounds with CDXs, i.e., (a) coprecipitation from the aqueous solution of both the host and the guest components;(b) coprecipitation by mixing the solution of the guest component in organic solvent with the aqueous solution of CDX; (c) kneading of an aqueous slurry or paste consisting of both the host and the guest components and finally drying.

In addition to these methods, some other ones have been developed. If the guest component is soluble in water and unstable on drying to get the final solid product, it is advisable to apply the freeze-drying method to the preparation. Such the preparation has a high solubility with advantages especially for pharmaceutical purposes. Nagai found that many drugs are easily included with α- and β-CDXs by freeze-drying method (7). Nakai reported that some drugs are included with β-CDX during cogrinding both the component (8). Both the freeze-drying and the cogrinding methods mentioned above gave the inclusion compounds in amorphous state.

IV. Characterization of CDX inclusion compounds

Many fundemental studies have been made on such as thermodynamics and kinetics of the inclusion process to understand the driving force for the formation, structure of the inclusion compounds, and determination methods for the stability constant. For these studies, various kinds of methods such as solubility determination, partition, dialysis, spectroscopy, electro-chemistry and reaction kinetics have been applied. Especially, the spectroscopy according to UV, fluorescence, ORD, CD, IR, NMR, ESR and Raman spectra has afforded a lot of useful means.

Uekama applied the liquid chromatography to the determination of stability constant of inclusion compounds of prostaglandins with α- and β-CDXs (9). This method has several advantages over the other ones, for example, using only a small amount of sample with a wide applicability to various systems and especially to chemically unstable ones.

X-ray structural analysis was done by Harada on the inclusion compounds of α-CDX-p-iodoaniline (10), α-CDX-DMSO-MeOH (1:1:2) dihydrate (11), and α-CDX-m-nitrophenol (1:2) (12). Tomita analyzed the crystal structures of the inclusion compounds of γ-aminobutyric acid (13), ethanol (14) and p-ethylaniline with α- or β-CDX (15). Kuge demonstrated the X-ray diffraction patterns of n-propyl alcohol/γ-CDX depended on the extent of hydration, i.e., on the drying state (16). Nagai showed according to X-ray diffractometry that the inclusion compounds of antiinflammatory drugs/β-CDX are obtained in amorphous state by freeze-drying (7).

Nakai studied a change in IR spectra due to the formation of inclusion compound by cogrinding of aspirin or benzoic acid with β-CDX, concluding that an interaction takes place between the drug and β-CDX in the ground mixture in the same way as in the inclusion compound (8).

Uekama studied CD (circular dichroism) of prostaglandin A_1 or tolbutamide/α- or β-CDX and also NMR of prostaglandin $F_{2\alpha}$/ α- or β-CDX, and proposed the possible structure of the inclusion compounds (17). Hatano demonstrated that ICD (induced CD) was quite useful for elucidating the structure, especially the orientation of the guest molecule, of β-CDX inclusion compounds of aromatic molecules (18).

Nagai applied the solid-state NMR and Raman spectroscopy to elucidate which molecular moiety of the guest molecule is included in the solid state of tolbutamide/β-CDX (19).

Ohara has applied ESR to confirm the formation of inclusion compounds of stable nitroxide radicals with CDX (20).

V. Researches of CDX in pharmaceutical field

It can be said that the most active and promising investigations of CDX inclusion compounds are carried out in the pharmaceutical field in Japan, i.e., at university laboratories or companies. Usually, more than ten papers of these studies have been presented every year at the Annual Meeting of the Pharmaceutical Society of Japan. In this connection, Teijin Ltd. and Ono Pharmaceutical Co. are leading in applying for the related patents. At the present, the products of injections and tablets of prostaglandins, which are stabilized by CDXs, are on market, i.e., PGE_1-α-CDX soluble powder for injections and PGE_2-β-CDX tablets (21).

Table I summarizes the examples of an improvement of pharmaceutical characteristics by means of CDX inclusion compounds.

Koizumi reported that the bioavailability of barbiturates after oral administration was enhanced in rabbits by β-CDX inclusion compounds in comparison with intact one (22). Nagai showed ketoprofen/β-CDX inclusion compound has a high bioavailability compared with ketoprofen alone (23).

Nagai showed also that the inclusion compound was effective for reducing the injury of stomach in the case of phenylbutazone (24). Uekama reported that acetohexamide/β-CDX lowered blood glucose level more than acetohexamide alone (25). He reported also that the hemolytic effects of chlorpromazine and flufenamic acid on human erythrocytes was lowered in the presence of CDXs in isotonic phosphate buffer at 37° (26). Koizumi reported that barbiturate/β-CDX inclusion compounds shortened the sleeping lag and lengthened the sleeping time in mice (27).

Regarding the stabilization of drugs, there are found, for example, Nagai's report on azapropazone/β-CDX (28), Yamana's on benzylpenicilline/β-CDX (29) and Yonezawa's on guaiazulene/α- or β-CDX (30). Ono Pharmaceutical Co. has been successful in putting the products to practical use by stabilization with CDX inclusion compounds, as mentioned already.

There are so many reports concerning the enhancement of solubility of drugs by CDX. The solubility curve with the addition of CDX depended on the kind of CDX and drugs, i.e., the

Table I. Examples of Improvement of Pharmaceutical Characteristics by CDX Inclusion Compounds

Enhancement of solubility :
prostaglandins, nonsteroidal antiinflammatory drugs, digitalis cardiacglycoside, steroid hormones, barbiturates, chloramphenicol, sulfonamides, phenytoin, benzodiazepines, coumarin anticoagulants, diuretics

Stabilization :
hydrolysis(aspirin, atropine, procaine, digitalis cardiacglycoside, prostacycline), oxidation(chlorpromazine, levomepromazine, epinephrine, aldehydes), photodecomposition (phenothiazines, ubiquinones, vitamines), dehydration (prostaglandin E groups)

Enhancement of bioavailability :
aspirin, phenytoin, digoxin, nonsteroidal antiinflammatory drugs, barbiturates, sulfonamides, diuretics

Modification of liquid drugs to powders :
oil-soluble vitamines(A,D,K), clofibrate, ONO-802, phenols, benzaldehyde, nitroglycerin, methyl salicylate, essential oils

Prevention of evaporation :
naphthalene, iodide, l-menthol, chlorbutanol, d-camphor, salicylic acid

Improvement of ill smell and taste :
chloral hydrate, prostaglandins, naphthalene, nonsteroidal antiinflammatory drugs, thymol, chloramphenicol

Reducing of stomach injury :
indomethacin, aspirin

Inhibition of hemolysis :
cationic drugs(e.g., phenothiazines), anionic drugs(e.g., flufenamic acid), antibiotics, benzyl alcohol, menadione

---

stability constant (28,31).

Many patents have been applied concerning the modification of liquid or volatile drugs to powdered form, for example, as shown in Table I already.

VI. CDX in agricultural fields or foods

CDXs can be employed for molecular encapsulation and thus for a stabilization of unstable or very volatile insecticides and herbicides. Several patents have been applied by Japanese companies in this field, for example, concerning the stabilization by β-CDX against vaporization of O,O-dimethyl-O-2,2-dichlorovinylphosphate, as belongs to the patent of Teijin Ltd.

Table II. Cyclodextrin Derivatives

Derivatives	Uses
Amination	
Imidazol alkylamino derivatives	Catalyst
$CDX-(OC_2H_4NH_2)_n$	Paper sizing
Amino-deoxy-CDX	Pharmaceuticals
Amino-CDX	Intermediate
Esterification	
$CDX-(OH)_{21-x}(ONO_2)_x$	Molecular weight determination
$CDX-(OH)_{21-x}(OSO_3H)_x$	Pharmaceuticals
$CDX-(OH)_{21-x}(OPO_3H_2)_x$	Structure studies
$CDX-(OH)_{21-x}(OCOC_6H_5)_x$	Structure studies
$CDX-(OH)_{21-x}(OCOR)_x$	Gas liquid chromatography, Detergent, Plasticizer
$CDX-O-C(=O)\sim COOH$	Plasticizer
Etherification	
$CDX-(OCH_3)_x$	Molecular weight determination
$CDX-(OH)_{14}(O-CH_2-\text{imidazolyl (HN, N)})_7$	Catalyst
CDX-Epichlorohydrin resin	Separation of isomer
$CDX-OCH_2-CHOH-CH_2OH$	Cross linking, Sizing
$CDX-O-(CH_2-CH-O)_nR$	Surface treatment
$CDX-O\sim SO_3^{\ominus}$	Treatment of sewage
Cross Linking	
CDX-Crosslinked resin	Separation of isomer
$CDX-(OCONHC_6H_5)_{21}$	Study of optical rotation

(32). Teijin Ltd. has also a patent concerning the chemical stabilization of allethrin with β-CDX inclusion compound (33).

Since Japanese government gave recently a licence of production of the powdered alcoholic beverages which are included in dextrin, the inclusion with β-CDX has been investigated finally to find a possibility of the production of the powdered alcoholic beverages, for which a patent has been applied already by Suntory Ltd. (34).

VII. CDX or its derivatives as model of enzyme

Many derivatives of CDX, as summarized in Table II, have been synthesized and subjected to investigations of application.

Additionally, various kinds of investigations have been done concerning these derivatives as a model of such enzymes as oxidoreductase, transferase, hydrolase, lyase, isomerase and lygase. In this field, very active studies have been made by polymer scientists, for example, Tabushi (35), Toda (36), Komiyama (37), Nozakura (38), Ohara (39), Osa (40), and so on. Among them, Tabushi synthesized many derivatives of CDX such as capped CDX and duplex CDX, investigating the sensitive enzyme-like action of these compounds. He is also intending an application of the cavity of CDX possibly acting as the field of synthesis of some bioactive substances.

"Literature Cited"

1. Okada, S., Tsuyama, N., Kitabatake, S.: Proceedings of the Symposium on Amylases (Osaka) 7 61,/1972/
2. Shiosaka, M., Bunya, H.: Proceedings of the Symposium on Amylases (Osaka) 8 43,/1973/
3. (a) Horikoshi, K.: Proceedings of the Symposium on Amylases (Osaka) 8 37,/1973/ (b) Nakamura, N., Horikoshi, K.: Agric. Biol. Chem. 40 753,/1976/ (c) Idem: ibid. 40 935, /1976/ (d) Idem: ibid. 40 1647,/1976/ (e) Idem: ibid. 40 1785,/1976/
4. Mifune, A., J. Synth. Org. Chem. Jap. 32 889,/1974/
5. Makita, T., Kojima, N., Hashimoto, Y., Ide, H., Tsuji, M., Fujisaki, Y.: Oyo Yakuri 10 449,/1975/
6. FDSC Report No. 55-093-1/2,/1980/: Private communication from Nihon Shokuhin Kako Co. Ltd., 8Fl, Shinkokusai Bldg., 3-4-1 Marunouchi, Chiyoda-ku, Tokyo.
7. Kurozumi, M., Nambu, N., Nagai, T.: Chem. Pharm. Bull. 23 3062,/1975/
8. (a) Nakai, Y., Nakajima, S., Yamamoto, K., Terada, K., Konno, T.: Chem. Pharm. Bull. 26 3419,/1978/ (b) Idem: ibid. 28 652,/1980/
9. Uekama, K., Hirayama, F., Irie, T.: Chem. Lett., 1978; p.661./

10. Harata, K., Uedaira, H.: Nature 253 190,/1975/
11. Harata, K.: Bull. Chem. Soc. Jpn.: 51 1644,/1978/
12. Harata, K., Uedaira, H., Tanaka, J.: Bull. Chem. Soc. Jpn.: 51 1627,/1978/
13. Tokuoka, R., Abe, M., Matsumoto, K., Fujiwara, T., Tomita, K.: Acta Cryst. B37 445,/1981/
14. Tokuoka, R., Abe, M., Fujiwara, T., Tomita, K., Saenger, W.: Chem. Lett., 1980; p.491./
15. Tokuoka, R., Fujiwara, T., Tomita, K.: Acta Cryst.: B37 1158,/1981/
16. Takeo, K., Kuge, T.: Agric. Biol. Chem.: 34 568,/1970/
17. (a) Uekama, K., Hirayama, F., Yamasaki, S., Otagiri, M., Ikeda, K.: Chem. Lett., 1977; p.1389./ (b) Uekama, K., Hirayama, F., Matsuo, N., Koinuma, H.: ibid., 1978; p.703./ (c) Uekama, K., Hirayama, F.: Chem. Pharm. Bull. 26 1195, /1978/
18. (a) Shimizu, H., Kaito, A., Hatano, M.: Bull. Chem. Soc. Jpn.: 52 2678,/1979/ (b) Idem: ibid. 54 513,/1981/
19. (a) Ueda, H., Nagai, T.: Chem. Pharm. Bull. 28 1415,/1980/ (b) Idem: ibid. 28 (in press),/1981/ (c) Nagai, T., Ueda, H., Nambu, N.: Abstract of F.I.P. Congress, P-XIV, No. 199/1981/
20. Ohara, M., Hettler, H., Gauss, D., Cramer, F.: Bioorg. Chem.: 8 211,/1979/
21. (a) Ono Pharmaceutical Co.: Japan Kokai Tokkyo Koho, 47-39057,/1972/ (b) Idem: ibid. 50-3362,/1975/ (c) Idem: ibid. 50-35319,/1975/ (d) Idem: ibid. 50-35324,/1975/
22. Koizumi, K., Kidera, Y.: Yakugaku Zasshi 97 705,/1977/
23. Nambu, N., Shimoda, M., Takahashi, Y., Ueda, H., Nagai, T.: Chem. Pharm. Bull. 26 2952,/1978/
24. Nambu, N., Kikuchi, K., Kikuchi, T., Takahashi, Y., Ueda, H., Nagai, T.: Chem. Pharm. Bull. 26 3609,/1978/
25. Uekama, K., Matsuo, N., Hirayama, F., Yamaguchi, T., Imamura, Y., Ichibagase, H.: Chem. Pharm. Bull. 27 398, /1979/
26. Uekama, K., Irie, T., Sunada, M., Otagiri, M., Tsubaki, K.:

J. Pharm. Dyn. 4 142,/1981/
27. Koizumi, K., Miki, H., Kubota, Y.: Chem. Pharm. Bull. 28 319,/1980/
28. Hamada, Y., Nambu, N., Nagai, T.: Chem. Pharm. Bull. 23 1205,/1975/
29. Minakami, Y., Ichimura, F., Yamana, T.: Yakuzaigaku 38 45, /1978/
30. Yonezawa, Y., Maruyama, S., Takagi, K.: Agric. Biol. Chem. 45 505,/1981/
31. Koizumi, K., Mitsui, K., Higuchi, K.: Yakugaku Zasshi 94 1515,/1974/
32. Teijin Ltd., SS Pharmaceutical Co. Ltd.,: Japan Kokai Tokkyo Koho, 49-134844,/1974/
33. (a) Teijin Ltd., Dai Nippon Jochugiku Co.,: U.S.P., 3,846,551,/1974/ (b) Teijin Ltd.,: Japan Kokai Tokkyo Koho, 49-26416,/1974/
34. Suntory Ltd., Japan Maize Products Co., Ltd.,: Japan Kokai Tokkyo Koho, 80-114283,/1980/
35. (a) Tabushi, I., Yamamura, K., Fujita, K., Kawakubo, H.: J. Am. Chem. Soc. 101 1019,/1979/ (b) Tabushi, I., Kuroda, Y., Mochizuki, A.: ibid. 102 1152,/1980/
36. (a) Iwakura, Y., Uno, K., Toda, F., Onozuka, S., Hattori, K., Bender, M.L.: J. Am. Chem. Soc. 97 4432,/1975/ (b) Kojima, M., Hattori, K., Toda, F.: Tetrahedron Lett. 21 2721,/1980/
37. (a) Komiyama, M., Inoue, S.: Chem. Lett., 1979; p.1101./ (b) Idem: Bull. Chem. Soc. Jpn. 53 2330, 3266,/1980/
38. (a) Harada, A., Furue, M., Nozakura, S.: Macromolecules 9 701,/1976/ (b) Idem: J. Polym. Sci., Polym. Chem. Ed. 16 189,/1978/
39. (a) Ohara, M., Watanabe, K.: Angew. Chem. Intern. Ed. 14 820,/1975/ (b) Ohara, M., Fukuda, J.: Pharmazie 33 467, /1978/
40. Ueno, A., Yoshimura, H., Saka, R., Osa, T.: J. Am. Chem. Soc. 101 2779,/1979/

INDUSTRIAL PRODUCTION OF CYCLODEXTRINS

K. HORIKOSHI,[1] N. NAKAMURA,[1] N. MATSUZAWA,[1] M. YAMAMOTO[2]

[1] The Institute of Physical and Chemical Research
Wako-shi Saitama 351, Japan
[2] Nihon Shokuhin Kako Co.
Tokyo 100, Japan

SUMMARY

We isolated three alkalophilic bacteria producing large amount of cyclodextrin glycosyltransferases (CGTases). Two of the enzymes, CGTases of *Bacillus* No. 38-2 and *Bacillus* No. 17-1 were purified by starch adsorption and DEAE-cellulose chromatography followed by gel filtration. The enzymes were very thermostable, especially *Bacillus* No. 38-2 enzyme was not inactivated up to 70 C. The enzymes isolated converted about 65-80 % of potato starch to cyclodextrins at 1 % (w/v) substrate concentration. Through the CGTase of *Bacillus* No. 38-2, crystalline β-cyclodextrin, γ-cyclodextrin and a mixture of cyclic and acyclic dextrins were produced on an industrial scale plant without using any organic solvent. One ton of potato starch suspension (15 % w/v) containing 10 mM $CaCl_2$ was liquefied by the CGTase at 90 C (pH 8.5) and then cooled to 60 C. The liquefite was readjusted to pH 8.5 with $CaCO_3$ and CGTase was added again. The cyclodextrin formation was maintained at 60 C for 45 h. After the reaction, the enzyme was inactivated at 80 C. The appreciable amount of bacterial α-amylase was added (pH 6.0) to hydrolyse saccharide which were not converted to cyclodextrin. The digest was decolorised and passed through ion exchangers. The refined digest was concentrated to 45 % (w/v) and transfered to a crystallizer. The crystalline β-cyclodextrin was collected by centrifugation. The filtrate containing cyclodextrins is available as CH for food additives. From this CH, γ-cyclodextrin was produced. The CH was treated with glucoamylase and passed through a Toyo Pearl HW-40 column to separate γ-cyclodextrin from other saccharides. The γ-cyclodextrin thus prepared was crystallised directly from water.

In 1969, the authors started the investigation of alkalophilic microorganisms which grew well at high alkaline pH values (pH 10-11). The microorganisms were isolated from soil by using alkaline media containing 1 % Na_2CO_3 (1). About 20 enzymes which are entirely new have been found by our method, which is called " Alkaline Fermentation " (2). The properties of the microorganisms isolated have been reported elsewhere. This paper deals with the isolation and purification of the new cyclodextrin glycosyltransferases (CGTases) produced by alkalophilic bacteria. Through one of these enzymes, production of cyclodextrins and a mixture of cyclic and acyclic dextrins on an industrial scale plant without using any kind of organic precipitant will be presented.

MATERIALS AND METHODS

Microorganisms and medium. Isolation method of alkalophilic bacteria was reported previously (3). Three strains No. 38-2, No. 13, and No. 17-1 were selected as the best enzyme producer from about 1,000 strains. These bacteria were aerobic, sporeforming, motile and rod-shaped bacteria. It is clear that all strains should belong to the genus *Bacillus*. The characteristic point of the bacteria was that the growth was especially good in alkaline media, the optimum pH for growth being about 10.

Preparation of crude enzymes. The organisms were grown aerobically for 3 days at 37 C in the II-medium (3). The cells were removed by centrifugation at 0 C. To 1 volume of the supernatant fluid were added 3 volumes of cold ethanol (-20 C). The mixture was stored in a cold room overnight. The precipitate formed was collected by centrifugation, washed with cold ethanol and dried in vacuo. Unless stated otherwise, all operations were carried out at room temperature. The crude enzymes thus prepared were stored in a refrigerator.

Dextrinizing activity. Dextrinizing activity was determined by the measurement of dextrinizing power using amylose as a substrate. The reaction mixture containing 0.2 ml of the enzyme solution (about 2 u/ml) and 0.3 ml of 0.05 % amylose in 0.1 M phosphate buffer (pH 7.0) was incubated at 40 C for 10 min. The reaction was stopped with 4 ml of 0.2 N HCl and 0.5 ml of iodine reagent (0.02 % I_2 and 0.2 % KI). Distilled water (4 ml) was added and absorbance at 700 nm was measured. One unit of the enzyme is defined as the amount of enzyme which reduces 10 % of absorbance of starch iodine complex per minute under the described conditions.

Analytical methods. Cyclodextrins were analysed by high performance liquid chromatography with a Shimadzu model LC-3A equipped a Shodex RI detector model SE-11. The column used was a Shodex NH pak J-411 which was operated at 40 C. The solvent system was acetonitrile: water (72.5:17.5 v/v) at a flow rate of 1.0 ml/min (4). Ampholine electrofocusing was done by the method of Vesterberg and Svensson (5). Molecular weight of the enzyme was estimated by the dodecylsulfate-polyacrylamide gel electrophoresis method (6).

Buffer systems. The following buffer systems were used: pH 2.5-6, 0.2 M acetate buffer; pH 5-8, 0.2 M phosphate buffer; pH 7-9, 0.2 M Tris-HCl buffer; pH 8.5-13, 0.2 M glycine-NaCl-NaOH buffer.

RESULTS AND DISCUSSION

Partial purification of the enzymes. The enzymes produced by *Bacillus* No. 38-2, *Bacillus* No. 13 and *Bacillus* No. 17-1 were purifed by the almost same manners.

1) Starch adsorption. The crude enzyme powder was dissolved in tap water and dialysed against running tap water for 2 days. About 3 % (w/v) corn starch and 20 % of ammonium sulfate were added to the dialysate and stirred for 60 min at 4 C. The corn starch which adsorbed the enzyme was collected by filtration and washed thoroughly with 10 mM phosphate buffer (pH 7.0) containing 20 % (w/v) of ammonium sulfate and 1 N NaCl. The enzyme was eluted from the starch with the same buffer containing 3 N NaCl and 0.1 M maltose, and then dialysed overnight against tap water followed by further dialysis against 10 mM glycine-NaOH-NaCl buffer (pH 8.5).

2) DEAE-cellulose chromatography. The dialysate was loaded onto a DEAE-cellulose column equilibrated with the same buffer described above. The enzyme was eluted with a linear gradient of NaCl (0-0.3 N). Active fractions were collected and concentrated by Ficoll 400.

3) Sephadex G-100 gel filtration. The concentrate was passed through a Sephadex G-100 column equilibrated with the same buffer containing 0.1 N NaCl. Enzymatically active fractions were collected, concentrated by Ficoll 400 and used as a partially purified enzyme preparation. The enzymes produced by Bacillus No. 38-2 and Bacillus No. 17-1 were purified further by DEAE-cellulose and preparative polyacrylamide gel electrophoresis. The crude enzyme of Bacillus No. 38-2 was a mixture of three enzymes: acid CGTase having optimum pH for enzyme action at 4.6; neutral CGTase, pH 7 and alkaline CGTase, pH 8.5. Other properties were almost identical. The crude enzyme of Bacillus No. 17-1 was a mixture of two enzymes: acid CGTase, optimum pH 4.5 and alkaline CGTase, optimum pH 9.5. Other properties were almost identical. Further properties will be presented elsewhere. In the following experiments, the partially purified enzyme preparations described above were used, because no difference was observed between purified and partially purified enzymes, especially in thermal stability and yield of cyclodextrins.

Physical properties of the enzymes. The molecular weights were 8.5-8.8 X 10^4 for the CGTases of Bacillus No. 38-2, and 5-6 X 10^4 for the CGTases of Bacillus No. 17-1 and Bacillus No. 13. Isoelectric points were 5.4-5.3 for Bacillus No. 38-2 CGTases and 4.5 for Bacillus No. 17-1 CGTases. The properties of the CGTases were summarized in Table 1.

Table 1. Properties of CGTases

	Optimum pH	Stable pH	Stability +Ca	Stability -Ca	Yield
Bacillus No. 38-2	4.5-9.0	5-10.5	70 C	60 C	75 %
Bacillus No. 17-1	9-9.5	6.5-10.0	60	55	65
Bacillus No. 13	9-9.5	6.5-10.0	60	55	60
Bacillus macerans	5.0-5.7	8.0-10.0	50	-	50
Bacillus megaterium	5.0-5.7	7.0-10.0	-	55	62

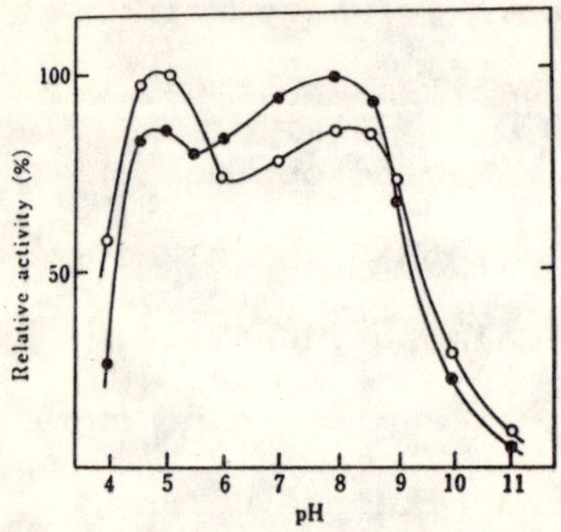

FIG. 1. Effect of pH on the Enzyme Activity.

The pH *vs* activity of the crude enzyme of an alkalophilic *Bacillus* sp. (ATCC 21783) was measured as described in the text. The pH was adjusted with the following buffer systems; acetate (pH 4~5.5), phosphate (pH 6~8) and glycine-NaOH-NaCl (pH 8.5~11). Symbols express that CGTase activity (●) and dextrinizing activity (○).

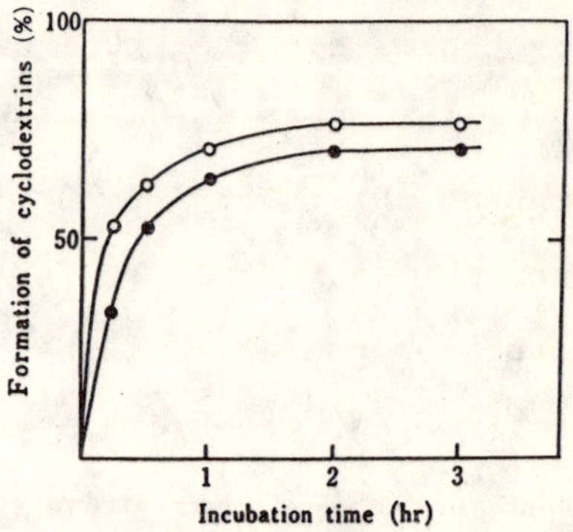

FIG. 2. Formation of Cyclodextrins from Potato Starch by the CGTase

Table 2. Yields of Cyclodextrins

Enzymes	Yield (%) α	β	γ
Bacillus No. 38-2	4	54	16
Bacillus No. 17-1	0	45	15
Bacillus No. 13	1	44	13
Bacilus macerans	12	11	4

Effects of pH on the enzyme action and stability. The CGTase activity was assayed by the following method. The reaction mixture containing 0.01 ml of the enzyme suitably diluted with distilled water, 5 mM α-cyclodextrin and 25 mM sucrose in 0.3 ml of 50 mM various buffer systems was incubated at 40 C for 10 min. Aliquot (0.1 ml) of the reaction mixture was incubated with 0.2 ml of 5 mg/ml of glucoamylase in 0.5 M acetate buffer (pH 5.5) for 5 h. Reducing sugar thus formed was determined by dinitrosalicylic acid method. As a reference, dextrinizing activity was also determined. The results are shown in Fig. 1. The optimum pH of the enzyme action are 4.5-5 and 7.0-9 for the CGTases of Bacillus No. 38-2, and 9.0-9.5 for the enzymes of Bacillus No. 17-1 and Bacillus No. 13. The enzyme was dissolved in various buffer solutions (pH 3-13) , and heated at 55 C and 60 C in the presence or absence of 10 mM $CaCl_2$. The residual activity was measured at pH 10.0. Bacillus No. 38-2 CGTase is the most stable at the range of pH 5-10.5 (Table 1).

Thermal stability. The enzyme dissolved in 0.1 M glycine-NaOH-NaCl buffer (pH 9.0) was heated at various temperatures for 30 min, and the residual activity was measured at pH 9.5. As shown in Table 1, Bacillus No. 38-2 CGTase is the most thermostable and stabilized by the addition of Ca^{2+}.

Formation of cyclodextrins from potato starch. Potato starch solution (1 % w/v) was incubated with the CGTases at pH 8.5 and 4.7 at 40 C and cyclodextrins formed were analysed by

HPLC method described above. Fig. 2 shows formation of cyclodextrins from potato starch by the CGTase of Bacillus No. 38-2. Table 2 summarizes the formation of α-, β- and γ-cyclodextrins by three CGTases at pH 8.5. It is quite striking that the yields of cyclodextrins are very high and that the ratio of α-, β- and γ-cyclodextrin is entirely different from those of Bacillus macerans CGTase. The yield of cyclodextrins from various carbohydrates were also tested. About 85-90 % of amylose, 65-70 % of amylopectin, 55-60 % of glycogen, 45-60 % of amylopectin β-limit dextrin, 20-25 % of maltotriose and 10-15 % of maltose were converted to cyclodextrins by Bacillus No. 38-2 CGTase.

Effect of substrate concentration on cyclodextrin formation. Unless stated otherwise, the following experiments were carried by using the CGTase of Bacillus No. 38-2. The optimum starch concentration represents a compromise of several factors. From theoretical consideration, the best yiels are expected from the most dilute solutions, because the enzyme is a type of glycosyltransferase. However, starch is a cheap raw material and the use of high concentration of starch is significant economic advantage for the production on an industrial scale, especially to minimize the volume of reaction vessel and make easily concentration of the hydrolysates. In these experiments, about 1 to 30 % (w/v) of potato starch was examined for the preparation of cyclodextrins. Potato starch suspension (about 30 % w/v) containing 10 mM $CaCl_2$ was liquefied by the CGTase of _Bacillus_ No. 38-2 (20 unit/g starch) at 85 C and pH 8.5 for 30 min. By this treatment the dextrose equivalent (D.E.) was less than 2 %. During liquefaction process, the D.E. value of liquefied starch must be lower than 2, because higher D.E. value makes lower yield of cyclodextrins as shown in Fig. 3. The liquefied potato starch which varied the concentration was incubated with CGTase (55 unit/g starch) at 65 C and pH 8.5 for 48 h. As shown in Fig. 4, high yield of cyclodextrins was observed with low substrate concentration. The optimum substrate concentration, however, was considered to be about 15 % for the preparation on an industrial scale.

Composition of cyclodextrins during production process. The liqefied starch (15 % w/v) was incubated with the CGTase (55 unit/g starch) at 65 C and pH 8.5 and the composition of cyclodextrins was analyzed by HPLC. The characteristic point is that the content of α-cyclodextrin is very low after 100 h-incubation, although gradually increasing. Total cyclodextrin content reached maximum after 48 h -incubation and kept constant up to 100 h (Fig. 5).

Industrial production of β-cyclodextrin and CH. From the results described above, the CGTase of _Bacillus_ No. 38-2 is the best enzyme to produce cyclodextrins on an industrial scale plant. The completely different and new process was established: One ton of potato starch suspension (15 % w/v) containing 10 mM $CaCl_2$ was liquefied by the crude CGTase of _Bacillus_ No. 38-2 (30 unit/g starch) at 85 C to 90 C and

Table 3. Specifications

	β-cyclodextrin	CH CH-20	 CH-30
Appearance	Crystal	Viscous liquid	
Moisture (%)	10	25	25
Cyclodextrin (%)	98	20	20
α-	0	2.7	2.7
β-	98	7.4	7.3
γ-	0	7.9	7.8
Glucose	0	3.6	11.3
Maltose	0	11.1	20.3
Oligsaccharide	1	65.3	48.4
Sweetness	Slight	25	36
D.E.*	1	20	30

* D.E.=Dextrose equivalent=The ratio of reducing sugars against dry substances

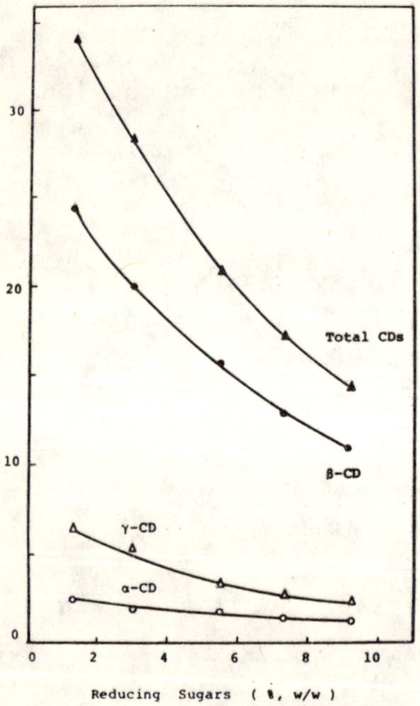

Fig. 3. Effect of Reducing sugars in the liquefite on cyclodextrins formation

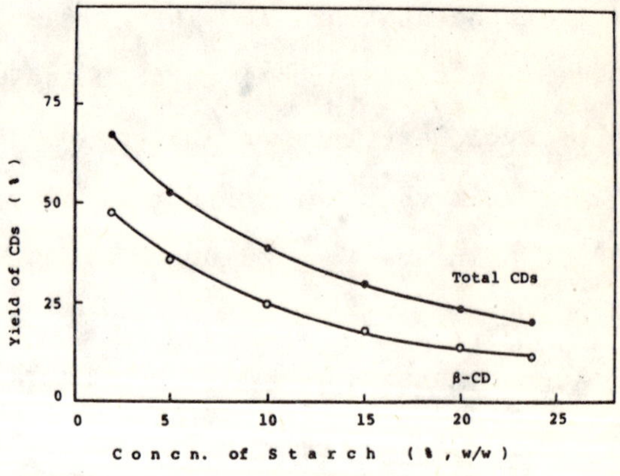

Fig. 4. Effect of substrate concentration on yield of cyclodextrin

pH 8.5 for 30 min with vigorous stirring and then cooled to 60 C. The liquefite was readjusted to pH 8.5 with $CaCO_3$ and the CGTase (55 unit/g starch) was added. The cyclodextrin formation was carried out at 60 C with stirring for further 45 h. After the reaction, the enzyme was inactivated by heating at 80 C. Then the pH of the reaction mixture was brought to 6.0 and appreciable amount of bacterial α-amylase was added to hydrolyse saccharides which were not converted to cyclodextrin. This process was maintained at 80 C for 10 to 12 h. The digest was refined by passing through an active charcol and ion exchangers. The refined digest was concentrated to 45 % (w/v) under reduced pressure at 60 C and transfered to a crystallizer. Beta-cyclodextrin was crystallized by lowering the temperature gradually. The crystalline material was separated by a basket type centrifuge, and washed with a small amount of water. The commercially available β-cyclodextrin (CELDEX N) was thus prepared. This preparation was dissolved with hot water and recrystallized, if necessary. The mother liquor containing cyclodextrins (20 % w/w dry matter base) was refined further by passing through a refining system and concentrated to 75 % (w/w). This product is now commercially available as cyclodextrin for food additives (CH-20 and CH-30), because the CHs can make inclusion compounds. Figure 6 shows new production process of β-cyclodextrin and CHs. Specifications and analytical data our products are shown in Table 3. The characteristic

Fig. 6.
Production Process of β-Cyclodextrin

Starch suspension
↓
Liquefaction
↓
Cooling
↓
Formation of CDs
— CGTase of Alkalophilic *Bacillus* sp.
↓
Inactivation by heating
↓
Degradation of acyclic dextrin
— α-Amylase
↓
Refining
↓
Crystallization

β-CD	Mother liquor
Drying	Refining
	Concentration
Crystalline β-cyclodextrin (CELDEX N)	CELDEX CH-20, CH-30

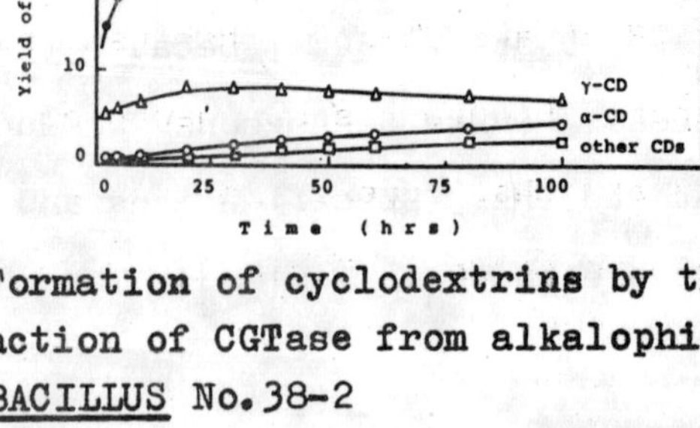

Fig. 5. Formation of cyclodextrins by the action of CGTase from alkalophilic BACILLUS No. 38-2

and most advantageous point is that our process does not use any organic solvent such as trichloroethylene, toluene, ethanol tec. The Ministry of Health and Public Welfare in Japan has permitted the use of our products in food.

Production of γ-cyclodextrin without using organic solvent.

About 8 % (w/w) of the CH is γ-cyclodextrin as shown in Table 3. Recently, we found Toyo Pearl HW-40 could separate γ-cyclodextrin from other cyclodextrins and glucose as shown in Fig. 7. By using this column, industrial perparation of γ-cyclodextrin was developed. The CH suitably diluted was treated with glucoamylase to hydrolyse acyclic dextrins at 55 C for 40 h (pH 5.5), then the enzyme added was denatured by heating at 85 C. The mixture of cyclodextrins and glucose was passed through a refining system, and concentrated to 70 to 75 % (w/w). The concentrate was passed through a Mitsubishi Diaion FRK-01 column followed by Toyo Pearl HW-40 gel filtration to isolate γ-cyclodextrin. The fraction of γ-cyclodextrin was collected, concentrated to 45-50 % (w/w), passed through a refining system and γ-cyclodextrin was crystallized directly from water. Figure 8 shows the flow sheet for the production of γ-cyclodextrin from CH. About 14.1 kg of pure γ-cyclodextrin were produced from 1 ton of CH by our process and the purity was higher than 98.5 %.

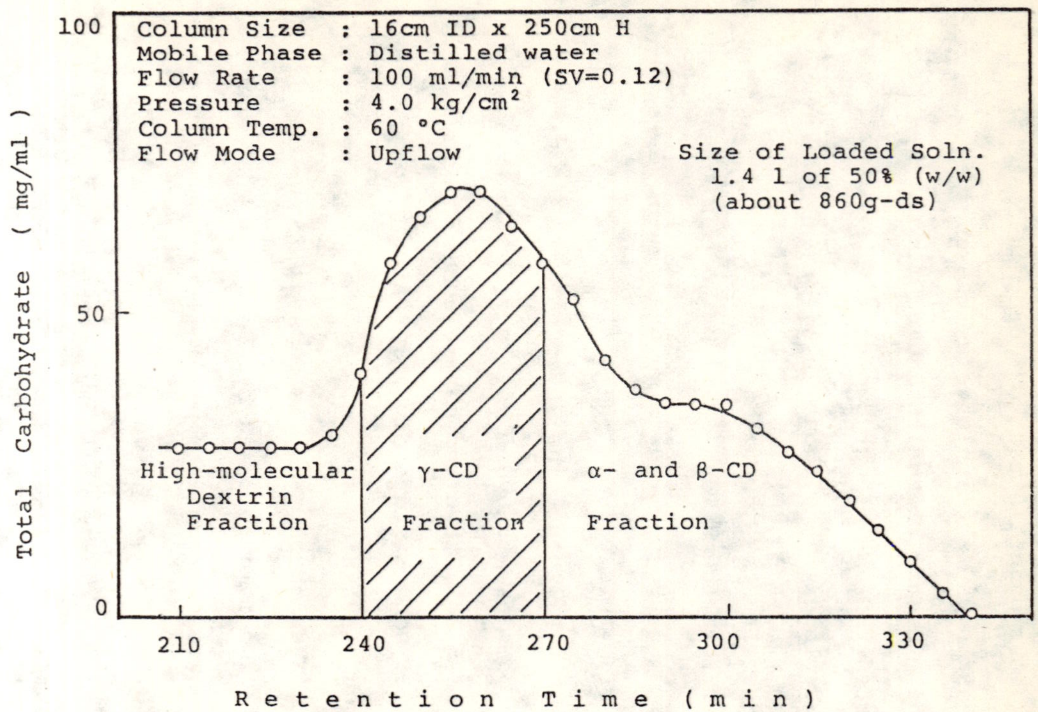

Fig. 7. ELUTION PROFILE of CYCLODEXTRIN MIXTURE on a COLUMN of TOYO PEARL HW-40

Fig. 8.

Preparation of Crystalline γ-Cyclodextrin

Celdex CH

Glucoamylase

DIAION FRK-01 ion exchanger column

CD fraction | Glucose fraction

Crystallization

Mother liquor | β-CD

Refining

CD mixture

Gel filtration with TOYO PEARL HW-40

High-molecular dextrin fraction | γ-CD fraction | α- and β-CD fraction

Refining

Crystallization

γ-CD | Filtrate

Drying

Crystalline γ-CD

Literature Cited

1. Horikoshi, K.: Agri.Biol.Chem. 35 /1971/
2. Horikoshi, K. and Akiba, T.: Alkalophilic Microorganisms, in press.
3. Nakamura, N. and Horikoshi, K.: Agri.Biol.Chem. 40 1647 /1976/
4. Yamamoto, M. and Horikoshi, K.: Stärke 33 244 /1981/
5. Vesterberg, O. and Svensson, H.: Acta Chem.Scand. 20 830 /1966/
6. Neville, Jr., D.M.: J.Biol.Chem. 246 6328 /1971/

I.Int.Symp. on Cyclodextrins 1.4.
Budapest, 1981

PROCESS DEVELOPMENT FOR THE PRODUCTION OF α-CYCLODEXTRIN

E. FLASCHEL, J.-P. LANDERT, A. RENKEN

Institute of Chemical Engineering, Swiss Federal Institute of Technology
CH-1015 Lausanne, Switzerland

SUMMARY

A process has been developed to utilize the highly active cyclodextrin-glycosyltransferase (CGT) from *Klebsiella pneumoniae* M5 al for α-cyclodextrin (α-CD) production. To maximize the α-CD yield, decanol has been used to precipitate α-CD during the reaction. Thus, yields of 50-60% are obtained. Since the crystalline complex is easily separated and decanol can be stripped off efficiently by steam distillation, this process seems to be economically feasible. The α-CD produced can be used in the food industry.

INTRODUCTION

Cyclodextrins are gaining industrial interest owing to their particular structure. These cyclic amyloses are commonly composed of 6, 7 or 8 (named α-, β-, γ-CD) glucose units. They undergo complexation with quite a large number of substances by molecular inclusion. Their potential for industrial application is substantiated by their ability to protect fragile substances from oxygen and UV, to delay the evolution of volatiles and to aid the ancillary processing of dangerous or volatile substances in powdered form.

Though α-, β- and γ-CD exhibit different inclusion specificity, β-CD has found more attention than the others. This is due to the fact that most of the cyclodextrin producing enzymes known produce predominantly β-CD. Furthermore β-CD is by far the least soluble form, leading to a relatively simple production and purification. Consequently β-CD is already in production on an industrial scale [1, 2] while α- and β-CD are still more or less regarded as specialities.

A further CD-producing enzyme has been found in the medium of *K. pneumoniae* [3], Since this CGT appeared to produce predominantly α-CD [4, 5], it has been chosen to develop a process for the production of α-CD.

MATERIALS AND METHODS

Analysis of cyclodextrins: HPLC on Lichrosorb-NH_2. 10 μm, with acetonitrile water 65/35 per volume.

Test for CGT activity: as described by Landert elsewhere in this Volume.

Starch: potato starch as obtained from Blattmann, Wädenswil, Switzerland.

CGT: concentrated preparation from *K. pneumoniae* M 5 al according to Bender [3].

Gelification of starch: heating of starch solution to 90°C.

Reaction media: normally without buffer, $CaCl_2$ (5 mmol/l).

CHARACTERIZATION OF CGT

For enzymatic reactions, the optimal conditions are in the first approximation provided by the pH and temperature activity and stability profiles. The pH-profiles are shown in Fig. 1. Though the CGT is stable over a wide pH range (6 to 10), the activity profile is rather narrow and optimal at pH 6 to 7. The operational pH is therefore fixed to 6.8.

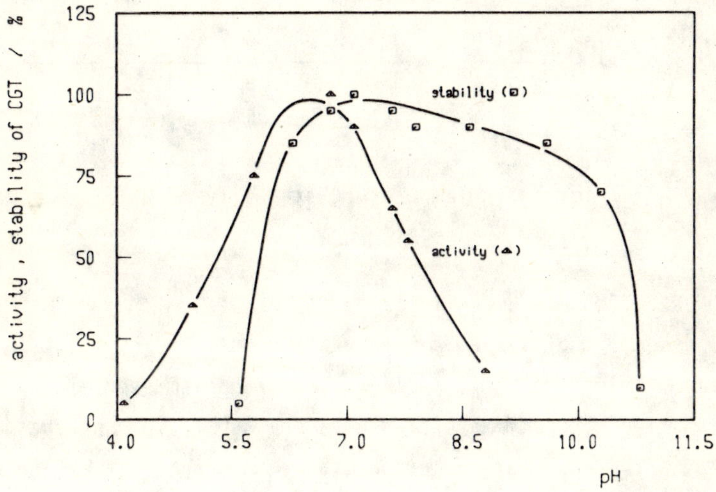

Fig. 1 pH activity and stability profiles of CGT

Figure 2 illustrates the temperature stability of CGT. Up to 40°C, it is sufficiently stable for use in the batch mode of operation, so this temperature has been chosen as standard.

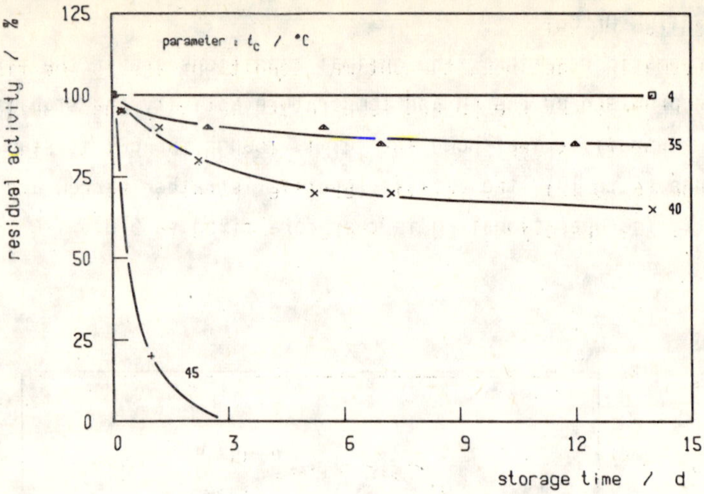

Fig. 2 Temperature stability of CGT

CYCLISATION REACTION and SELECTIVITY of CGT

It is assumed that all CGTs catalyze not only the cyclisation of amylose but also their disproportionation. The mechanism seems to be quite the same: cyclisation can be interpreted as a one-substrate- and the disproportionation as a two-substrate-reaction. There is also great evidence for a hydrolytic activity. Since the reverse reactions are also catalyzed, the reaction media are normally such multi-component systems that they cannot be analyzed and described completely.

Figure 3 shows the CD-production in batch reaction as a function of the normalized reaction time ($E_0 . t$). Since the product concentration is a distinct function of this combined variable, it can be concluded that the CGT acts with a steady state kinetics. The more astonishing fact is that the formation of α-CD (CD_6) is kinetically favoured compared to that of β-CD (CD_7) by orders of magnitude. The yields for longer reaction times are shown in Fig. 4. α-CD reaches rapidly a maximum concentration and declines. β-CD

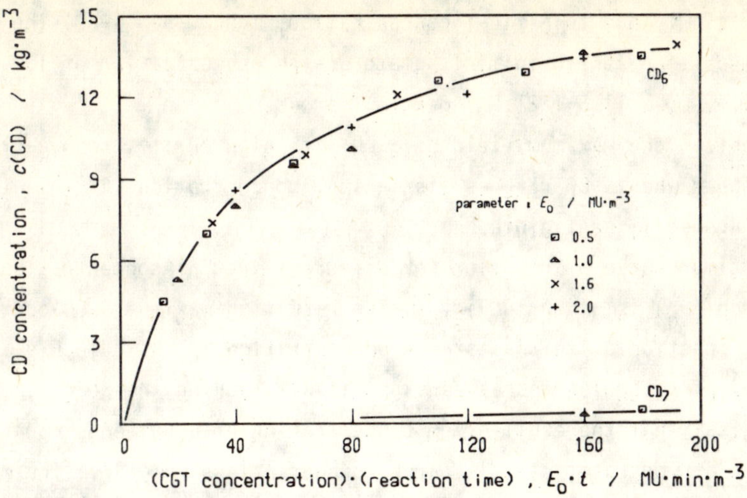

Fig. 3 Cyclodextrin concentration as a function of normalized reaction time

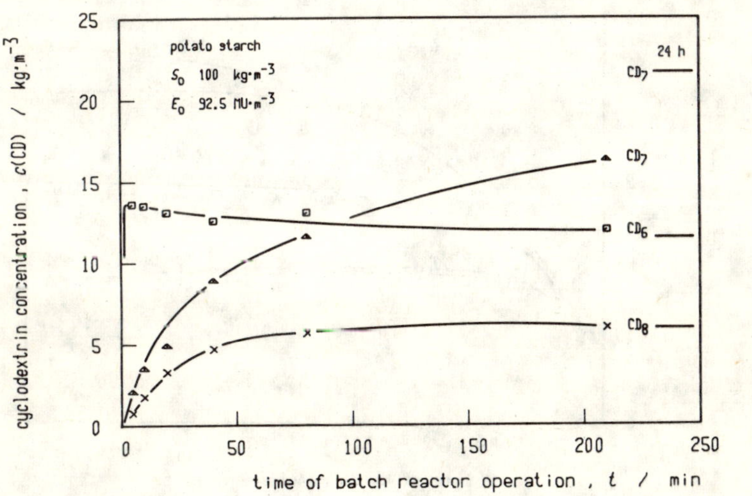

Fig. 4 Product distribution near equilibrium

is formed slowly but to a final concentration higher than α-CD. The equilibrium composition for an initial potato starch concentration of 100 kg·m^{-3} is α-: β-: γ-CD = 11.5 : 21.5 : 6 kg·m^{-3} at 40°C.

To obtain the maximum yield of α-CD with high purity, the reaction has to be stopped when α-CD reaches its maximum concentration. At this time, β- and γ-CD are still negligible.

The time course of reaction for different initial concentrations of potato starch (S_0) is given in Fig. 5. The resulting maximum α-CD concentration as a function of initial starch concentration is shown in Fig. 6. Since the number of molecules increases significantly during the reaction, it is quite logical that the equilibrium concentration of α-CD tends to a constant value even at moderate initial starch concentrations (Fig. 6). The maximum achievable α-CD yield is limited for potato starch to 13.5 kg·m^{-3}. This yield is too small to be useful for production on an industrial scale.

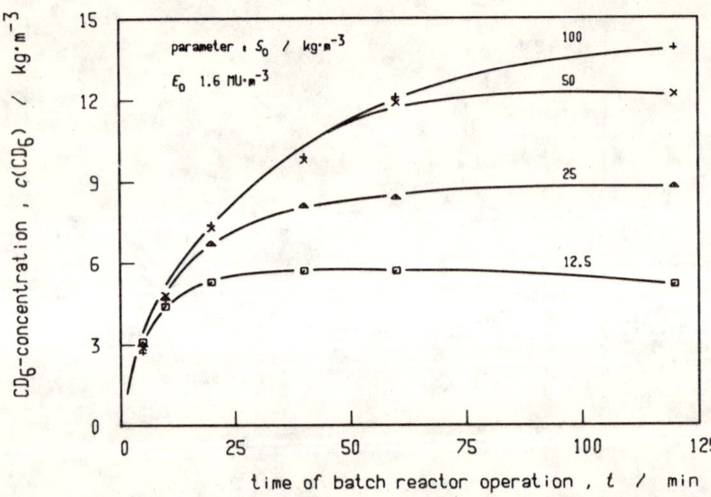

Fig. 5 Reaction characteristics for different initial starch concentrations.

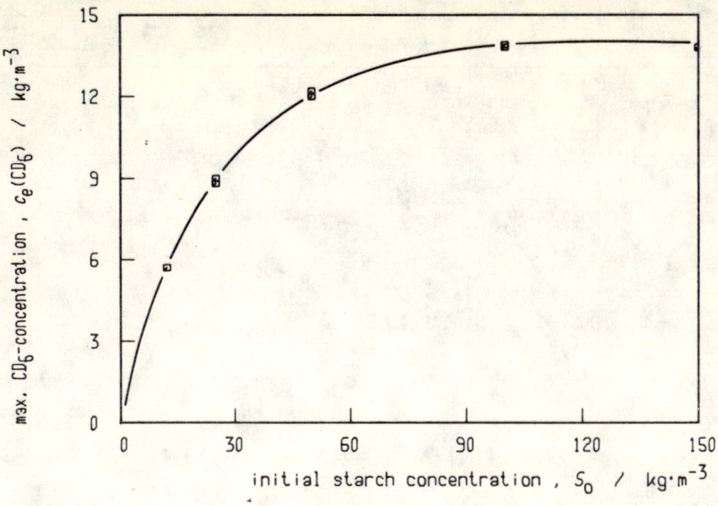

Fig. 6 Maximum α-CD yield as a function of initial starch concentration

DECANOL PROCESS

The relatively low equilibrium concentration of α-CD requires means by which the α-CD yield can be increased. The simplest idea is to utilize the natural ability of CDs to form complexes with hydrophobic substances. If such a complex is sparely soluble, the equilibrium of cyclisation could be guided to higher CD-yields, since only the soluble CDs are relevant for the reaction equilibrium.

Such a complexing agent should be non-toxic, FDA-approved and compatible with CGT. An agent with these characteristics is decanol, which has been described to undergo complexation preferentially with α-CD [6]. α-CD in saturated decanol solution is only soluble to 6.5 kg·m^{-3} at 40°C. This should be the main parameter, limiting the overall α-CD yield. The influence of decanol on the CGT reaction is illustrated in Fig. 7. The α-CD yield and the reaction rate increases with increasing decanol concentration. Unter the specified conditions, a decanol concentration of 5 kg·m^{-3} is sufficient for

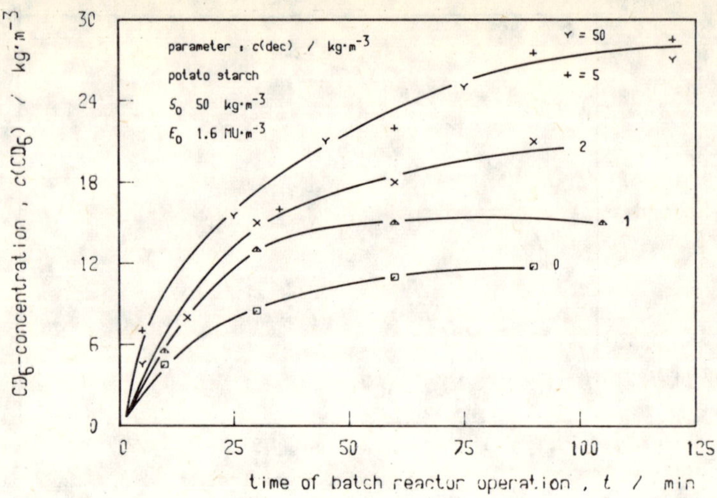

Fig. 7 Influence of the decanol concentration on the α-CD yield

maximum yield, indicating that the formed complex might have a 1:1 composition. With potato starch, an α-CD yield of 60% is obtained. Since the reaction rate is enhanced by decanol, the occurence of β-CD and γ-CD can be totally neglected. On a technical point of view, it is only possible to work with potato starch up to a concentration of 50 $kg \cdot m^{-3}$. It would be more favourable to operate at higher initial starch concentrations. Therefore starch with a lower mean molar mass will be employed.

Since the decanol-α-CD complex can be filtered off after reaction, the purification steps are quite simple. Decanol is stripped off by steam distillation under pressure. This leads to residual decanol in the crystalline α-CD well below 5 ppm, the concentration which is admissible for food industrial applications.

ACKNOWLEDGEMENT

The aid of Dr. H. Bender, University of Freiburg, W.-Germany, is greatly appreciated. By virtue of his valuable help, a considerable amount of CGT is at our disposal. Our thanks are also due to Dr. M.-R. Kula, GBF, Braunschweig - Stöckheim, W.- Germany, who kindly provided us with an initial quantity of CGT and particularly for her advice to enter this field of research.

SYMBOLS and UNITS

c	$kg \cdot m^{-3}$	concentration
c_e	$kg \cdot m^{-3}$	pseudoequilibrium concentration
E_0	$MU \cdot m^{-3}$	initial enzyme concentration
S_0	$kg \cdot m^{-3}$	initial substrate (starch) concentration
t	min	time

LITERATURE CITED

1. Matzuzawa, M., Kawano, M., Nakamura, N., and Horikoshi, K., Die Stärke, 27(12), 410-431, 1975
2. Horikoshi, K., Proc. Biochem., May 1979, p. 26-30
3. Bender, H.,Arch. Microbiol. 111, 271-282, 1977
4. Bender, H., Carbohydrate Res., 78, 133-145, 1980
5. Bender, H., Carbohydrate Res., 78, 147-162, 1980
6. Armbruster, F.C., U.S. Patent 3,640,847, Feb. 8, 1972

I.Int.Symp. on Cyclodextrins
Budapest, 1981 1.5.

THE PREPARATION AND SOME PROPERTIES OF GLUCOSYL-CYCLODEXTRINS

SHOICHI KOBAYASHI,[1] KEIJI KAINUMA,[1] DEXTER FRENCH[2]

[1] National Food Research Institute
Yatabe, Tsukuba, Ibaraki 305, Japan
[2] Iowa State University, Department of Biochemistry
Ames, Iowa, 50011, U.S.A.

SUMMARY

Glucosyl-cyclodextrins were prepared by the action of *Bacillus macerans* cyclomaltodextrin glucanotransferase (BME)on branched dextrin of waxy corn starch prepared by removing the cyclodextrins with bromobenzene and tetrachloroethane from the BME-starch digest.

Branched cyclodextrins were effectively produced by the action of BME on branched dextrin in the presence of SDS and were further degraded to glucosyl-cyclodextrins by the combined action of *Aspergillus oryzae* α-amylase and glucoamylase. These enzymes also degrade β and γ -cyclodextrins, which interfere with the separation of glucosyl-cyclodextrins by paper-chromatography, to glucose.

Glucosyl-cyclodextrins were separated by paperchromatography. Glucosyl-α-cyclodextrin was crystallized from water and made into a crystal complex with iodine. Glucosyl-cyclodextrins have extremely high solubility compared to the original cyclodextrins, solubilyze various kinds of oily substances in them and are extraordinary resistant to the action of *Aspergillus oryzae* α-amylase.

INTRODUCTION

Cyclodextrins (CDs) are produced by the action of cyclomaltodextrin glucanotransferase on starch (1, 2). The main cyclic products are α-CD, β-CD and γ-CD. In addition to these purely 1,4-linked cyclic molecules, various branched CDs have been reported (3, 4, 5). Among these branched CDs, glucosylcyclodextrins (G_1-CDs) have been shown to be useful for the diagnosis of glycogen storage disease type III (4, 6). However, the yields of G_1-CDs have been so low that it has not been practical to prepare them in a pure state in sufficient quantity and their properties have not been investigated. It is supposed that the properties are different from those of original CDs, because of the presence of branches attached by α-1, 6 linkages (Fig. 1). Branched CDs could possibly be used in the pharmaceutical, food and cosmetic industries.

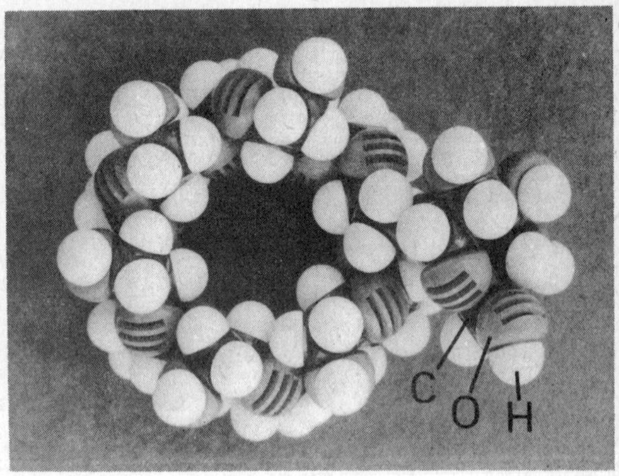

Fig.1 The molecular model of G_1-α-CD
C: Carbon atom, O: Oxygen atom
H: Hydrogen atom

In our previous report (2) it was shown that the addition of sodium dodecyl sulfate (SDS) during the enzymolysis greatly enhanced the formation of α-CD and repressed the formation of β-CD and γ-CD. With branched or chemically modified substrate, SDS enhanced the formation of the corresponding branched or modified α-CD. Action of glucoamylase on the mixed CDs (mixture of α, β and branched CDs) converted the various branched CDs present into corresponding

G_1-CDs. However, purification of G_1-CDs from such a mixture was complicated because this preparation contained significant ammounts of α-CD and β-CD, which interfered with the chromatographic separation.

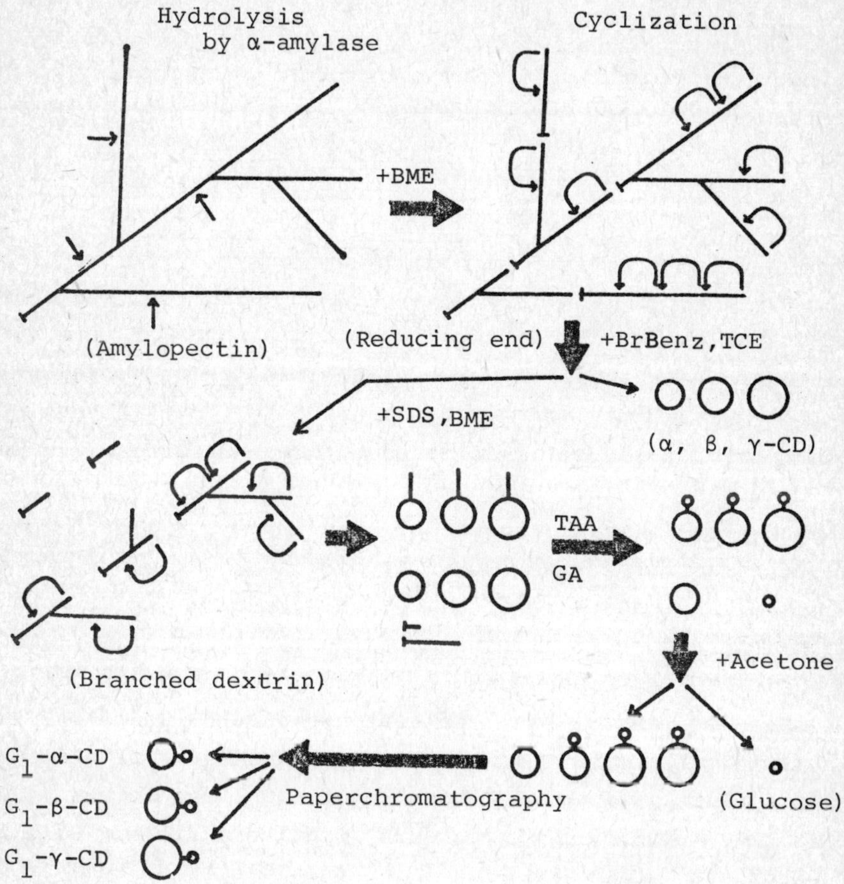

Fig.2 Schematic explanation of the preparation of G_1-CDs (Refer to EXPERIMENTAL in the text for details)

The present paper reports a method for the large scale preparation of G_1-CDs using a highly branched dextrin as the substrate for BME. The various glycosyl-CDs present were converted to G_1-CDs by the glucoamylase action, and the β-CD was degraded by the joint action of *Aspergillus oryzae* α-amylase and glucoamylase. Chromatographic techniques enabled us to obtain the G_1-CDs in a pure or crystalline state. The principle of this method is shown schematically in Fig. 2.

EXPERIMENTAL

Materials. BME was prepared and partially purified through the starch adsorption step as previously described (8). The lyophilized enzyme had an activity of 15 THU (Tilden and Hudson units (1)) per mg. Bacterial liquefying α-amylase, a crude preparation (BLA) was kindly donated by Hankyu Kyoei Bussan Co., Ltd., Osaka, Japan. Glucozyme, a crude preparation of glucoamylase (GA) was kindly donated by Nagase Sangyo Co., Ltd., Osaka, Japan. *Aspergillus oryzae* α-amylase, a crude preparation (TAA) was kindly donated by Sankyo Co., Ltd., Tokyo, Japan. Waxy corn starch was kindly donated by Nihon Shokuhin Kako Co., Ltd., Shizuoka, Japan. Glucoamylase, pure grade (36 IU/mg) was purchased from Seikagaku Kogyo Co., Ltd., Tokyo, Japan. SDS, 1,1,2,2-tetrachloroethane (TCE), bromobenzene (Br Benz) and the other chemicals used were of analytical grade.

HPLC (High pressure liquid chromatography). The chromatograph used was TRIROTOR, Nihon Bunko Co., Ltd., Tokyo, Japan and the conditions were as follows: solvent, 60 and 65% acetonitrile; flow rate, 2 mL/min; attenuation, 8X; detection, RI; and column, precolumn (4.6 mm ID X 5 cm long), analysis column (4.6 mm ID X 25 cm long) were prepared by packing a filler, Fine SIL NH_2 (purchased from Nihon Bunko) in stainless steel tubes. Injected samples (50 μL) conained 0.5-1.5 mg of carbohydrate in water.

Glucoamylase treatment of samples for analysing CDs. 400 μL of a sample (total carbohydrate, 5-10%) was mixed with 200 μL of water and 200 μL of pure glucoamylase solution containing 100 IU of the enzyme activity in 1 mL of 0.1 M acetate buffer (pH 5.20), and the mixture was incubated at 50°C for 1 hr to degrade acyclic oligosaccharides and branches of branched CDs which interfere in the HPLC analysis of CDs. The reaction mixture was heated in a boiling water bath for 5 min and centrifuged at 3500 rpm for 10 min. 50 µl of the clear supernatant was used for HPLC analysis.

Paper chromatography. Sample solutions were spotted on Toyoroshi No. 50 filter paper, 20 cm high and 20 or 25 cm wide, and the spotted paper was irrigated three times with 1-propanol-1-butanol-water(5:3:4) at 70°C, each irrigation requiring about 4 hr. CDs were revealed by dipping the chromatogram in a freshly prepared solution of 0.5% iodine in 90% acetone, and drying in a draft. Spots of appropriate intensities are obtained with 25 µg of α-CD, 100 µg of β-CD, 200 µg of γ-CD, and 100 to 400 µg of branched CD. Preparative paper chromatography was carried out by streaking the sample on a paper of size 35 X 35 cm, irrigating three times (each irrigation requiring 15 hr) and developing as described. The individual colored bands were cut out and eluted with water. By this procedure, it was possible to purify 60 mg of G_1-α and G_1-γ-CD on a single sheet of paper.

Preparation of G_1-CDs. Waxy corn starch (400 g) was suspended in 2 L of water, BLA (500 mg) was added, liquefied in a boiling water bath and autoclaved at 120°C for 15 min. The starch-iodine reaction of the solution showed a purple colour. The solution was added to 20 mL of 1 M acetate buffer (pH 6.0) and the mixture was incubated with 270 mg of BME at 40°C for 24 hr with stirring. The reaction mixture was mixed with 30 mL of TCE and 30 mL of Brbenz at 10°C and agitated overnight followed by centrifugation at 8000 X G for 10 min. The supernatant was allowed to stand in a cold room (4°C) for 2 days and was centrifuged again to remove the lees appearing during storage. The supernatant was mixed with 12 g of SDS and 800 mg of BME and incubated at 40°C for 48 hr with stirring followed by autoclaving at 120°C for 15 min. The reaction mixture was stood in a cold room overnight and centrifuged at 8000 X G for 10 min to remove the precipitated SDS.

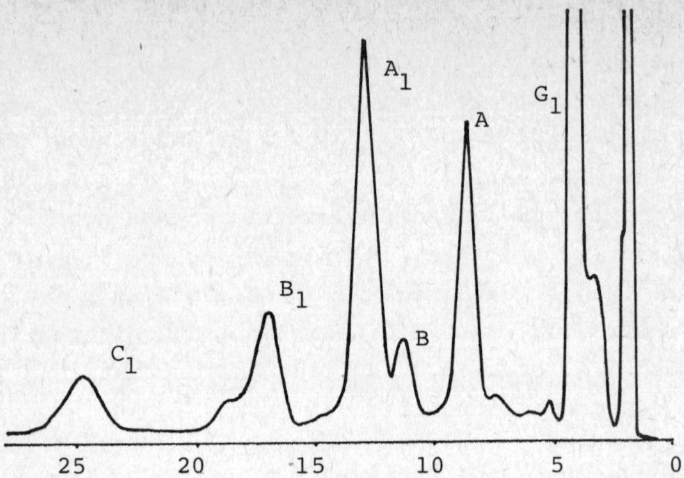

Fig.3 HPLC profile of branched dextrin-SDS-BME digest(after glucoamylase treatment)

G_1: Glucose, A: α-CD, B: β-CD

A_1: G_1-α-CD, B_1: G_1-β-CD, C_1: G_1-γ-CD

The supernatant contained 26% G_1-α-CD, 11.5% G_1-β-CD, 6.6% G_1-γ-CD, 13% α-CD, 3.5% β-CD and 39.4% glucose after glucoamylase treatment as shown in Fig. 3. The supernatant was mixed with 0.5 g of TAA and 2.5 g of GA and digested at 40°C for 24 hr with rotary shaking. The treatment was extremely effective for the elimination of β-CD which interferes with the separation of G_1-α-CD by paperchromatography. The separation of α, β, γ, -CDs and G_1-α, β, γ-CDs is shown in Fig. 4. From this figure it is seen that the Rfs of β-CD and G_1-α-CD are almost the same.

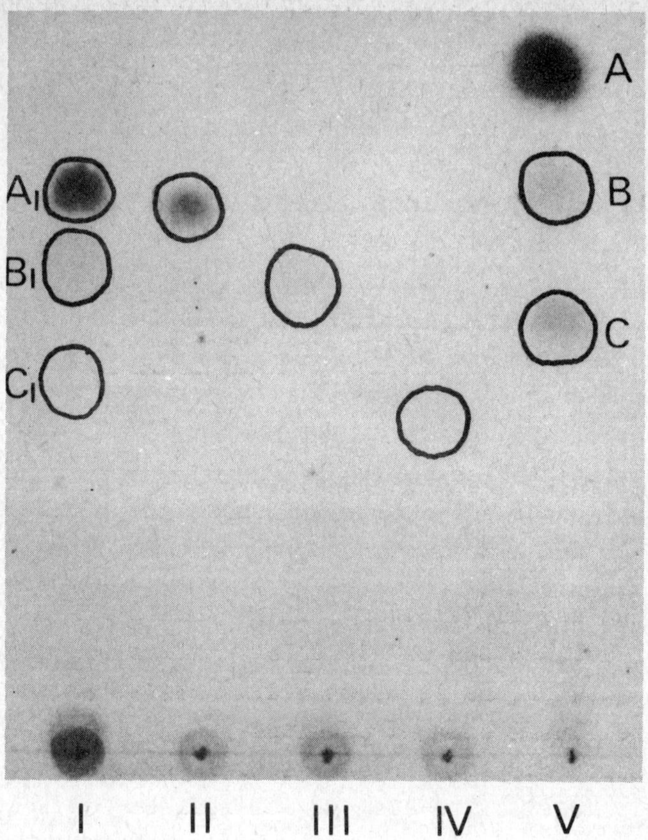

Fig.4 The paperchromatographic separation of various CDs

I: Mixture of G_1-CDs

II: G_1-α-CD(purple spot)

III: G_1-β-CD(yellow spot)

IV: G_1-γ-CD(yellow spot)

V: Mixture of α(A_1, blue), β(B_1, yellow) and γ-CD(C_1, brown spot)

To remove the residual SDS, the digest containing 25.8% G_1-α-CD, 10.6% G_1-β-CD, 6.6% G_1-γ-CD, 11.7% α-CD and 45.3% glucose, was heated in a boiling water bath for 15 min and then applied onto a mixed bed resin column (5.6 X 62 cm, Amberlite IRA-411 700 mL and IR-120 350 mL). With this procedure 40 g of SDS was removed. The eluate, concentrated to 250 mL, was added to 750 mL of acetone to precipitate CDs. The precipitate, in which a small amount of glucose was present, was collected by standing in a cold room overnight followed by decantation. The yield of the precipitate (G_1-CDs mixture) was 73 g (18.3% of the starting starch substrate).

The precipitate was dissolved in water to obtain a 10% concentraion, and the solution was streaked on a sheet of chromatography paper and chromatographed to separate the G_1-α-CD. G_1-β-CD was also prepared by paperchromatography, but this preparation showed two peaks by HPLC analysis. One of the peaks may correspond to the single branched glucosyl-β-CD (G_1-β-CD) and the other one possibly to the doubly branched glucosyl-β-CD ($2G_1$-β-CD). G_1-γ-CD was prepared in a pure state showing one peak with HPLC analysis.

The properties of G_1-CDs. 50 mg of G_1-α-CD was dissolved in 50 μL of water in a small V type vial. One drop of the solution was placed on a hole slide glass, covered with cover glass and stood for 2-3 days for crystallization. The crystals observed under polarized light is shown in Fig. 5. The crystals of G_1-α-CD-iodine complex was also formed. It was exceedingly difficult to crystallize G_1-β-CD, and rod shaped crystals have been formed rarely. The reason for this maybe because the preparation was a mixture of G_1-β-CDs. G_1-γ-CD has not been crystallized so far.

The susceptibility of G_1-CDs to *Asp. oryzae* α-amylase is considerably lower than that of original CDs (Fig. 6). The rate of degradation of G_1-CDs is about one tenth that of the original CDs. Details of these results will be reported elsewhere.

Fig.5 Crystals of G_1-α-CD from water, crossed polarizers. (X35)

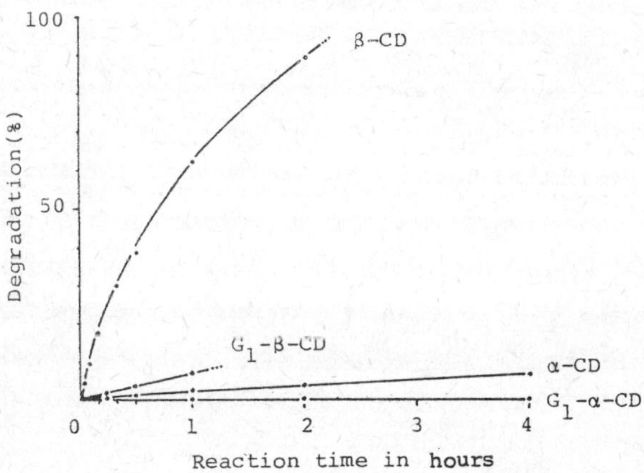

Fig.6 Degradation of various CDs by Aspergillus oryzae α-amylase

Reaction mixture: Substrate 1%, enzyme (Crystalline 3X, from Sankyo Co., Ltd., 120IU/mg) 600IU/g of substrate, 10mM acetate buffer containing 1mM $CaCl_2$,pH5.5

Determination of degradation was by HPLC analysis

The solubility of G_1-CDs was about ten times as high as that of original CDs, and powdered G_1-α-CD was instantly soluble in water. G_1-CDs solubilized various kinds of oily materials in them. (Table 1). 20 μL of each of the oily samples was added to 10 mL of 10^{-2}M G_1-α-CD solution and shaken to observe the formation of precipitation. As shown in the table, G_1-α-CD solubilizes tetrachloroethane, decanol, lemon oil, orange oil, which form precipitates with α-CD. Bromobenzene which does not form complex with α-CD in also solubilized by G_1-α-CD.

Table 1 Solubilization of oily compounds by G_1-α-CD

	α-CD	G_1-α-CD
Tetrachloroethane	+++	-
1-Decanol	+++	±
Lemon oil	+	-
Orange oil	+	-
Fluorobenzene	-	-
Cyclohexane	-	-
Bromobenzene	*	-

-~+++: Degree of precipitate formation
*: no complex formation

LITERATURE CITED

1. French, D.: Adv. Carbohydr. Chem. 12 189, (1957)
2. Thoma, J.A. and Stewart, L., in Whistler, R.L. and Paschall, E.F. (Eds.): Starch, Chemistry and Technology, Vol. 1, Academic Press Inc., New York and London, 1965; pp. 209-249
3. French, D., Pulley, A.O., Effenberger, J.A., Rougvie, M.A. and Abdullah, M.: Arch. Biochem. Biophys. 111 153,(1965)
4. Taylor, P.M. and Whelan, W.J.: Arch Biochem. Biophys. 113 500, (1966)
5. Stark, J.R.: Biochem. J. 102 27, (1967)
6. Taylor, P.M. and Whelan, W.J.: Proceedings of 4th Meeting of the Fed. European Biochem. Soc. p.101, (1967)
7. Kobayashi, S., Kainuma, K. and Suzuki, S.: Bull. Agric. Chem. Soc. Jpn. 51 691, (1977)
8. Kobayashi, S., Kainuma, K. and Suzuki, S.: Carbohydr. Res. 61 229, (1978)

SEPARATION OF CYCLODEXTRINS WITH GELCHROMATOGRAPHY AND HPLC

M. SZILASI,[1] K. OTTA,[1] B. ZSADON,[1] J. SZEJTLI,[2] F. TÜDŐS[1]

[1] Department of Chemical Technology, Eötvös Loránd University
Múzeum krt. 4/B
H-1088 Budapest, Hungary
[2] CHINOIN Pharmaceutical-Chemical Works, Tó u. 1-5
H-1045 Budapest, Hungary

Previously, several chromatographic methods have been described for the separation of cyclodextrins. We elaborated a chromatographic method for analytical and for preparative separation of α-β- and γ-cyclodextrin on dextran gel (Sephadex) columns [1].

However, there was no method for rapid chromatographic analysis of samples containing α-, β- and γ-cyclodextrin. We found high performance liquid chromatography on μ Bondapak--carbohydrate column with acetonitril-water mixtures as eluents to be very suitable for rapid qualitative and quantitative analysis [2].

I. SEPARATION ON DEXTRAN GEL COLUMNS.

The chromatographic behaviour of α-, β- and γ-ciklodextrin on strongly cross-linked dextran gels was found to be very characteristic.
The α-, β- and γ-ciklodextrins were the products of the Chinoin Pharmaceutical and Chemical Works (Budapest, Hungary). The investigations were carried out with crystalline samples of chromatographic purity, identified with chemical and physical methods.

CHROMATOGRAPHIC METHOD AND APPARATUS

Figure 1 illustrates the scheme of the chromatograph used for these investigations. The dextran gels were swelled in distilled water, the samples were introduced in the form of aqueous solution and eluted with distilled water. The columns Pharmacia K 16/100 (1,6 x 100 cm) were used with termostat jackets, flow adaptors and 1 mm i.d. capillary connecter tubing (Pharmacia Fine Chemicals AB, Uppsala, Sweden). One column or more connected in series were used, according to need. In every case the columns were thermostated.

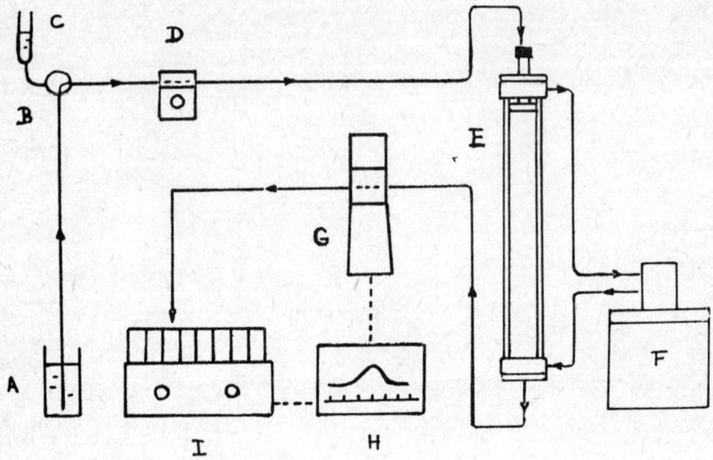

Figure. 1.Scheme of the chromatograph. A Eluent; B distribution valve; C Sample; D Peristaltic pump; E Column with flow adaptor and thermostat jacket; F Thermostat; G polarimeter with flowthrough type sample cell; A Recorder; I Fraction collector.

The other parts of the equipment: Vario Perpex Pump LKB 12000, Ultra Rac Fraction collector LKB 7000 and Flat Bed Recorder LKB 2066 (LKB Producter AB, Stockholm, Sweden).

RESULTS

Figure 2. shows the chromatogramm of a mixture of α-β- and γ-cyclodextrin (α=A, β=B, γ=G) on Sephadex G-15 at room temperature.

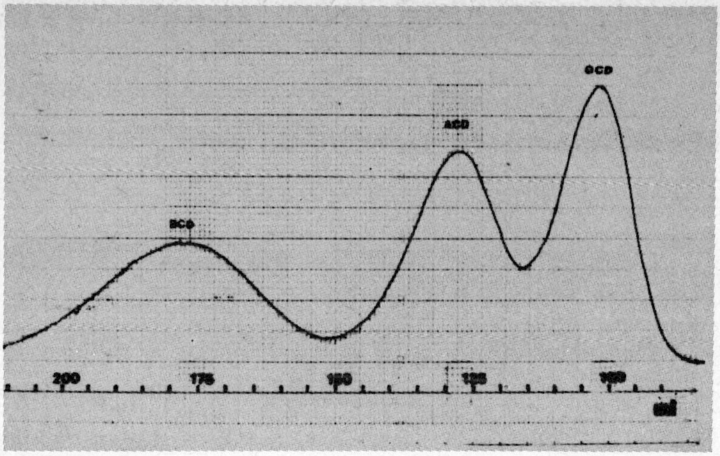

Figure 2. Chromatography of α-CD (20 mg) + β-CD (20 mg) + γ-CD (20 mg) on Sephadex G-15, at 23°C. K 16/100 column, total bed volume V_t = 195 ml, flow velocity 20 ml/h, chart speed 4 cm/h.

On dextran gels of type Sephadex G-25 and G-10 we found an elution sequence identical with that shown on the Fig.2. although the separation of cyclodextrins was not so effective as on the type G-15. On Sephadex G-25 γ CD can be well separated, but the peaks of α- and β-CD do not come apart, as it is apparent from Fig. 3.

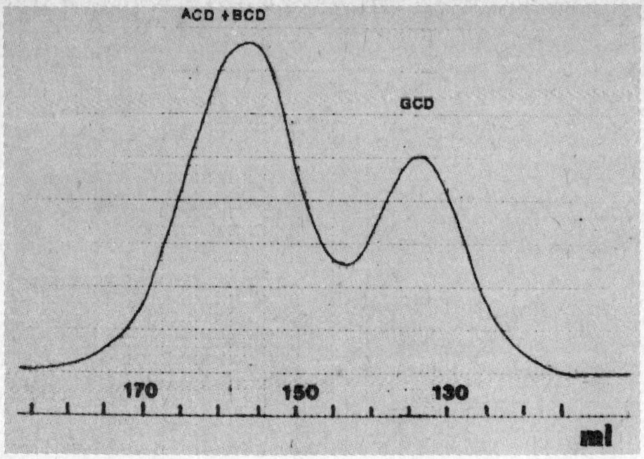

Figure 3. Chromatography of α-CD (20 mg) + β-CD (20 mg) + γ-CD (20 mg) on Sephadex G-25, at 23 °C. K 16/100 column, V_t= 195 ml flow velocity 20 ml/h, chart speed 4 cm/h.

On Sephadex G-10 all the three peaks are practically melting together, as it is apparent from Fig. 4.

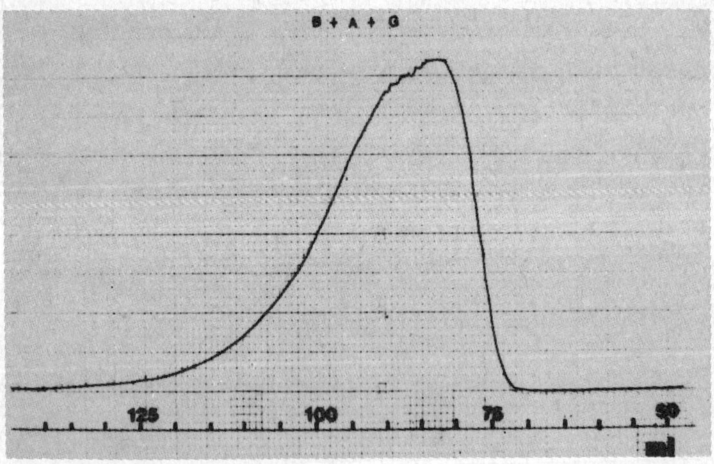

Figure 4. Chromatography of α-CD (20 mg)+ β-CD (2 mg)+ γ-CD (20 mg) on Sephadex G-10, at 23 °C. K 16/100 column, V_t= 195 ml, flow velocity 20 ml/h, chart speed 4 cm/h.

The combinations of gels of Sephadex G-15 and G-25 may be used with good result for the separation of α-, β- and γ, cyclodextrin. In the example showed in Fig. 5 two colums each of 1 m were used, which were filled with Sephadex G-15 and G-25, respectively. With this combined system complete separation of α-, β- and γ-cyclodextrin were possible.

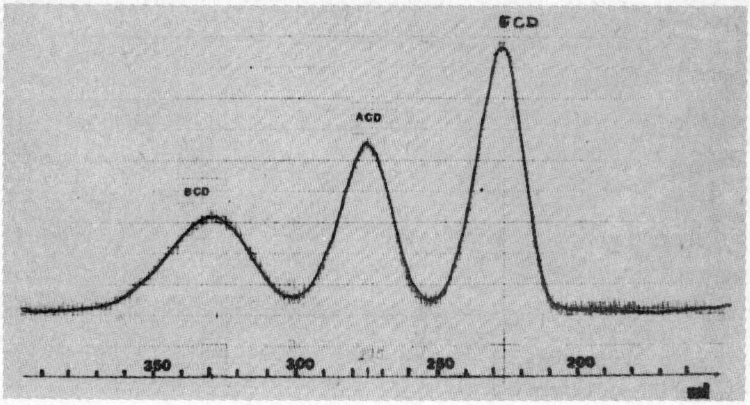

Figure 5. Effective separation of α-CD (20 mg) + β-CD (20 mg)+ γ-CD (2o mg) on a combined system of a Sephadex G-25 and a Sephadex G-15 column (2x K 16/100), at 23°C. Total bed volume 390 ml, flow velocity 20 ml/h, chart speed 2 cm/h.

II. SEPARATION OF CYCLODEXTRINS BY HPLC

We found high-performance liquid chromatography (HPLC) on μ Bondapak-carbohydrate column packing to be very suitable for rapid chromatographic separation of samples containing α-CD, β-CD γ-CD, glucose and malto-olygomers.

CHROMATOGRAPHIC METHOD AND APPARATUS

Liquid chromatography was performed using a Waters Model ALC/GPC 201 chromatograph, with U6K injector a R 400 differential refractometer and a column μ Bondapak-carbohydrate (30 cm x 4 mm I.D; packed with μ silica with an amine functional group) and thermostatted at 25°C. The sample solution contained α-CD, β-CD, γ-CD and glucose or malto-olygomers in distilled water. Acetonitrile-water mixtures were used for the elution. (Acetonitrile was HPLC grade from Fisher Scientific Company, water was twice distilled in glass apparatus)

RESULTS

Fig. 6 illustrating the separation of cyclodextrins (A = α-CD. B = β-CD, G = γ-CD) and glucose (Gl) by HPLC

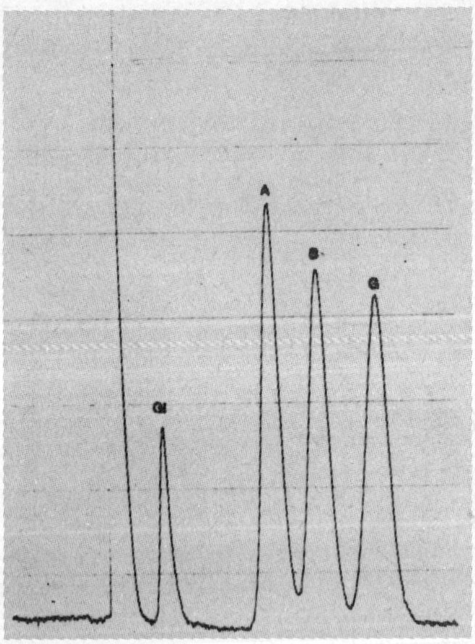

Figure 6. Chromatography of α-CD(A), β-CD(B), γ-CD(G) and glucose (Gl) Column: μ Bondapak/carbohydrate 4 mm x 30 cm, eluent: acetonitrile-water (70:30, v/v), flow-rate: 2ml/min, detector: differential refractometer 4 x.

Fig. 7. Shows the separation of malto-olygomers in the presence of α-cyclodextrin.

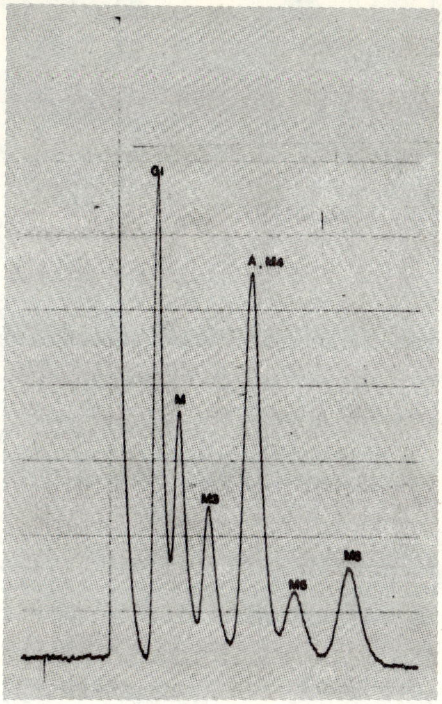

Fig. 7. Separation of malto-olygomers (Gl-glucose, M-maltose, M3-maltotriose, M4-maltotetraose, M5-maltopentaose, M6-maltohexaose) in the presence of α-cyclodextrin (A). Column: μ Bondapak/carbohydrate 4 mm x 30 cm, eluent: acetonitril-water (70.30, v/v), flow-rate: 2 ml/min, detector: differential refractometer 4 x.

We have found that the selectivity can be altered (increased or reduced) by changing the composition of the acetonitril-water eluent. When we used acetonitrile-water 80:20 v/v as eluent, the separation of α cyclodextrin (A) and malto-tetraose (M4) was succesful too. (see 8. Fig.)

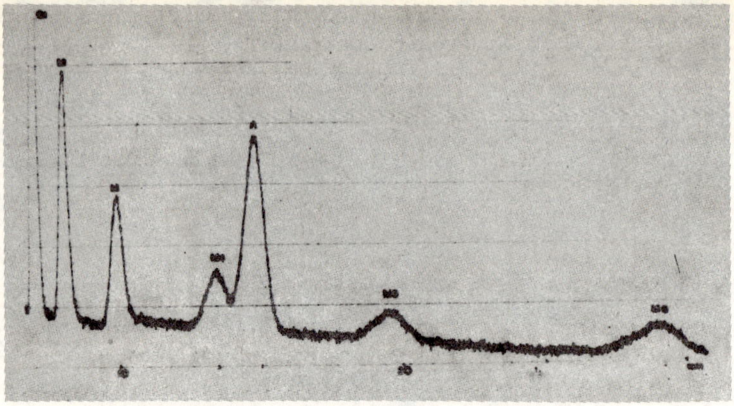

Figure 8. Separation of α-cyclodextrin (A) and maltotetraose (M4) in the presence of other malto-olygomers. Column: μ Bondapak/carbohydrate 4 mm x 30 cm, eluent: acetonitril--water (80:20, v/v), flow rate: 2 ml/min, detector: differential refractometer 4 x.

Both chromatographic methods are suitable for qualitative and quantitative analyses of cyclodextrins in mixtures.

On the bases of our experiments we propose to use the gelchromatography for preparative purposes and the high-performance method for rapid chromatographic analysis.

REFERENCES:

1. B. Zsadon, M. Szilasi, J. Szejtli, G. Seres and F. Tüdős, Stärke, 3o 276 (1978)

2. B. Zsadon, K.H. Otta, F. Tüdős, J. Szejtli, Journal of Chromatography 172, 190 (1979).

RAPID AND SIMPLE SPECTROPHOTOMETRIC METHOD FOR DETERMINATION OF MICRO-AMOUNTS OF CYCLODEXTRINS

M. VIKMON

CHINOIN Pharmaceutical-Chemical Works, Tó u. 1-5
H-1045 Budapest, Hungary

SUMMARY

Cyclodextrins are known to form inclusion complexes with many kinds of organic dye molecules.

Phenolphtalein, the acid-base indicator forms stable colourless inclusion complexes with α-, β- and γ-cyclodextrins in aqueous solutions at pH 10,5. Complex stability constants /K_c/ of the different CD-complexes are:

$$K_{\beta CD} = 2,16 \cdot 10^4; \quad K_{\gamma CD} = 1,92 \cdot 10^3; \quad K_{\alpha CD} = 1,50 \cdot 10^2$$

The molar ratio of phenolphtalein to any of the CDs is 1:1. The concentration range where the decrease of color intensity of phenolphtalein is proportional to CD concentrations was determined:

0 - 50 µg/ml for β CD
0 - 300 µg/ml for γ CD
0 - 3000 µg/ml for α CD

This method is also easily applicable to assort different batches of CD comparing the samples with standards.

Cyclodextrins are known to form inclusion complexes with many kinds or organic dye molecules (1,2,3).

One of the most frequently used acid-base indicator, phenolphtalein forms a very stable colourless inclusion complex with β-cyclodextrin in $4 \cdot 10^{-3}$M Na_2CO_3 solution at pH 10,5. Stability constant /K_C/ was found to be relatively high, $K_{\beta CD} = 2,16 \cdot 10^4\ M^{-1}$ and can be easily measured spectrophotometrically (4). γ- and α-cyclodextrin also form inclusion complexes with phenolphtalein under similar conditions but complex stability constants are very different from the one of β-cyclodextrin complex:

$$K_{\gamma CD} = 1,92 \cdot 10^3\ M^{-1}$$
$$K_{\alpha CD} = 1,50 \cdot 10^2\ M^{-1}$$

It must be mentioned that these stability constants were calculated using the commonly accepted and simple Benesi--Hildebrand method, because the concentrations of the components were not comparable, γ-cyclodextrin and α-cyclodextrin were taken in high excess.

Buvári and Barcza's (4) calculation was based on the generally known relations of coordination chemistry because the concentrations of phenolphtalein and β-cyclodextrin were nearly identical. Their method delivered an indirect proof for the 1:1 ratio of components in the complex. The same 1:1 complex formation was assumed with γ-and α-cyclodextrins too.

In our measurements the concentration of phenolphtalein was $3 \cdot 10^{-5}$ M, concentrations of cyclodextrins were varied between 0-$4 \cdot 10^{-5}$M for β-cyclodextrin, 0-$3 \cdot 10^{-4}$M for γ--cyclodextrin and 0-$4 \cdot 10^{-3}$M for α-cyclodextrin. The pH was supported by $4 \cdot 10^{-3}$M Na_2CO_3.

<u>Experimental</u>

Materials: β-cyclodextrins used were samples of Nihon Shokuhin Kako Ltd. Japan and Chinoin. γ-cyclodextrin standard was obtained from SENN Chem. Ltd., and the samples of CHINOIN Res. Lab. were examined.

α-cyclodextrin used was product of Nihon Shokuhin Kako Ltd. Japan. Other materials used were of reagent grade.

Cyclodextrin stock-solutions:

β-cyclodextrin: $5 \cdot 10^{-4}$M
γ-cyclodextrin: $2{,}5 \cdot 10^{-3}$M
α-cyclodextrin: $2{,}5 \cdot 10^{-2}$M

/Water content of cyclodextrins was taken into account./

Phenolphtalein solution /I/ $3{,}75 \cdot 10^{-3}$M /it can be stored for a long time/ 0,1194 g phenolphtalein was dissolved in 100 ml of 96 v/v % ethanol.

$3{,}75 \cdot 10^{-4}$M phenolphtalein stock-solution /II/ /diluting 10 ml of solution /I/ to 100 ml with water/ must always be prepared freshly, because phenolphtalein easily precipitates from aqueous solutions.

$4 \cdot 10^{-2}$M Na_2CO_3 solution /CO_2 free/.

Analytical procedure

An appropriate volume /0,5, 1,0, 1,5, 2,0, 2,5, 5,0 ml/ of cyclodextrin solution was added to 2,0 ml of phenolphtalein stock-solution /II/ in a 25 ml volumetric flask. After addition of 2,5 ml $4 \cdot 10^{-2}$M Na_2CO_3 solution the volume was brought to 25 ml with freshly distilled water.

Absorbances were measured within 10 minutes at 550 nm with UV-Vis Spectrophotometer in 1 cm cells at 25 ± 2 °C.

Results and discussion

The decrease of colour intensity of phenolphtalein was found to be fairly proportional to the cyclodextrin concentration in the following ranges:

0-50 µg/ml for β-cyclodextrin
0-300 µg/ml for γ-cyclodextrin
0-3000 µg/ml for α-cyclodextrin

Figures 1, 2 and 3 are the calibration curves for determination of cyclodextrins.

E_F is the absorbance of 3. 10^{-5}M phenolphtalein solution at pH 10,5.

$$\Delta E = E_F - E_{DF}$$

E_{DF} is the absorbance in the presence of cyclodextrin.

This method proved to be useful e.g. at studying the permeation of CDs through different membranes.

Through absorbances are approximately linear in the above mentioned concentration ranges ΔE values measured at higher concentrations are also characteristic and can be well reproduced. Indifferent, non-complexing agents /sugars or linear dextrins/ have no influence on the absorbance, therefore ΔE values measured, obviously decreases in the presence of non-complexing impurities of the cyclodextrin samples. When the impurities are other cyclodextrins, the considerable differences in the K_C values assure a ΔE different from that of the standard.

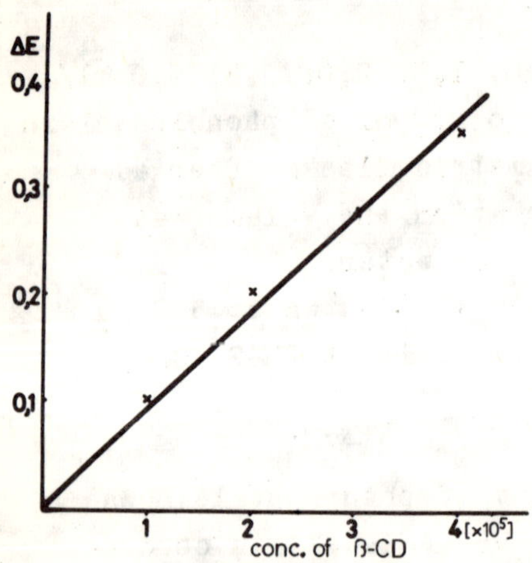

Fig. 1.
β-CD concentration
0-4 . 10^{-5} M

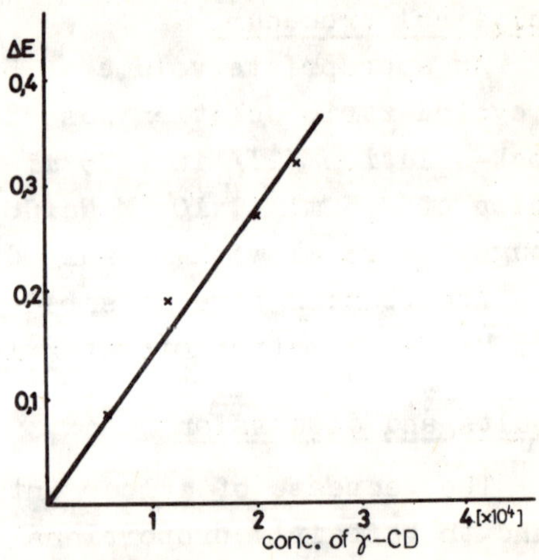

Fig. 2.
γ-CD concentration
0-3 . 10^{-4} M

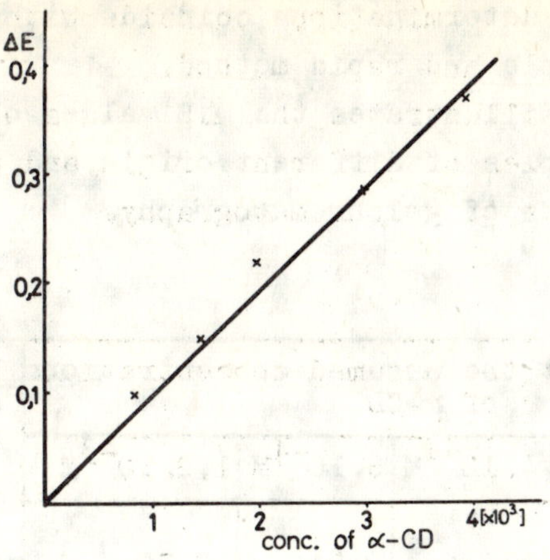

Fig. 3.
α-CD concentration
0-4 . 10^{-3} M

Figure 4. illustrates the method on two different β-CD samples.

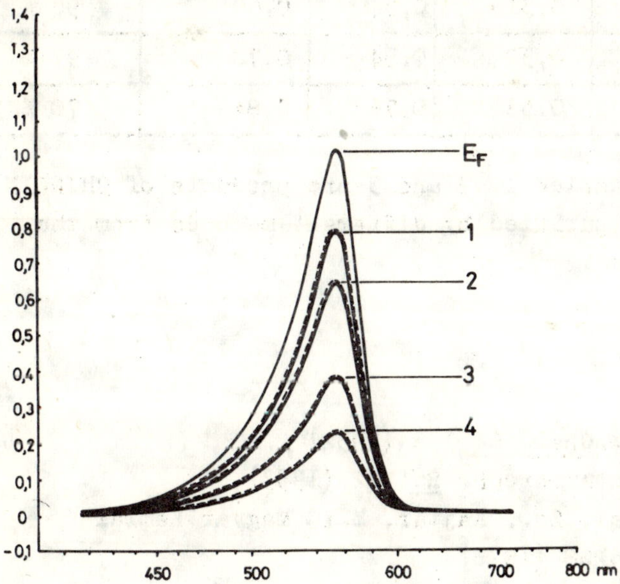

β-cyclodextrin concentrations /M/
1./ 2 . 10^{-5}
2./ 4 . 10^{-5}
3./ 10^{-4}
4./ 2 . 10^{-4}
—— CD of CHINOIN batch number 770306
--- CD of Nihon Shokuhin Kako Ltd. Japan

Fig. 4. Visible spectrum of 3 . 10^{-5}M phenolphtalein at pH 10,5 in various concentrations of β--cyclodextrin in 4 . 10^{-3}M Na_2CO_3 solution at 25±2 °C.

The results of HPLC determinations coincide with the results of the above simple and rapid method.

The following table illustrates the ΔE values obtained with γ-cyclodextrin samples of different origin and purity as compared to the results of gelchromatography.

Table I.

Sample	ΔE values at the assumed concentrations of γ-CD				Purity according gelchr.
	$1,2.10^{-4}$M	$2,4.10^{-4}$M	6.10^{-4}M	$1,2.10^{-3}$M	
SENN CHEM. standard	0,20	0,32	0,54	0,70	min. 97 %
γ-CD 1.	0,19	0,30	0,52	0,68	min. 97 %
γ-CD 2.	0,19	0,31	0,54	0,69	96 %
γ-CD 3.	0,20	0,32	0,54	0,70	97 %
crude γ-CD	0,36	0,53	0,76	0,84	70 %

γ-cyclodextrin samples 1, 2 and 3 are products of CHINOIN and the samples were purified by different methods from the crude γ-cyclodextrin.

Literature Cited

1. Cramer, F.: Angew. Chem. 64 136 (1952)
2. Broser, W.: Z. Naturforsch. 86 722 (1953)
3. Szejtli, J., Budai, Zs., Kajtár, M.: Magyar Kémiai Folyóirat 84 68 (1978)
4. Buvári, Á., Barcza, L.: Inorg. Chim. Acta 33 L179-L180.

ENZYMOLOGY, TOXICOLOGY AND METABOLISM

I.Int.Symp. on Cyclodextrins
Budapest, 1981

ENZYMOLOGY OF THE CYCLODEXTRINS

H. BENDER

Chemisches Laboratorium der Universität, Albertstr. 21
D-7800 Freiburg i. Br., FRG

SUMMARY

The cyclization reaction is a special type of the 4-α-D-glucopyranosyltransfer reactions typical of the cyclodextrin glycosyltransferases (CGTs). The enzyme from K.pneumoniae M 5 al simultaneously catalyzes cyclization and an acceptor-dependent transfer of linear fragments of 4-α-D-glucopyranosyl chains (disproportionation). Maximum cyclization rates are obtained with chain lengths of glc_{16}-glc_{80}, indicating the dependence upon the helical conformation of the substrate. αCD is formed initially by most of the CGTs. The rates of ßCD- and γCD-formation are extremely low. In contrast to αCD the higher CDs scarcely participate in reverse reactions and, therefore, accumulate. Only the incubation time determines which CD is obtained as the main cyclic product. By shifting the reaction equilibrium in the cyclization direction the yields of CDs can be enlarged significantly. Using an ultra-/diafiltration reactor, 70% of starch (enzyme-substrate ratio w/w) 1:100,000) can be converted into CDs.

Essentially 6 bacterial strains (4 bacilli) are known for producing extracellular CGTs. Depending upon the strain and the cultural conditions, 5-430mg of enzyme per l of culture filtrate were obtained, and purified up to 140fold (overall yields 10-78%). Because of the different assay methods the CGTs of the various bacterial sources cannot be compared at present. It cannot be decided, therefore, which enzyme is the most suited for the production of CDs on a technical scale.

The history of cyclodextrin glycosyltransferase -research is a very strange one for an enzymologist by several reasons: a) For 70 years only one organism (Bacillus macerans) has been known for producing this type of enzyme. b) Quantitative investigations of the chemistry of the cyclodextrin glycosyltransferases (1,4-α-D-glucan:[1,4-α-D-glucopyranosyl] transferases (cyclizing) (EC 2.4.1.19), CGTs) have been severely handicapped by the lack of a specific assay method. c) As indicated by more than 600 publications and patents (1) the chemists have been fascinated by the clathrate-forming properties of the characteristic cyclic transfer products. In spite of the fact that the CGTs are very interesting enzymes, the investigations of both the enzyme chemistry and the mechanism of the reactions have been neglected.

According to D.French, one of the pioneers in CGT-research, "the Schardinger dextrins could be manufactured on a very large scale, if there were a suitable market for them" (2). Up to now, the cyclodextrins (CDs) have been much too expensive for application on a technical scale. Production of CDs of low costs depends upon both an easily available CGT of wellknown characteristics and economic methods for manufacturing the cyclic compounds.

1. Enzyme chemistry. Some molecular characteristics of the CGTs from three bacterial strains are summarized in Table I.

Table I

CGT of	Molecular weight	Quarternary structure
B.macerans	145,000	dimeric
K.pneumoniae M5 al	68,000	monomeric
alk. Bacillus	88,000	?

Evidently the enzymes from the various bacterial sources differ at least in their sizes and their protein structures. The amino acid analysis of the CGT from Klebsiella pneumoniae M5 al has not been completed yet. SΔQ calculations for the CGT and for various α-amylases from pro- and eukaryotes point at surprising similarities between the two types of enzyme.

The CGTs are known to catalyze the reversible 1,4-α-D-glucopyranosyltransfer reactions:

A)

$$G_n \underset{\text{"coupling"}}{\overset{\text{cyclization}}{\rightleftharpoons}} G_{(n-x)} + x$$

(x=6, αCD; x=7, βCD; x=8, γCD; G_n, 1,4-α-D-glucopyranosyl chain with n glucopyranosyl residues)

B)

$$G_n + G_m \underset{\text{disproportionation}}{\rightleftharpoons} G_{(m-x)} + G_{(n+x)}$$

(G_m, G_n, 1,4-α-D-glucopyranosyl chains with m(n) glucopyranosyl residues; x, linear part of a chain). Generally, the part of a 1,4-α-D-glucopyranosyl chain containing the non-reducing end is transferred via the hydroxyl group of C_1 to the hydroxyl group of mainly C_4 of the acceptor. Reaction A) is only a special type of B): The non-reducing end of one chain itself serves as acceptor. It should be mentioned that the acceptor binding sites of the enzyme are not absolutely specific for glucose or maltooligosaccharides: methyl-α-D-glucopyranoside, 4-nitrophenyl-α-D-glucopyranoside, D-galactose, kojibiose, L-sorbose, D-xylose have been shown to have acceptor properties.

The CGTs have been supposed to attack suited substrates from their non-reducing chain ends exclusively (exo-fashion of attack). To understand the reactions A) and B) it is necessary to know the nature of G_n. As shown for the CGT from Klebsiella pneumoniae M 5 al, 1,4-α-D-glucopyranosyl chains with lengths of 16 - 80 glucopyranosyl residues are necessary for maximum cyclization rates, indicating the dependence upon helical conformation of the substrate. No matter, whether the chains are free or "bound", as is the case with the branched α-D-glucans amylopectin or glycogen. Chains $< G_{14}$ cannot be cyclized directly, chains $> G_{100}$ are poor substrates for the cyclization reaction (the reactions catalyzed by the CGT with long-chain amyloses will be discussed below).

αCD is formed initially by most of the CGTs. The reversible reaction A) approaches to a reaction equilibrium - the kinetics of cyclization essentially equal to a first order reaction. Further on the CGT is supposed to be inhibited by the transfer product αCD competitively(3): Independent from the initial substrate concentrations never more than the maximum concentration of 20mM αCD have been obtained with 1,4-α-D-glucopyranosyl chains $\overline{d.p.}$16 in the absence of clathrate-forming compounds.

Kinetic studies on the cyclization reaction have been scarcely possible because of the lack of a suitable assay method. A specific assay for the initial cyclization rate, however, is near at hand: A photometric assay method has been worked out, based on the marked increase in ultraviolet absorption of iodine by complexing with αCD (4). The initial rates of cyclization have been measured directly with suited substrates in the presence of iodine by the increase in optical density at 290nm. The specific activities of the CGT from Klebsiella pneumoniae M 5 al - given as the initial cyclization rates - with various substrates are displayed in Table II.

Table II.

Substrate	initial substrate concentration	kat x kg^{-1} of protein
1,4-α-D-glucopyranosyl-chains $\overline{d.p.}$16 a)	3mM	4.35
amylose (potato) $\overline{d.p.}$ 839 a)	0.18mM	0.036
amylopectin (waxy maize) mol.weight $> 10^6$	3.5%	2.67
native starch (potato)	3.5%	2.7
glycogen (E.coli) mol.weight $> 10^7$	0.6%	2.06

a) $\overline{d.p.}$, average degree of polymerization

Besides αCD, the CGTs are known to produce βCD and γCD. The rates of the formation of the higher CDs are extremely low. At least βCD scarcely participates in reverse reactions and, therefore, accumulates at the cost of αCD in the course of secondary transfer reactions. Accordingly, the time of incubation only determines which CD is obtained as the main product. It is impossible to give absolute ratios of the 3 main CDs (see 5).

The conditions for the reverse reaction of A), the "coupling" reaction, are quite different. $G_{(n-x)}$ may be glucose or maltooligosaccharides, and glucosides or some other sugars as mentioned above. The coupling rate of the Klebsiella-CGT has been determined using a coupled enzymic assay with maltodextrinphosphorylase, phosphoglucomutase, and glucose-6-phosphate-dehydrogenase. With the most effective acceptor maltose (initial concentration 0.4mM) 0.9mM of αCD (initial concentration 1mM) are coupled per g of protein and minute. Accordingly, the coupling activity amounts to only 0.5% of the initial cyclization rate with 1,4-α-D-glucopyranosyl chains $\overline{d.p.}$16 at the corresponding initial substrate concentration.

Once formed the linear chains are homologized, yielding shorter and longer disproportionation products. Chains long enough are substrates for the cyclization reaction. In the course of long-termed incubation, therefore, most of the αCD can be "transformed" into e.g. βCD.

What is the nature of C_n for the disproportionation reaction B)? Pure maltose proves to be a very poor (if any) substrate. Maltotriose and higher maltooligosaccharides are homologized with remarkable velocity. In the course of the disproportionation glucose is formed, which can be used as a mark for the reaction rate.

As shown for the CGT from Klebsiella pneumoniae M 5 al the cyclization- and the disproportionation reaction occur simultaneously with all the substrates assayed (3,7). The efficiency of cyclization (depending upon the probability of free acceptor binding sites of the enzyme) decreases with the higher substrate concentration at the cost of disproportionation. From French and co-workers the equilibrium constants for the main CDs have been reported. This is scarcely possible: For determination of the constants, the substrate G_n must be defined exactly. Since G_n always changes by disproportionation, its concentration can never be determined satisfactorily.

2. <u>1,4-α-D-Glucopyranosyltransfer reactions catalyzed with starch</u>. Because of the low costs native starches will be the substrates of choice for the production of CDs on a technical scale. Starches consist of the minor component amylose (long 1,4-α-D-glucopyranosyl chains, scarcely branched), and the major component amylopectin (short 1,4-α-D-glucopyranosyl chains, highly branched). Long chain amyloses prove to be poor substrates for the cyclization reaction. αCD is formed with maximum rate only after an initial degradation of the amylose into chains with lengths $< G_{100}$. The chain-shortening (evidenced by rapid retrogradation of the shortened chains) is enhanced significantly by the addition of maltooligosaccharides, maltose proves to be most effective (7,8). Evidently the chain-shortening is realized by a special kind of the disproportionation reaction resembling to an endo-fashion of attack. Only at low initial substrate concentrations (the retrogradation being retarded) the amyloses yield more than 90% of CDs.

The cluster model proposed by French (9) is generally accepted for the structure of the amylopectins at present. According to this model, both all the A-chains and the external parts of the B-chains, but also parts of the longer inner B-chains can serve for cyclization (and disproportionation). The initial chain-shortening reaction catalyzed by the CGT with amylose can be noted with the amylopectins and their ß-limit dextrins as well: The cleavage of the longer inner B-chains (presumedly with amylose-like conformation) yields fragments of markedly lower molecular weights. It is the cause of the rapid liquefaction of amylopectin- or starch gels. αCD is formed with maximum rate initially. The maximum yields of the cyclic compounds equal the degree of ß-amylolysis, in spite of the fact that the CGTs can by-pass part of the α-1,6-branching points. Without the use of debranching enzymes, however, the highly branched clusters of the amylopectin molecules are resistant to CGT-attack.

Using starch as the substrate, the A-chains of amylopectin serve as the acceptors for the chain-shortening of amylose. Caused by the disproportionation reactions both the shortened amylose chains and the amylopectin fragments with lengthened external chains retrograde at higher substrate concentrations. Just after a short incubation time all the soluble amylose has disappeared from the digests.

3. Production of CDs with regard to economy. Unfortunately the fine structures of amyloses and amylopectins, respectively, are not known in detail at present. Therefore it is scarcely possible to decide which starch is suited best for CD-production. Potato starch e.g. (easily soluble in hot water) proves to be a poorer substrate than is maize starch (nearly insoluble in hot water,must be pre-dissolved in dimethyl sulfoxide). From this finding that starch seems to be the better substrate for CD-production, which contains the more helical conformation (shorter amylose chains, helical regions of amylopectin). Some properties of potato- and maize starch are summarized in Table III.

Table III.

		% of total starch	$\overline{\text{d.p.}}$ (length of unit chain) a)	degree of ß-amylolysis (%)
maize	amylose	28	1,100	78
	amylopectin	62	23	58
potato	amylose	23	3,000	76
	amylopectin	67	24	56

a) $\overline{\text{d.p.}}$, average degree of polymerization

With regard to economy highly concentrated starch pastes should be used. As mentioned above, the efficiency of cyclization decreases at the cost of disproportionation with the higher substrate concentrations, yielding larger amounts of noncyclic, partly insoluble transfer products. This is the main problem for manufacturing the CDs on a technical scale.

As shown, the amounts of αCD formed by most of the CGTs initially are limited by both the reaction equilibrium, and a competitive inhibition mechanism. The limitation can be bypassed by shifting the reaction equilibrium in the cyclization direction using clathrate-forming compounds (αCD-clathrates are resistant to CGT-attack). Most of the clathrate formers are extremely toxic, and have to be removed by time consuming and very expensive procedures. Because of the high solubility of αCD in water, the preparation of pure compound is another problem. The reaction equilibrium can be shifted in the cyclization direction by use of an ultra/diafiltration reactor as well. With maize starch (initial substrate concentration 5%, substrate-enzyme ratio 100,000: 1) the overall yield of CDs was 70%. Most of the cyclic compounds (65%) were obtained as αCD with 98% purity from the ultra/diafiltrate in the course of 15 h incubation (30°C). Calculating the total costs it seems to be more desirable to obtain a nearly pure product, and not maximum yields of CDs.
ßCD is mostly used at present. Although produced slowly (except the CGT from Bacillus megaterium) it can be isolated from long-termed digests in good yields (50-60%, related to substrate) very easily and, therefore, is the cheapest. It is questionable, however, whether it is the most suited for technical use generally.

Because of the larger cavity and the high solubility in water, the γCD is of special interest (e.g. for antibiotica guests). Unfortunately, methods for economic preparations are unknown at present. Therefore, the γCD is much too expensive for technical use. Scaling up the conditions for γCD-production

Table IV. CGTs from four bacterial species. Yields of enzyme during cultivation and purification

Organism [a]		mode of cultivation	$mg \times l^{-1}$ of culture filtrate	purification	
				overall yield(%)	xfold purification necessary
B. macerans	(13)	batch	62	10-35	64-80
" "	(14)	"	26	12	140
" "	(15)	"	43	50	35
" " (mutant ?)	(16)	"	360-480	-	-
B. macerans	(11)	continuous	16	-	-
" "	(11)	continuous two stage	100	-	-
B. megaterium	(17)	batch	260	50	135
alk. Bacillus	(18)	"	430	35	74
K. pneumoniae	(12)	continuous	150	78	3

a) Numbers in parantheses indicate references

is a very important task for the future.

4. Enzyme sources. Which CGT is most suited for technical use ? As is the case with bacterial α-amylases a thermo-stable enzyme might be superior to others (10). All the CGTs attack the dissolved substrate only. Accordingly, the substrates have to be dissolved by heating (sterilizing the solution at least partly). As the CGTs liquefy e.g. starch gels rapidly, the start temperature is of minor interest. Enzyme properties other than thermostability will decide which CGT is preferred. Such properties may be the specific activities, and the costs of preparation of enzyme on a technical scale.

Unfortunately the CGTs from various bacterial sources have been assayed with a series of different methods, lacking the precision necessary for accurate research. It is impossible, therefore, to compare the specific activities of the various enzymes at present.

Essentially 6 bacteria (4 bacilli) have been shown to produce extracellular CGT. They all are inducible wild-type organisms producing the enzyme when growing with starch or starch products. Constitutive or hyper-producing mutants are unknown. Genetic investigations have never been made. The bacteria have been grown mainly batchwise. The synthesis of enzyme during continuous growth of Bacillus macerans was studied (11), but only continuously grown Klebsiella pneumoniae M5 al have been used as the source of CGT (12). The cultural conditions for maximum enzyme production have not been scaled up systematically. With the various bacteria 5 - 430mg enzyme per liter of culture filtrate have been obtained (Table IV). For the preparation of pure enzyme purifications up to 140fold were necessary, the overall yields were 10% - 78%. The preparation of pure CGT, however, is of theoretical interest only, e.g. for investigating the enzyme characteristics exactly. For technical use interfering enzyme impurities (diminishing the yields of cyclic transfer products) have to be removed only.

4. Literature cited

1. Saenger,W.: Angew. Chem. 92 343/1980/
2. French,D.: MTP Int.Rev.Sci.Org.Chem.Ser.5,Carbohydr. 267/1975/
3. Bender,H.: Carbohydr.Res. 78 133/1980/
4. Bender,H.: Analyt. Biochem. in press /1981/
5. Kitahata,S.,Okada,S.: Agr.Biol.Chem. 38 2413/1974/
6. Bender,H.: Carbohydr.Res. 65 85/1978/
7. Bender,H.: Carbohydr.Res. 78 147/1980/
8. Bender,H.: Carbohydr.Res. in press /1981/
9. French,D.: J.Animal Sci. 37 1048/1973/
10. Horikoshi,K.: Process Biochem. 26/1979/
11. Lane,A.G., Pirt,S.J.: J.appl.Chem.Biotechnol. 23 309/1973/
12. Bender,H.: Arch.Microbiol. 113 49/1977/
13. Schwimmer,S., Garibaldi,S.J.: Cereal Chem. 29 108/1952/
14. DePinto,J.A., Campbell,L.L.: Biochemistry 7 114/1968/
15. Kobayashi,S., Kainuma,K., Suzuki,S.: Carbohydr.Res. 61 229/1978/
16. Miskolci-Török,M.,Seres,G.,Vakaliu,H.,Szejtli,J.,Járai,M.: Hung. Teljes 17 926/1980/
17. Kitahata,S.,Tsuyama,N.,Okada,S.: Agr. Biol.Chem. 38 387/1974/
18. Nakamura,N.,Horikoshi,K.: Agr.Biol.Chem. 40 935/1976/

A PHOTOMETRIC TEST FOR THE CYCLISATION ACTIVITY OF CYCLODEXTRIN-GLYCOSYLTRANSFERASES

J.-P. LANDERT, E. FLASCHEL, A. RENKEN

Institute of Chemical Engineering, Swiss Federal Institute of Technology
CH-1015 Lausanne, Switzerland

Summary

A quick and reliable test has been developed to measure the cyclisation activity of cyclodextrin-glycosyltransferases (CGT). Since the reproducibility depends mostly on the nature of the substrate, maltotriose has been employed. The cyclodextrins (CD) formed are measured directly in situ by complexation with methylorange and by recording the change of absorbance photometrically. The test has successfully been used with CGT from *Klebsiella pneumoniae* M5 al and CGT from alkalophilic *Bacillus sp.*

Introduction

The activity tests for cyclodextrin-glycosyltransferases (CGT) previously described suffer from lack of specificity, tedious manipulations and unreliability. They can be classified as follows, according to the different activities of CGT:

- hydrolytic-activity tests

  The hydrolysis of starch is measured by means of its ability to undergo complexation with iodine. Although this method is not specific for CGT, it is used for its sensitivity and rapidity by several authors. (e.g. Nakamura N. and Horikoshi K. [1])

- coupling-activity tests

  These tests are employed in many variations (e.g. Nakamura N. and Horikoshi K. [2]). They are based on coupling of α-CD to an acceptor, thus forming linear dextrins. These are cut down by hydrolyzing enzymes and analysed by conventional methods. The manipulations however are too tedious for a quick routine test and the supposed linearity of response is often quite questionable.

- cyclisation-activity tests

  There are no quick and reliable tests known which are based on the measurement of CD-formation. The widespread Tilden-Hudson-test [3] certainly does not meet those requirements.

Summarizing these methods, it can be said that up to now, no specific and practicable test is known which meets the needs of routine analysis.

Materials and Reagents

α-CD was obtained from Serva, Heidelberg, W. Germany, β- and γ-CD as well as maltotriose (95%) from Sigma Chemicals, St. Louis, U.S.A., and methylorange from Fluka, Buchs, Switzerland. CGT from *Klebs. pneumoniae* was prepared according to Bender [5]. CGT from alkalophilic *Bac. sp.* was kindly supplied by N. Nakamura from Nihon Shokuhin Kako Co Ltd., Japan. The preparation contained 31'000 U/g according to [1]. Prior to the test, the solutions of latter CGT were filtered (0,45 μ-filter, Millipore Corp., Bedford, U.S.A.). Proteins were measured according to Bio-Rad, using albumine as standard. The photometer employed has been a Perkin-Elmer/Hitachi model 200.

Test principle

The test principle is based on the well known ability of CD to undergo complexation with common dyes, such as methylorange [4]. The absorbance of this dye decreases upon complexation proportionally to the CD-concentration in the range shown in fig. 1 (under test conditions). This allows to follow the formation of CD directly in a photometer. Methylorange has no harmful influence on the enzymatic reaction.

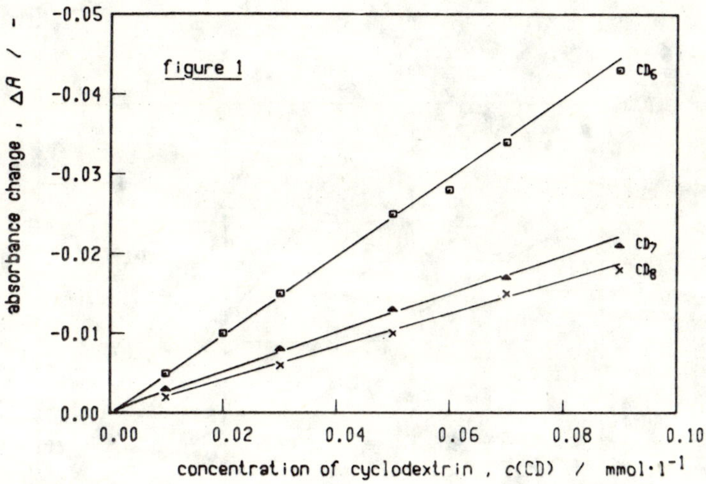

Fig. 1 Absorbance change of methylorange upon complexation with CD

If starch would be used as substrate, it would be impossible to follow such a change in absorbance accurately, due to its opacity. Since the CGT is able to build up its substrate for cyclisation out of oligosaccharides, maltotriose was chosen. Its commercial availability at high purities provides a reproducible basis for the test.

Test procedure

The following conditions allow the monitoring of CGT activity in culture filtrates of *Klebsiella pneumoniae* without preconcentration. Using other buffer systems, they may be easily adopted for other CGTs.

The reagent solutions are based on a 0.1 M maleate buffer of pH 6.8 containing 5 mM $CaCl_2$. 2 ml of a 0.15 mM methylorange solution containing 15 mg/ml maltotriose are placed in a quartz cuvette of 1 cm light path and thermostated at 25°C. 1 ml of adequately diluted enzyme solution is added and the change of absorbance measured at 546 nm. The test is monitored on a chart recorder (fig. 2) and the maximal rate of decrease of absorbance determined.

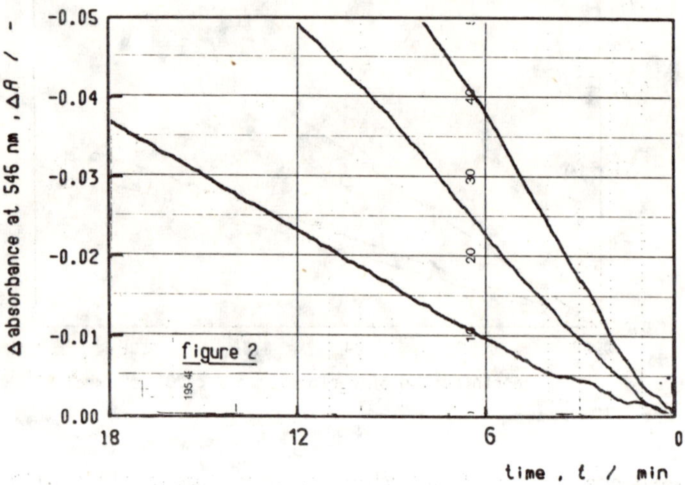

Fig. 2 Absorbance change vs. time for different concentrations of CGT from *Klebs. pneum.* (5, 10, 15, U/ml respectively)

This rate is linearly correlated to the concentration of CGT, as shown in fig. 3. CGT from *K. pneumoniae* produces predominantly α-CD. Therefore a unit is defined as a maximal decrease of absorbance of $0.5 \cdot 10^{-3} min^{-1}$, which corresponds under test conditions to a cyclisation rate of 1 μmol α-CD $l^{-1} min^{-1}$.

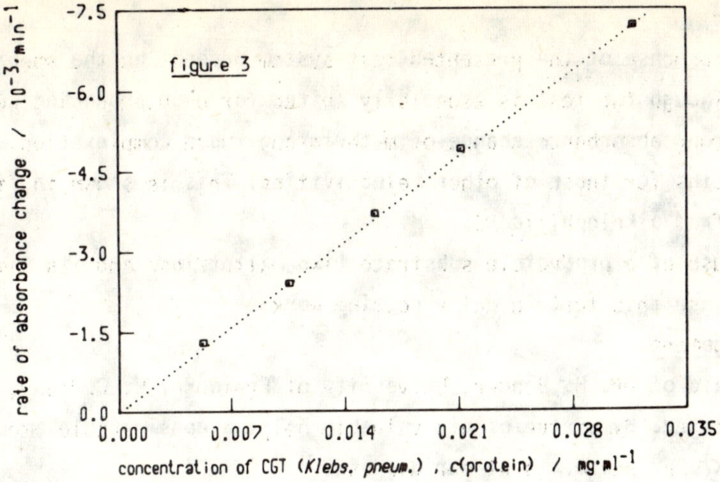

Fig. 3 Rate of absorbance change for serial dilutions of CGT from *Klebs. pneum.*

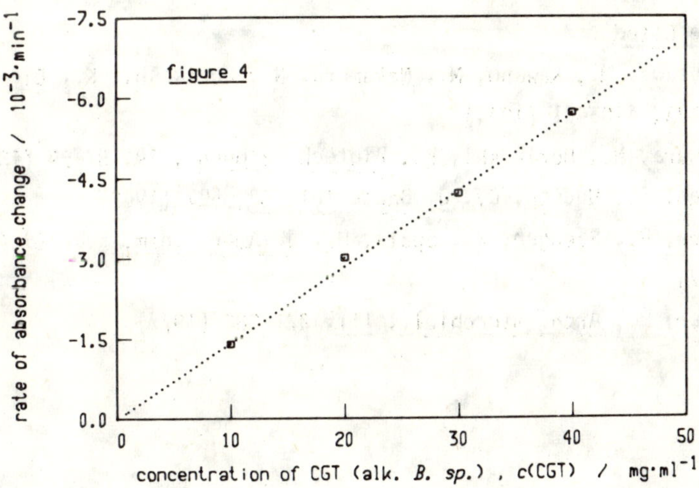

Fig. 4 Rate of absorbance change for serial dilutions of CGT from alk. *Bacillus sp.* at pH 8

Conclusion

The response of the presented test system depends on the specificity of the CGT. Though the test is especially suited for α-CD producing CGTs, due to the strong absorbance change of methylorange upon complexation, the linearity remains for those of other selectivities. This is shown in fig. 4 with CGT from alkalophilic *Bac. sp.*

The use of a producible substrate like maltotriose and its simplicity permit to use this test in daily routine work.

Acknowledgement

The aid of Dr. H. Bender, University of Freiburg, W. Germany, is greatly appreciated. By virtue of his valuable help, a considerable amount of CGT from *Klebs. pneum.* is at our disposal.

Our thanks are also due to Dr. N. Nakamura, Nihon Shokuhin Kako Co Ltd, Japan, who kindly provided us with a sample of CGT from alkalophilic *Bacillus sp.*

Literature Cited

1. Matzuzawa, M., Kawano, M., Nakamura, N., Horikoshi, K., Die Stärke, 27 (12), 410-431 (1975)
2. Nakamura, N., Horikoshi, K., Biotech. Bioeng., 19, 87-99 (1977)
3. Tilden, E., Hudson, C., J. Bacteriol. 43, 527 (1942)
4. Cramer, F., Saenger, W., Spatz, H., J. Amer. Chem. Soc. 89 (1), 14 (1967)
5. Bender, H., Arch. Microbiol., 111, 271-282 (1977)

I.Int.Symp. on Cyclodextrins
Budapest, 1981 2.3.

ACTION OF CGT ON BETA LIMIT DEXTRIN OF AMYLOPECTIN

RAINER SIEBERT

Chemisches Laboratorium der Universität, Albertstr. 21
D-7800 Freiburg, FRG

SUMMARY

The action of CGT from Klebsiella pneumoniae M 5 al on beta limit dextrin from amylopectin was investigated. It was shown that CGT is shortening the longer B-chains of the substrate under formation of products of low molecular weight.

AIM

The CGT from Klebsiella pneumoniae M 5 al causes rapid liquefaction of amylopectin gels which is probably due to the formation of products of considerably decreased molecular weight. Beta limit dextrin is a suitable substrate to study the mode of attack because in these molecules the A-chains and exterior parts of the B-chains are shortened to "stubs" while the framework of amylopectin molecules is preserved. In accordance with the cluster model of amylopectin proposed by Dexter French (1) the beta limit dextrin is drawn in Fig. 1.

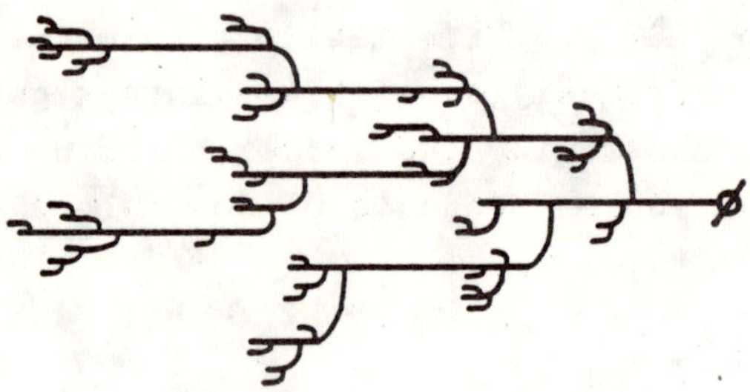

Fig.1 Cluster model of beta limit dextrin

RESULTS

The beta limit dextrin - 3.5 % (w/v) - was incubated in 1o mM Tris pH 7.2, 5 mM $CaCl_2$, with 14 ug/ml CGT for different times and aliquots of 6 mg carbohydrate were analyzed by GPC on a Sepharose CL-6B column (m.w. range $1o^4 - 1o^6$). Elution was in the descending direction at 9.8 ml/h with distilled water on a 2 x 9o cm column and fractions of 4.5 ml were collected. Total carbohydrate in each fraction was determined with anthrone. Results are shown in Fig. 2 and indicate the rapid formation of products of lower molecular weight. In the course of a 2o h incubation besides noncyclic products of low molecular weight 25 % of cyclodextrins (related to total carbohydrate) were formed as determined by the amyloglucosidase method.

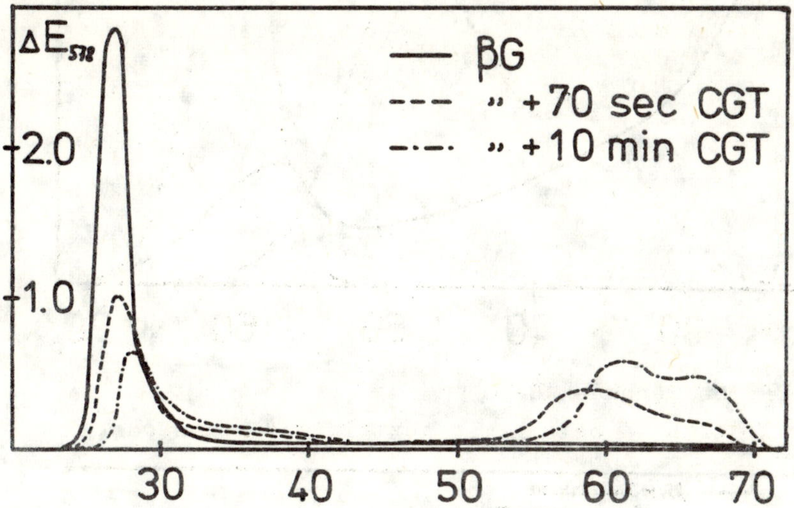

Fig. 2 GPC on Sepharose CL-6B
CGT-treated beta limit dextrin

In a different approach a digest was analyzed by GPC on Sephadex G-5o (m.w. range 5oo - 1o^4) and compared with amylopectin debranched by isoamylase and maltose as standards (Fig. 3). As can be seen products of very low molecular weight are formed. Elution was in the descending direction at 13.5 ml/h with 1o mM phosphate buffer pH 7.o on a 2 x 9o cm column and fractions of 4.5 ml were collected.

After the action and denaturation of CGT the whole digests were additionally debranched with pullulanase and subjected to chromatography on the same Sephadex column. The results (Fig. 4) reveal that the longer B-chains - about 53 % of total carbohydrate in beta limit dextrin - have been shortened.

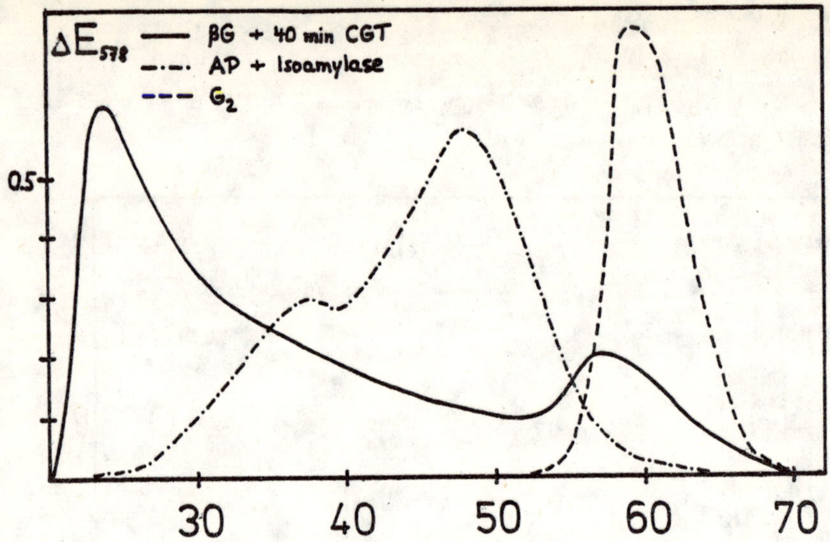

Fig. 3 GPC on Sephadex G-5o
CGT-treated beta limit dextrin and standards

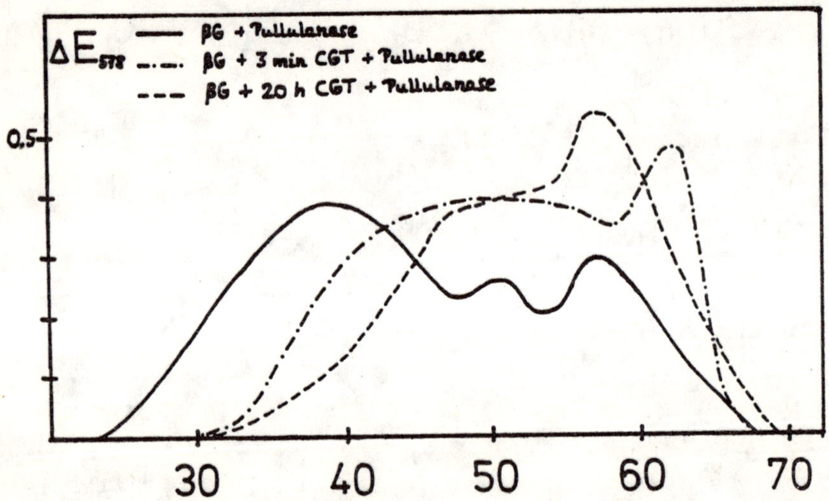

Fig. 4 GPC on Sephadex G-5o
Debranched CGT-treated beta limit dextrin

Fig. 5 demonstrates that CGT-treated beta limit dextrin is again a suitable substrate for the action of beta amylase. Beta limit dextrin treated in advance for 2o h with CGT exhibited 14 % beta amylolysis as calculated by integration and by determination of the formed reducing ends with Nelson reagent. This should be due to transfer reactions, in part it could be due to so-called "buried chains" which might be exposed after digestion with CGT. (Column dimensions in Fig. 5: 2 x 92 cm).

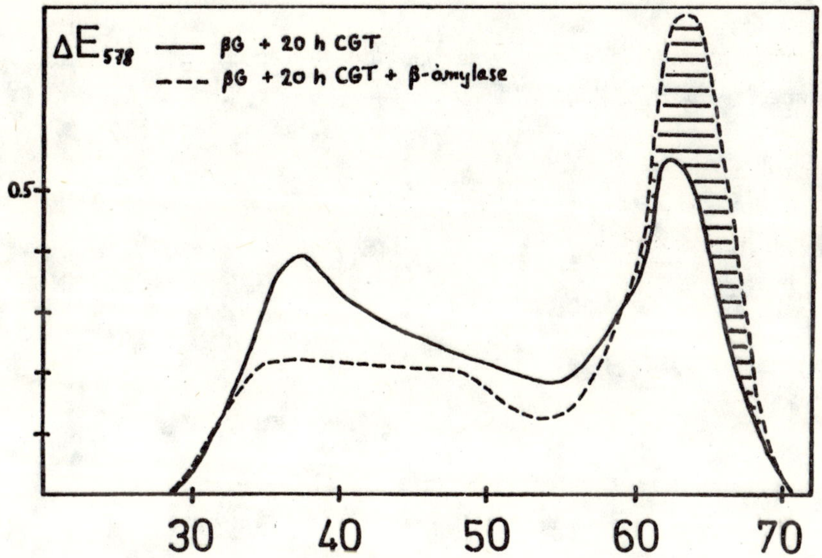

Fig. 5 GPC on Sephadex G-5o
CGT-treatment and beta amylolysis

CONCLUSIONS

The results show that CGT attack on beta limit dextrin leads to a rapid decrease in molecular weight. CGT is shortening the longer B-chains under formation of cyclic and noncyclic oligosaccharides. It has not yet been studied if branched cyclodextrins are among these products of low molecular weight. Further investigations are needed to elucidate the structure of the remaining branched glucan molecules of approximately d.p. 25 - 6o. Perhaps these molecules may lead us to some information about the fine structure of amylopectin molecules (for example the existence of highly branched clusters which are resistant to attack by CGT). Or inversely: They might tell us about the structural requirements for the action of CGT on branched glucans.

ACKNOWLEDGMENT

This research was supported by the Deutsche Forschungsgemeinschaft (Be 789/2).

Literature cited

1. French, D.: J.Animal Sci. 37 1o48,/1973/.

I.Int.Symp. on Cyclodextrins 2.4.
Budapest, 1981

ABSORPTION AND METABOLISM OF β-CYCLODEXTRIN BY RATS

ANDREA GERLÓCZY,[1] ANNA FÓNAGY,[2] JÓZSEF SZEJTLI[1]

[1] CHINOIN Pharmaceutical-Chemical Works, Tó u. 1-5
H-1045 Budapest, Hungary
[2] "Frédéric Joliot-Curie" National Research Institute
for Radiobiology and Radiohygiene, Pentz K. u. 5
H-1221 Budapest, Hungary

The oxygen consumption of small intestinal slices of rat was measured in Warburg apparatus using glucose, maltose, starch and beta-cyclodextrin as substrates. The oxygen consumption was increased in case of glucose, maltose and starch, while beta-cyclodextrin did not produce any change.

Universally labelled ^{14}C-beta-cyclodextrin or ^{14}C-glucose were given orally to rats and the radioactivity was measured in blood by liquid scintillation method. The peak of radioactivity appeared in blood within 10 minutes in the case of labelled glucose, while a rather protracted and low maximum was found between 2 and 10 hours in the case of labelled beta-cyclodextrin.

Rats were treated orally with ^{14}C-glucose or ^{14}C-beta-cyclodextrin and the radioactivity of the respired $^{14}CO_2$ was measured. The total respired radioactivity was approximately of the same quantity after ^{14}C-beta-cyclodextrin and ^{14}C-glucose treatment during the 24 hours period. Radioactivity maximum appeared during the first two hours after ^{14}C-glucose treatment, while it appeared between the 4th and 8th hours after ^{14}C-beta-cyclodextrin treatment. There was no significant difference in the tissue distribution of radioactivity after ^{14}C-glucose and ^{14}C-beta-cyclodextrin treatment. A larger amount of radioactivity was found in the cecum of animals treated with beta-cyclodextrin.

Respiration of $^{14}CO_2$ proves that beta-cyclodextrin is metabolized in rat.

β-cyclodextrin /βCD/ is to be widely used in drug and food industry, therefore, it is very important to study its metabolism in the mammalian organism.

Data available in the literature regarding the biological effect of βCD are contradictory enough. The intravenously administered βCD was toxic /LD_{50} in rats: 788 mg/kg/ /1/. According to French's monography, the orally administered βCD was also toxic /2/. This observation was one of the very hindering factors of application of βCD in the drug and food industry.

New results have been published concerning the oral administration of βCD during the last 6 years. Makita and co-workers /3/ did not found any toxic symptom in rats treated orally with βCD for 6 months. Similar results were obtained by Szejtli and Sebestyén /4/.

Cyclodextrins are known to be very resistant to the attack of the common starch-hydrolyzing enzymes, however, α-amylases slightly hydrolyze the cyclodextrin ring.

Taking into consideration the above mentioned data three kinds of pathways of βCD seem to be possible in the mammalian organism following oral administration. One of the possibilities is that βCD is not absorbed from the gastrointestinal system, it is excreted without any change by feces. The other one is that only a very few amount of βCD is absorbed and that is less than that induce toxic symptoms. Finally, βCD may be metabolized similar to other α 1→4 linkaged polysaccharides. This last possibility is supported by Andersen and co-workers /5/, who found that βCD was metabolized to a similar extent as starch.

Universally labelled ^{14}C-βCD gave an opportunity to study the absorption and metabolism of βCD. The labelled βCD was prepared by the team of the Technical University, Budapest.

Absorption of ^{14}C-βCD and ^{14}C-glucose as control was examined /6/. Rats were treated orally with the labelled

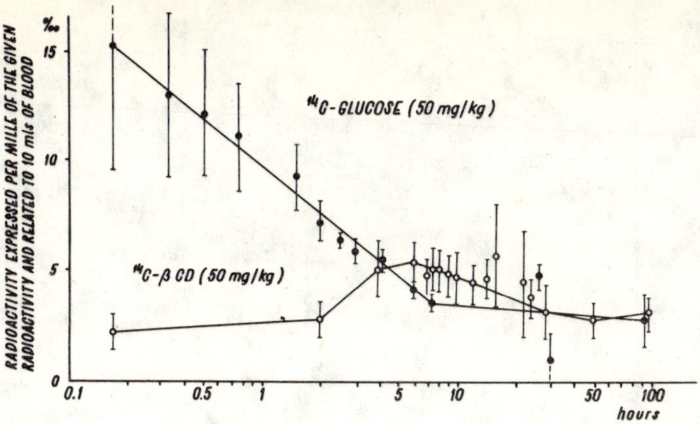

Fig.1

Absorption of universally labelled ^{14}C-βCD and ^{14}C-glucose in rat after oral administration

substances and the change of radioactivity of blood was determined by liquid scintillation method /Fig.1/.

The peak of radioactivity appeared in blood in a few minutes after ^{14}C-glucose treatment, and vanished from the blood practically within 8 hours. It is very likely that the residual radioactivity originated from carbon atoms incorporated into the various components of the blood. The quite protracted maximum of radioactivity appeared between the 2nd and 10th hours after ^{14}C-βCD treatment. The value of the maximum was about three times lower than in case of ^{14}C-glucose.

Since a small amount of radioactivity deriving from ^{14}C-βCD appeared in blood following oral administration, βCD is absorbed from the gastrointestinal system in some way. The fact that the same residual radioactivity can be observed in case of ^{14}C-βCD as that of ^{14}C-glucose suggests that βCD is metabolized in form of glucose.

The metabolism of βCD was investigated under <u>in vitro</u> circumstances at first.

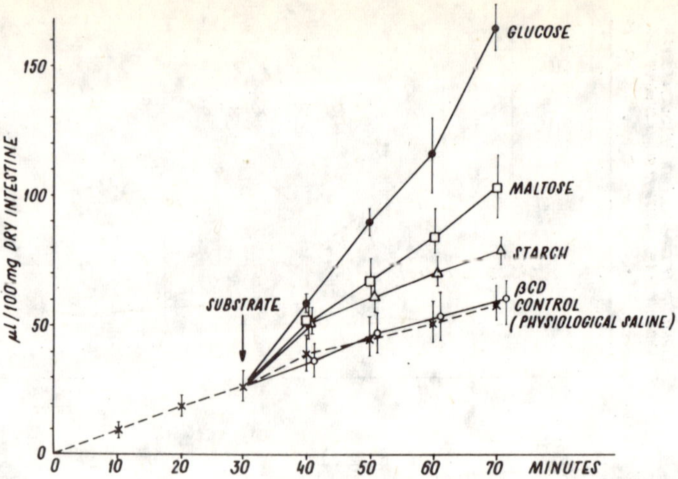

Fig.2

Oxygen consumption of small intestinal slices in Warburg apparatus

Oxygen consumption of rat small intestinal slices was measured in Warburg apparatus, using glucose, maltose, starch and βCD as substrates /Fig.2/.

The oxygen consumption was increased in case of glucose, maltose; for starch the increase was small enough. So they behaved like substrates. On the contrary, βCD did not increase the oxygen consumption of intestinal slices, therefore it is not metabolized under in vitro circumstances.

In order to gain further information about the metabolism of βCD a metabolism cage was constructed suitable for measuring the radioactivity of respired $^{14}CO_2$ as well as that of urine and feces at the end point of experiment. Radioactivity respired by rat was measured after oral addition of ^{14}C-βCD and ^{14}C-glucose by liquid scintillation method /Fig.3/.

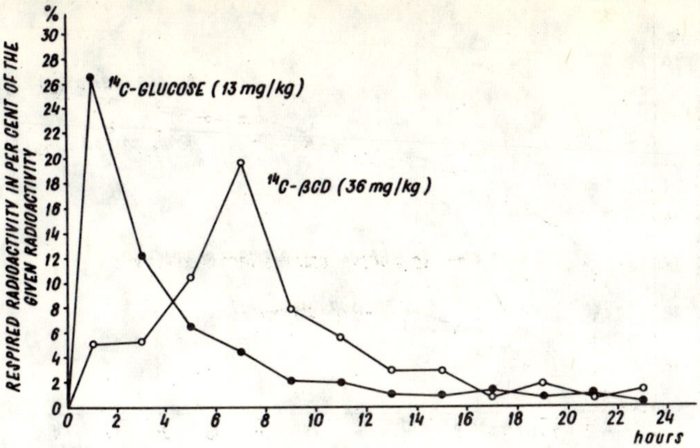

Fig.3

Radioactivity respired by rats after oral administration of ^{14}C-βCD and ^{14}C-glucose

The largest amount of radioactivity in the respired CO_2 was measured during the first two hours after ^{14}C-glucose treatment. Radioactivity appeared in the respired CO_2 after ^{14}C-βCD treatment, too. The largest amount of radioactivity was measured between the 6th and 8th hours after treatment. These results are in good correlation with the absorption curve.

Fig.4 demonstrates the radioactivity respired by rat after oral ^{14}C-βCD and ^{14}C-glucose treatment, expressed in cumulative form. The respiration curve of ^{14}C-glucose and 36 mg/kg dose ^{14}C-βCD represents the same experiment we have mentioned above. Respired radioactivity was measured after a large - 313 mg/kg - dose ^{14}C-βCD treatment, too.

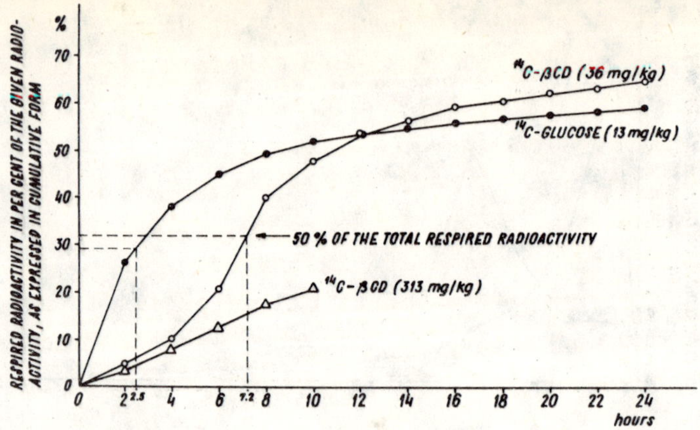

Fig.4

Radioactivity respired by rats after oral administration of ^{14}C-βCD and ^{14}C-glucose. Cumulative form

In case of 36 mg/kg dose ^{14}C-βCD the total respired radioactivity was approximately of the same quantity as after ^{14}C-glucose treatment. In case of 313 mg/kg dose ^{14}C-βCD the respired radioactivity was about half as much as in case of ^{14}C-glucose at the end point /at the 10th hour/. 50 % of the respired radioactivity appeared at 2.5 hours after addition of ^{14}C-glucose and at 7 hours after addition of 36 mg/kg dose ^{14}C-βCD.

As the production of $^{14}CO_2$ is one of the criterions to determine whether a ^{14}C-labelled substrate is metabolized or not, therefore, our results prove that βCD is metabolized in rats. It is very likely, that some part of βCD is excreted from the organism after large dose treatment.

At the end point of the respiration experiment rats were exterminated and the distribution of radioactivity was measured in the organs. No significant difference was found between rats treated with ^{14}C-βCD and ^{14}C-glucose, with the exception that there was a higher amount of radioactivity in the large intestine of ^{14}C-βCD treated animals. The quantity of radioactivity was about fivefold more in animals treated with 36 mg/kg dose ^{14}C-βCD /10 % of the given radioactivity/ than in ^{14}C-glucose treated animals /2 % of the given radioactivity/. 10 hours after the administration of the higher dose ^{14}C-βCD the 55 % of the orally given radioactivity was found in the large intestine.

Since only a very few amount of radioactivity appears in the respired CO_2 after the 24th hour, and the radioactivity decreases already after the 10th hour, it is very likely that βCD is not absorbed in a considerable extent from the large intestine.

Summarizing the results, the following conclusions may be drawn:

As radioactivity appears in blood after ^{14}C-βCD treatment, therefore βCD is absorbed from the gastrointestinal system in some way.

Appearance of the radioactivity in the respired CO_2 is a proof for <u>in vivo</u> metabolism of βCD in rats. During a 24 hours period βCD is metabolized to a similar extent as glucose in case of 36 mg/kg dose. This observation is in accordance with the results of Andersen et al. /5/.

Taking into consideration all these data and the fact that the intravenously toxic βCD does not induce toxic symptoms after oral administration it is very likely that βCD itself is not absorbed from the gastrointestinal system in a considerable amount. Only glucose derived from βCD on amylase action appears in blood.

However, it cannot be excluded that βCD is absorbed to a very few extent as an intact molecule. Attempts were made to determine βCD from the blood by high pressure liquid chromatography method. Due to difficulties of extraction procedure of βCD from blood the complete evaluation of the results is not possible at present. However, mention should be made of that 8 hours after oral administration of 313 mg/kg ^{14}C-βCD a very small amount, that is 3 μg of βCD was detected in 1000 μg of blood.

On the basis of investigations of radioactivity distribution it is very likely that some part of βCD is excreted by the feces after high dose of βCD treatment.

These results allow us to draw the conclusion that βCD might safely be applied in the drug and food industry.

ACKNOWLEDGMENT

Thanks are due to Dr.B.Bánky /Technical University, Budapest, Institute of Agricultural Chemical Technology/ for preparing the universally labelled ^{14}C-βCD.

Literature Cited

1. Frank, D.W., Gray, J.E., Weaver, R.N.: Am.J.Pathol. 83 367 /1976/
2. French, D.: Adv.Carbohydrate Chem. 12 189 /1957/
3. Makita, T., Ojima, N., Hashimoto, V., Ide, H., Tsuji, M., Fujisaki, Y.: Oyo Yakuri 10 449 /1975/
4. Szejtli, J., Sebestyén, G.: Die Stärke 31 385 /1979/
5. Andersen, G.M., Robbins, F.M., Domingues, F.J., Moores, R.G., Long, C.L.: Toxicol.Appl.Pharmacol. 5 257 /1963/
6. Szejtli, J., Gerlóczy, A., Fónagy, A.: Arzneim.-Forsch/ /Drug Res. 30 808 /1980/

I.Int.Symp. on Cyclodextrins
Budapest, 1981

TOXICITY STUDIES OF BETA-CYCLODEXTRIN

VERA GERGELY, G. SEBESTYÉN, S. VIRÁG

CHINOIN Pharmaceutical-Chemical Works, Tó u. 1-5
H-1045 Budapest, Hungary

SUMMARY

The acute oral and parenteral LD-5o of beta-cyclodextrin was established on rats, mice and dogs.

The long term toxicity studies were performed on rats and dogs following the two weeks dose finding studies given the test substance by oral route. All of the animals were autopsied and the tissue samples were examined histopathologically in both experiments.

Investigation of the effect of beta-cyclodextrin on hepatic drug metabolizing enzyme system of rats did not show any alteration in the activity.

According to our results the beta-cyclodextrin given orally to rats and dogs did not cause any side effect during the long term toxicity studies.

A wide range toxicity study of beta-cyclodextrin /beta-CD/ has been performed on mice, rats and dogs.

Acute toxicity studies

The acute LD-5o values were estimated after a single oral, intraperitoneal or subcutaneous administration. In case of the rodents ten males and ten females were treated in each dose group while the dogs were grouped by four in 1:1 sex ratio. The observation period durated for fortnight.

The oral LD-5o value of the compound is greater than 5000 mg/kg in dogs.

TABLE I. Acute LD-5o values of beta-CD in rat

Sex	Mode of administration	LD_{50} mg/kg
M	po.	>5000
F	po.	>5000
M	sc.	>1000
F	sc.	>1000
M	ip.	372.79 /297.05 - 467.84/
F	ip.	356.10 /267.83 - 473.45/

TABLE II. Acute LD-50 values of beta-CD in mouse

Sex	Mode of administration	LD_{50} mg/kg
M	po.	> 3000
F	po.	> 3000
M	sc.	419.4 /310.71 - 566.1/
F	sc.	412.31 /3o1.72 - 563.45/
M	ip.	372.03 /251.92 - 549.39/
F	ip.	330.98 /259.41 - 422.29/

Three months oral toxicity study on rats

The study was performed on Long-Evans rats given the test substance in the following oral doses: 200, 400 and 600 mg/kg.

Laboratory investigations

Hematology: Hemoglobin, hematocrit, RBC, WBC, MCH, MCV, MCHC, differential cell count

Blood chemistry: SGOT, SGPT, Alkaline phosphatase, BUN, glucose, creatinine, bilirubine, total protein and protein electrophoresis

Urinanalysis: pH, protein, blood, glucose, ubg, bilirubin, ketons and sediment /qualitative analysis/

Terminal investigations

All of the animals were autopsied and the following organs were weighed: heart, lung, liver, kidneys, spleen.

The bodyweight and food consumption were measured weekly

during the treatment period and daily clinical observations has been performed.

Results: There were no clinical signs during the treatment. None of the measured laboratory parameters showed any significant difference in the treated groups compared to the controls. There were no macroscopic changes in the rats and even the histological study could not improve any pathological change.

Six months oral toxicity study on rats

Long-Evans rats were used and they were treated with the following doses of beta-CD: 2oo, 4oo and 6oo mg/kg po.

The same clinical and laboratory parameters were investigated and measured as in the subacute toxicity study.

After the six months treatment we have used a two months recovery period without any treatment in order to investigate the disappearence of the toxic signs if there is any.

According to the laboratorical and histopathological results the test substance proved to be non-toxic in rats using the above mentioned oral doses.

Six months oral toxicity study on dogs

The Beagle dogs were treated as follows: 1oo, 25o and 5oo mg/kg. Out of the investigations used on the rats the following clinical observations were performed: pulse rate, respiratory rate and body temperature.

At the autopsy the following organs were weighed: heart, lung, liver, kidneys, spleen, brain and gonads.

Taking into consideration all of the laboratory, clinical, pathological and histological results the beta-CD did not proved to be toxic in Beagle dogs.

Mutagene study

The Long-Evans rats were treated with 2oo, 4oo and 6oo mg/kg oral dose of beta-CD. The animals were then exterminated two hours after the 8 mg/kg ip. colchicin administration. It has been established that the incidence of spontaneous aberration did not increased and there were no sign of mutation in the treated animals.

Teratological study

The investigation was made on Long-Evans rats given the beta-CD in 2oo, 4oo and 6oo mg/kg oral doses. There were 25 females per group and they were treated between the 7th and 16th days of pregnancy. The mother animals were sectioned one day before the delivery and the following parameters has been investigated: bodyweight and sex ratio of pups, resorptions, visceral and sceletal malformations. The latter was examined after Alizarin Red-S staining.

The compound proved to be non-toxic neither on the pregnant females nor on the development of the pups.

Effect on rat liver microsomal enzymes

The effect on the microsomal enzyme system was investigated on Sprague-Dawley rats given the test substance in 2oo and 6oo mg/kg oral dose for three or seven days.

We could not observe any difference in the Hexobarbital sleeping time of treated male or female rats compared to the controls.

The microsomal protein and cytochrome P-45o content did not show any alteration among the treated and control groups.

In an other experiment 75 mg/kg ip. Antipyrine was given to the pretreated and control males and females. The serum Antipyrine level was measured two hours later and no significant difference was detectable between the pretreated and control groups.

The compound has also not shown any metabolic interaction with the anticoagulant Syncumar in male and female rats.

According to these results the investigated compound has neither inductive nor inhibiting effect on rat liver micros omal enzymes.

Carcinogenecity study

The one year long carcinogenecity study is on work now.

The Long-Evans rats are treated with 2oo, 4oo and 6oo mg/kg beta-CD orally.

ABSORPTION AND ELIMINATION OF CYCLODEXTRIN DERIVATIVES BY RABBITS AND RATS

P. SZABÓ,[1] T. FERENCZY,[1] J. SERFŐZŐ,[1] J. SZEJTLI,[2] A. LIPTÁK[3]

[1] Department of Zoology and Anthropology, Kossuth Lajos University, Egyetem tér 1
H-4010 Debrecen, Hungary
[2] CHINOIN Pharmaceutical-Chemical Works, Tó u. 1-5
H-1045 Budapest, Hungary
[3] Institute of Biochemistry, Kossuth Lajos University, Egyetem tér 1
H-4010 Debrecen, Hungary

SUMMARY

The intestinal absorption, digestibility by the colon bacterial flora and urinary elimination of the β-cyclodextrin and its methylated derivatives were studied. The in vitro absorption of the methylated derivatives of cyclodextrin proved to be slow, and could not be inhibited by phloretin. The absorption is concentration dependent, and shows no saturation limit. The results point to a passive transport mechanism. The β-cyclodextrin and its methylated derivative could be digested by a culture of colonic symbionts. The results render quite probable that these compounds may be converted into glucose by the colon bacteria. Following oral administration the urinary elimination was rather low, while following the intramuscular injection the heptakis(2,6-di-O-methyl)-β-cyclodextrin was excreted from the organism quantitatively in 24 hours. The plasma concentration of intravenously injected heptekis-(2,6-di-O-methyl)-β-cyclodextrin decreased abruptly in 1-2 hours, and after 6 hours, it could be detected in blood plasma only in traces.

Andersen et al. (1), Szejtli et al. (2) and Gerlóczy et al. (3) published data on absorption and metabolism of cyclodextrins. Methyl-cyclodextrin-derivatives are characterized with considerably improved solubility, and because their complexes are also readily soluble in water (4,5), they can be considered as potential drug carriers for poorly soluble drugs. This was the motivation of the investigation of intestinal absorption and elimination of β-cyclodextrin, 2,6-di-O--methyl-β-cyclodextrin and 2,3,6-tri-O-methyl-β-cyclodextrin (further on: -CD, DM- -CD and TM- -CD, respectively) in rabbits and rats.

MATERIALS and METHODS

The experimental animals were rabbits of 2 to 2,5 kg weight and rats of 140 to 180 g of the stock CFY. They were fed with standard granulated food-preparation (LATI, Gödöllő), water was given them ad libitum.

The compound studied were chemically pure; the β-cyclodextrin is a product of CHINOIN, Budapest, the methylated derivatives were prepared by the Biochemical Institute of the Kossuth L. University, Debrecen.

In vitro absorption studies: the technique "everted sacs" of Wilson-Wiseman (6) was used. The incubating medium was a Krebs-Ringer bicarbonate buffer solution (pH = 7,4), as volume marker, inulin was applied.

The changes in the concentration of the inulin applied as volume marker was determined with Kulka's method (7).

In vivo absorption studies were performed in rats by the technique "ligated loop" as described by Musacchia (8).

The colonic bacteria are presumably able to decompose the cyclodextrins. For examining the symbionts the modified method of Griffiths-Smith (9) and Doyle-Griffiths (10) was used. The inoculum was prepared from the coecal content of rats. The applied culture medium was that of type Thioglycollate U.S.P. (Oxoid). The medium contained β-CD and DM-β-CD in a concentration of 1,66 and 1,00 mg/ml, respectively. The incubation was carried out at 37 °C under anaerobic circumstances. Samples containing inoculum + medium and medium + substrate, respectively, served as control. The samples taken at different time intervals were treated with 10 % trichloracetic acid, this was followed by centrifugation, and then the carbohydrate content was determined from the supernatant. The result were referred to 100 mg dry matter content of the samples.

Elimination studies: the urinary elimination of single doses of 150 mg/kg DM-β-CD and TM-β-CD administered orally to rats was determined from 24, 48 and 72 hour specimens of urine. In other experiments, the urinary elimination of 150 mg/kg dosis administered intramuscularly was studied in observation periods of 24, 48 and 72 hours.

The determination of the β-CD was carried out by spectrophotometry with Roe's antron reaction (11) from the incubatum deproteinized in 1:1 ratio with 10 % trichloracetic acid. The determination of DM-β-CD and TM-β-CD from the liquid phase was carried out as follows: 1 ml of blood plasma and urine, respectively, having been deproteinized, was shaken in 3x3 ml of chloroform; after the evaporation of the organic solvent, the residue was dissolved in distilled water, and from the aliquot parts the glucose was determined quantitatively with Roe's method (11).

The statistical evaluation of the results was carried out by Student's "t" test (12).

RESULTS

Table I. demonstrates the results of the mucosa-serosa flux (in jejuno-ileal segments of rats) of β-cyclodextrin as well as of several methylated derivatives. No significant difference was found between the transport rate of TM-α-CD and that of TM-β-CD ($P > 0,05$). The fact, that 0,2 M/l phloretin - which inhibits strongly the active transport of D-glucose (13) - has no influence on the transport of β-CD and its methylated derivatives suggests that these compounds are absorbed by a passive transport.

In "ligated loop" in vivo intestinal absorption experiments the absorption of DM-β-CD showed to be unambiguously concentration dependent (Fig.1.).

Within the studied concentration range the absorption of DM-β-CD did not indicate saturation limit.

Table II. illustrates the effect of rat bacterial flora on β-CD and DM-β-CD. Both compounds were utilized by the bacteria however

Table I.

Transport activity of cyclodextrins in "everted-sacs" preparation			
	nM/100 mg fresh weight	SD ±	Significant
β-CD	1,74	0,71	--
β-CD + phloretin	2,48	0,71	P > 0,05
DM-β-CD	1,72	0,27	P > 0,05
DM-β-CD + phloretin	1,94	0,52	P > 0,05
TM-α-CD	1,54	0,28	P > 0,05
TM-α-CD + phloretin	1,44	0,29	P > 0,05
TM-β-CD	2,80	0,60	P > 0,05
TM-β-CD + phloretin	2,90	0,27	P > 0,05

Table II.

The digestive effect of coecal bacteria on β-CD and DM-β-CD			
incubation	uptake mg/100 mg dry weight		number of experiments
	7 days	14 days	
Medium thioglycollate + 1,66 mg/ml β-CD	27,707 ±3,25	29,609 ±6,60	20
Medium thioglycollate + 1,0 mg/ml DM-β-CD	6,131 ±3,74	7,107 ±4,70	6

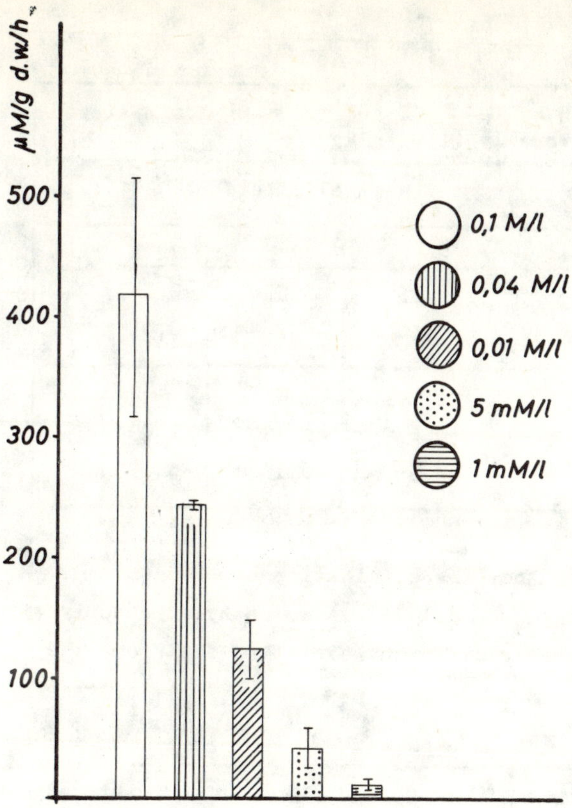

Fig.1.
Intestinal absorption of DM-β-CD in rats

at a different rate. The bacterial decomposition of the DM-β-CD proved to be considerably slower than that of the β-CD. These data support the assumption, that the cyclodextrins administered orally may be utilized presumably by being converted into glucose in the colon due to the symbionts.

The data of Table III. show the urinary elimination of the orally administered DM-β-CD and TM-β-CD. 2,71 % of the DM-β-CD was found to be eliminated in the 24 hour urine, and additional 1,59 % and 0,43 % on the following two days. With TM-β-CD the value of elimination was recorded to be 1,22 % on the first, 0,35 % on the second and 0,17 % on the third day. It is not clear whether the methylated cyclodextrins on account of their extreme solubility were not absorbed from the intestine or following the absorption were utilized by the organism; it can be said only with certainly, that they are not excreted with the urine.

Table III.

Urinary elimination of DM-β-CD and TM-β-CD administered orally (150 mg/kg)			
administration	elimination		
DM-β-CD	24 h	48 h	72 h
36,410 ±4,68 mg	0,99 ±0,40 mg	0,528 ±0,26 mg	0,163 ±0,01 mg
TM-β-CD 54,07 ±6,76 mg	0,677 ±0,59 mg	0,197 ±0,18 mg	0,094 ±0,05 mg

The data of Table IV. show the elimination of DM-β-CD and TM-β-CD administered intramuscularly. The two methylated derivatives were found to have essentially different excretion features. The full

Table IV.

Urinary elimination of DM-β-CD and TM-β-CD administered intramuscularly (150 mg/kg)			
administration	elimination		
DM-β-CD	24 h	48 h	72 h
33,875 ±4,88 mg	33,654 ±7,87 mg	0,671 ±0,36 mg	0,222 ±0,31 mg
TM-β-CD 30,33 ±4,87 mg	4,06 ±3,61 mg	0,257 ±0,33 mg	0,275 ±0,15 mg

quantity of the DM-β-CD injected was eliminated with the urine within 24 hours. Contrary to this, the TM-β-CD was excreted in a minimum amount even during an observation period of 72 hours. The cause of this essential difference is not known, but the solubility of TM-β-CD is inferior to that of the DM-β-CD.

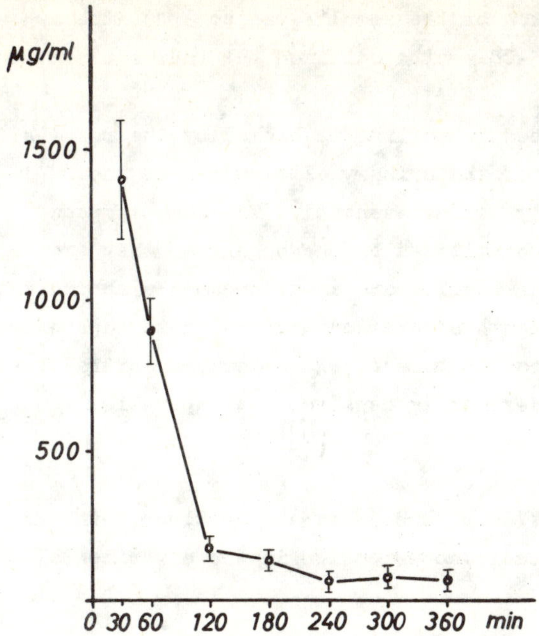

Fig.2.
Changes of DM-β-CD concentration in the plasma of rabbits

Figure 2. illustrates the blood plasma concentration of DM-β-CD administered intravenously. In 1 to 2 hours after the injection the concentration fell abruptly, and after 6 hours, the compound may have been detected only in traces. The increase of the concentration of DM-β-CD in the urine examined simultaneously refers to that the kidneys eliminated the compound with a remarkable activity.

DISCUSSION OF THE RESULTS

The intestinal absorption of β-CD and its methylated derivatives seems to be a passive transport, it does not indicate any saturation maximum and may not be inhibited by phloretin. The concentration of intact molecules of β-CD and DM-β-CD in the blood - after oral administration - is very low, the decomposition into glucose units within the intestine may be assumed.

The microsymbionts of the colon may have an important part in the metabolism of cyclodextrins. Our experiments in vitro demonstrate the active bacterial decomposition of the β-CD and the DM-β-CD. It may be assumed that also this part of the intestines has a great role in the absorption and utilization of glucose converted from the cyclodextrins. From the comparison of the results one comes to the assumption that the intact β-CD may be utilized but indirectly by the organism of the rat.

The hypothesis described in outline is backed by the results, too, obtained in the study of the urinary elimination of the DM-β-CD administered intramuscularly or intravenously. The same compounds given parenterally cannot be utilized by the organism, they are eliminated with the urine. These results are in accordance with the data that in chronic oral treatments alterations referring to nephrosis may not be found, whereas in case of short-term parenteral treatments the intact β-cyclodextrin induces serious nephrosis.

Acknowledgement

Thanks are due to Mrs.Zádor, Mrs.Preczner, Mr.Tibor Tóth and Miss Gabriella Molnár for their contribution in the experimental work.

Literature Cited

1. Anderson, G.H., Robbins, F.M., Domingues, F.J., Moores, R.G. and Long, C.L.: Toxicol.Appl.Pharmacol. 5 257 (1963)
2. Szejtli, J., Gerlóczy, A., Fónagy, A.: Arzneim.Forsch., 30 808 (1980)
3. Gerlóczy, A., Fónagy, A., Szejtli, J.: This Symposium (1981)
4. Szejtli, J., Bolla-Pusztai, É.,Stadler-Szőke, Á.: Hung.Patent Appl. 1141/80 (1980)
5. Pitha, J.: Life Sci., 29 307 (1981)
6. Wilson, T.H. and Wiseman, G.: J.Physiol. 123 116 (1954)
7. Kulka, R.G.: Biochem.J. 63 542 (1956)
8. Musacchia, X.J.: Exp.Physiol.and Biochem. 1 379 (1968)
9. Griffiths, L.H. and Smith, G.E.: Biochem.J. 128 901 (1972)
10. Doyle, B. and Griffiths, L.A.: Xenobiotica 10 247 (1980)
11. Roe, J.H.: J.Biol.Chem. 212 335 (1955)
12. Weber, E.: Grundriss der Biologischen Statistik, Springer Verlag, Jena 1964
13. Randles, J. and Kimmich, G.A.: Am.Jour.Physiol. 234 C64-C72 (1978)

I.Int.Symp. on Cyclodextrins 2.7.
Budapest, 1981

RENAL EFFECTS OF PARENTERALLY ADMINISTERED METHYLATED CYCLODEXTRINS ON RABBITS

J. SERFŐZŐ, P. SZABÓ, T. FERENCZY, A. TÓTH-JAKAB

Department of Zoology and Anthropology, Kossuth Lajos University, Egyetem tér 1
H-4010 Debrecen, Hungary

SUMMARY

The renal effects of a 12 days intramuscular treatment with 50 mg/kg/day doses of β-CD, heptakis-monomethyl-β-CD /= MM-β-CD/, heptakis-dimethyl-β-CD /DM-β-CD/ and heptakis-trimethyl-β-CD /= TM-β-CD/ were studied on rabbits, with emphasis on the glomerular circulation of the nephron, and in the ultrastructural changes of the epithelial cells of the proximal tubule.

At this relatively high dosis MM-β-CD and TM-β-CD caused an increased necrosis while the effect of β-CD and DM-β-CD were more moderate.

MM-β-CD and TM-β-CD reduce the blood supply of the nephron, the concomitant hypoxia increases the lysosomal activity in the epithel cells. This state leads to the development of progrediated necrosis. β-CD and DM-β-CD increase the blood supply of the nephron. In this case the development of necrosis and activation of lysosomal system are caused by the recirculation of these substances connected with their enhanced reabsorption.

Detailed toxicity studies proved that orally administered β-CD is not toxic. However this is not true for parenterally administered cyclodextrins or cyclodextrin-derivatives. Frank et al. (2) observed kidney damages following parenteral application of cyclodextrins.

The renal effect of a 12 days treatment with 50 mg/kg/day β-CD, MM-β-CD, DM-β-CD and TM-β-CD were studied on rabbits, with emphasis on the glomerural circulation of the nephron, and in the ultrastructural changes of the epithelial cells of the proximal tubule.

MATERIALS and METHODS

The β-CD was a product of CHINOIN, Budapest /weight loss on drying was 14 %/. The purity of the dry substance was of analytical grade. The substituted cyclodextrins:

MM-β-CD: heptakis-6-O-monomethyl-β-CD
DM-β-CD: heptakis-2,6-di-O-methyl-β-CD
TM-β-CD: heptakis-2,3,6-tri-O-methyl-β-CD

were prepared by dr. A. Lipták /Institute of Biochemistry of the Kossuth University, Debrecen/. These substances were at least of 95 % homogenity /ascertained by TLC and NMR investigations/.

The examinations were carried out in ten week old rabbits, Oryctologus cuniculus /domesticus/, of 1,700 to 2,100 g weight. The treatment lasted twelve days, during this time doses of 50 mg/kg of β-CD, MM-β-CD, DM-β-CD and TM-β-CD, respectively, were administered every day intramulcularly.

The blood supply and the structural state of the glomeruli of the nephron were studied by light microscopy. The material excised from the renal cortex was fixed in 80 % ethanol, then embedded in paraffin. The five μm thick sections were stained with Delafield's hematoxylin and Chromotrop 2R and studied by a Zeiss Amplival microscop.

The number of the open and closed glomeruli, i.e. those supplied with blood, and those excluded from the circulation, respectively, was determined. From the changes in the number of the open glomeruli charted in a diagram, it may be concluded on the blood supply of the renal cortex.

For the studies by electron microscopy, the material was cut to the size required by a McIlwain tissue chopper, then fixed at 4 °C with 3 % glutaraldehyde, postfixed with 1 % OsO_4 solution in 0,1 M collidine buffer at pH 7,2 (3). The preparations were embedded in Araldite, and sliced by LKB Ultratome III ultramicrotome. The ultrathin sections were stained with lead citrate (6). The fine structure of the epithelial cells of the proximal tubule was studied by TESLA BS 540 transmission electron microscop.

RESULTS

Light Microscopy

It is a structural characteristic of the renal cortex in rabbit that in the nephron population about half of the glomeruli of the Malpighian body are open, the capillaries in these being dilated and icluding only a few blood elements, whereas in the other half part of cases, the glomeruli are closed, i.e. the capillaries become narrower, include a number of blood elements, and get in part or entirely choked /Fig.1./. The phenomenon points to the intermittent functioning of the glomeruli. Under the influence of the β-CD and its derivatives, the proportion between the open and the closed glomeruli shifts, involving a modification in the renal hemodynamics, too. The β-CD and the DM-β-CD increase the number of the open glomeruli, and also the intensity of the renal circulation grows hereby, while the MM-β-CD and the TM-β-CD take effect in the opposite direction /Fig.2./. The β-CD and its methylated derivatives induce in certain cases degeneration of the Malpighian body, too./The number of the capillaries in the glomerulus decreases, the latter turns into being lobularly structured, the wall of the afferent arteriole thickens /Fig.2./.

Electron Microscopy

The ultrastructure of the epithelial cells in the proximal tubule of the intact kidney shows the characteristic signs of a transport in the apicobasal direction: an active vesiculation of substances in the cytoplasma extensions in the basal zone. The vesicotubular system of the cytoplasma has a normal structure, and intact limiting membranes /Fig.3./.

The β-CD modifies the state of the epithelial cell. The transport in the apicobasal direction increases as it is indicated by the great number of vesicles to be found in the apical zone of the cell, by a quantity of larger-sized lysosomes, and an increased content of deposits in the basal extensions of cytoplasma /Fig.4./. In the intermediate and basal part of the cell necrotizing mitochondria can be observed.

The effect of the MM-β-CD manifests itself in causing a progressive cellular necrosis. Figure 5. shows a number of lysosomes, also giant ones among them, degenerating mitochondria as well as cytosegresomes isolating the necrotizing areas in the cytoplasma. The active transport is indicated by the dense material filling the interstice of the basal zone of the cell.

The DM-β-CD has a similar effect as the unsubstituted molecule, however the structural changes and the influence on function of epithelial cells is somewhat more expressed. This presents itself in developing numerous lysosomes and in filling the interstice with the dense material /Fig.6./. In addition, the vesicles in the apical zone as well as the mitochondria in the intermediate and basal areas are swelling and necrotized.

The effect of the TM-β-CD results in breaking down the basal structure of the epithelial cell. The vesicotubular system becomes rearranged in columns reaching across the cell in the apicobasal direction, forming agglomerated dark, amorphous material /Fig.7./. Giant lysosomes including acicular crystals appear in the cytoplasma.

DISCUSSION

No fundamental difference was found between the renal toxicity of parenterally administered β-cyclodextrin and its methylated derivatives. This is illustrated clearly by the modified ratio of the open and closed glomeruli, and by the structural changes of the vesicotubular system of the epithelial cells.

Necrosis in the epithelial cells is assumed to be connected with increased activity and developing of lysosomes (2). Like β-CD and its methylated derivatives, mannitol, sucrose and dextran induce, too, an increase in the lysosomal activity (1,4,5), however, there is an essential difference in the final results of the processes: in the latter cases, it does not come to the development of necrosis. Due to the slow recirculation, the cyclodextrins are continuously reabsorbed from the glomerular filtrate and accumulate in the epithelial cell of the proximal tubule. These substances induce there in a way not yet known exactly lytic processes producing cytosegresomes and multivesicular bodies, initiating hereby the cell necrosis. The final effect of various cyclodextrin derivatives differ from each other in whether the blood supply of the nephron population becomes increased or reduced under the influence of the changes induced by them in the renal hemodynamics. In cases of open glomeruli, the cyclodextrins get from the glomerular filtrate into the epithelial cells, otherwise, however, it does not come to their reabsorption. Thus, at least in first approach, the nephrotoxic effect of the cyclodextrins depend on whether the substances in question get or do not get to be filtered in the glomeruli. It is easy to see that, after all, this depends on the blood supply of the glomeruli. It is noteworthy, that the solubility of the mentioned substances is very different: the solubility of MM-β-CD is about 3 fold, of the TM-β-CD 10 fold, while of DM-β-CD is at least 25 fold higher than that of the unsubstituted β-CD.

Attention should be paid to that observation, that the β-CD and the DM-β-CD exert their effect immediately on the cell whereas the MM-β-CD and the TM-β-CD do not or, at least, essentially not so because of reducing the blood supply of the nephron, and herewith the filtration, too. Despite of this, the most marked necrosis of epithelial cells may be seen in these latter cases. The phenomenon seems to be connected with the hypoxia developing as a result of the decline of blood supply of the nephron. The epithelial cells are sensible to hypoxia as it is indicated otherwise by the swelling and necrosis of their mitochondria, too. Giant lysosomes including acicular crystals appear under such circumstances as an indication of the irreversible changes taking place in the cell.

In intact rabbits, the renal glomeruli are approximately half-and-half open and closed, respectively. Relying upon this fact, it may be established that the nephrons function intermittently. The β-CD and its methylated derivatives effect a change in this state by making possible an increase or a decline in the circulatory activity. In whatever direction this process may happen, it excercise an influence on the renal circulation.

Apart from influencing the blood supply of the glomeruli of the nephron, the β-CD and its methylated derivatives have a necrotizing effect on the glomerular capillaries and the afferent arterioles. Therefore parenteral application of β-CD and its derivatives will be limited to relatively low doses. No attempt has been yet made to find a non-toxic level, newertheless the studied 50 mg/kg/day dosis is considered to be a provocative, toxic level if administered parenterally.

Literature Cited

1. Diomi, P., Ericsson, J.L.E., Matheson, N.A., Shearer, J.R.: Lab.Invest. 22 355 (1970)
2. Frank, D.W., Gray, J.E., Weaver, R.N.: Am.J.Pathol. 83 367 (1976)
3. Karnovsky, M.J.: J.Cell.Biol. 27 137A (1965)
4. Maunsbach, A.B., Madden, S.C., Latta, H.: Lab.Invest. 11 421 (1962)
5. Trump, B.F., Janigan, B.T.: Lab.Invest. 11 395 (1962)
6. Venable, J.H., Goggeshall, R.: J.Cell Biol. 25 407(1964)

FIGURES

Figure 1. Light microscopical structure of the open /A/, closed /B/ and necrotizing glomeruli /C/ in the renal cortex. x 380.

Figure 2. Effects of the cyclodextrines on the glomerular intermission in the rabbit kidney.
A = β-CD, B = MM-β-CD, C = DM-β-CD, D = TM-β-CD
■ = control, □ = therapy

Figure 3. Fine structure of the epithelial cell of the proximal tubule in the intact rabbit kidney. x 40,000.

Figure 4. The effect of β-CD on the epithelial cell: the activity of apicobasal transport and the lysosome formation increases. x 40,000.

Figure 5. The effect of MM-β-CD on the epithelial cell: Giant lysosomes, cytosegresomes and necrotizing mitochondria appear in the cytoplasm. x 40,000.

Figure 6. The effect of DM-β-CD on the epithelial cell: the activity of apicobasal transport and the lysosome formation are increased. x 40,000.

Figure 7. The effect of TM-β-CD on the epithelial cell: acicular crystal appears in the giant lysosome, and the necrotizing cell organelles accumulate in the interstice. x 40,000.

ABBREVIATIONS

AA = afferent arteriole, AC = acicular crystal, AP = apical pit, AV = apical vesicle, BC = Bowman's capsule, BM = basal membrane, C = capillary lumen, CS = cytosegresome, ER = endoplasmic reticulum, G = Golgi apparatus, I = interstice, L = lysosome, LAC = lysosome with acicular crystal, M = mitochondrium, MV = microvillus, MW = multivesicular body, N = nucleus.

Note: The light and electron microscopical figures are drawing illustrations were prepared after original photo demonstration!

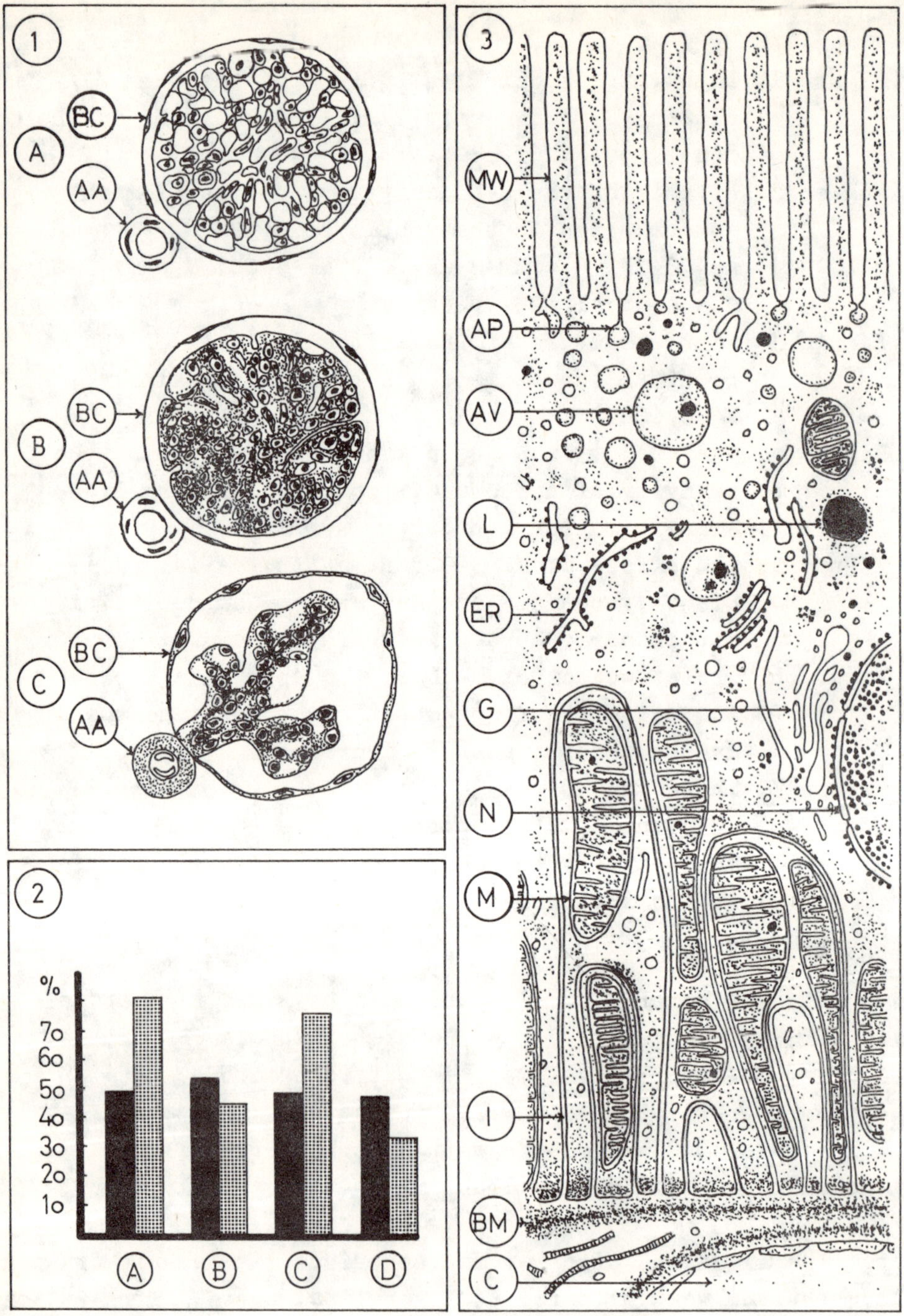
1
BC
A
AA
B
BC
AA
C
BC
AA
2
%
70
60
50
40
30
20
10
A
B
C
D
3
MW
AP
AV
L
ER
G
N
M
I
BM
C

Figs 1-3

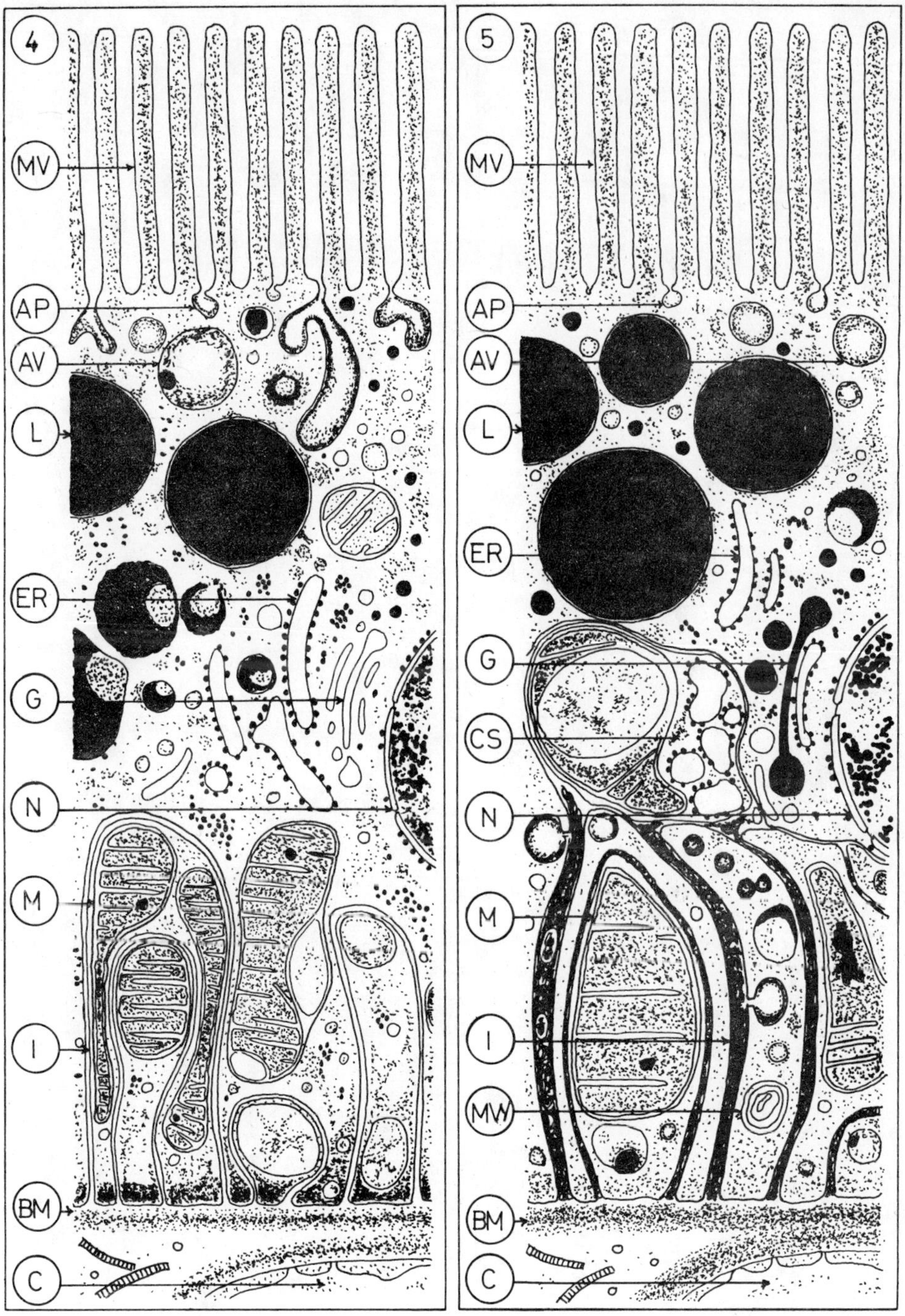

Figs 4-5

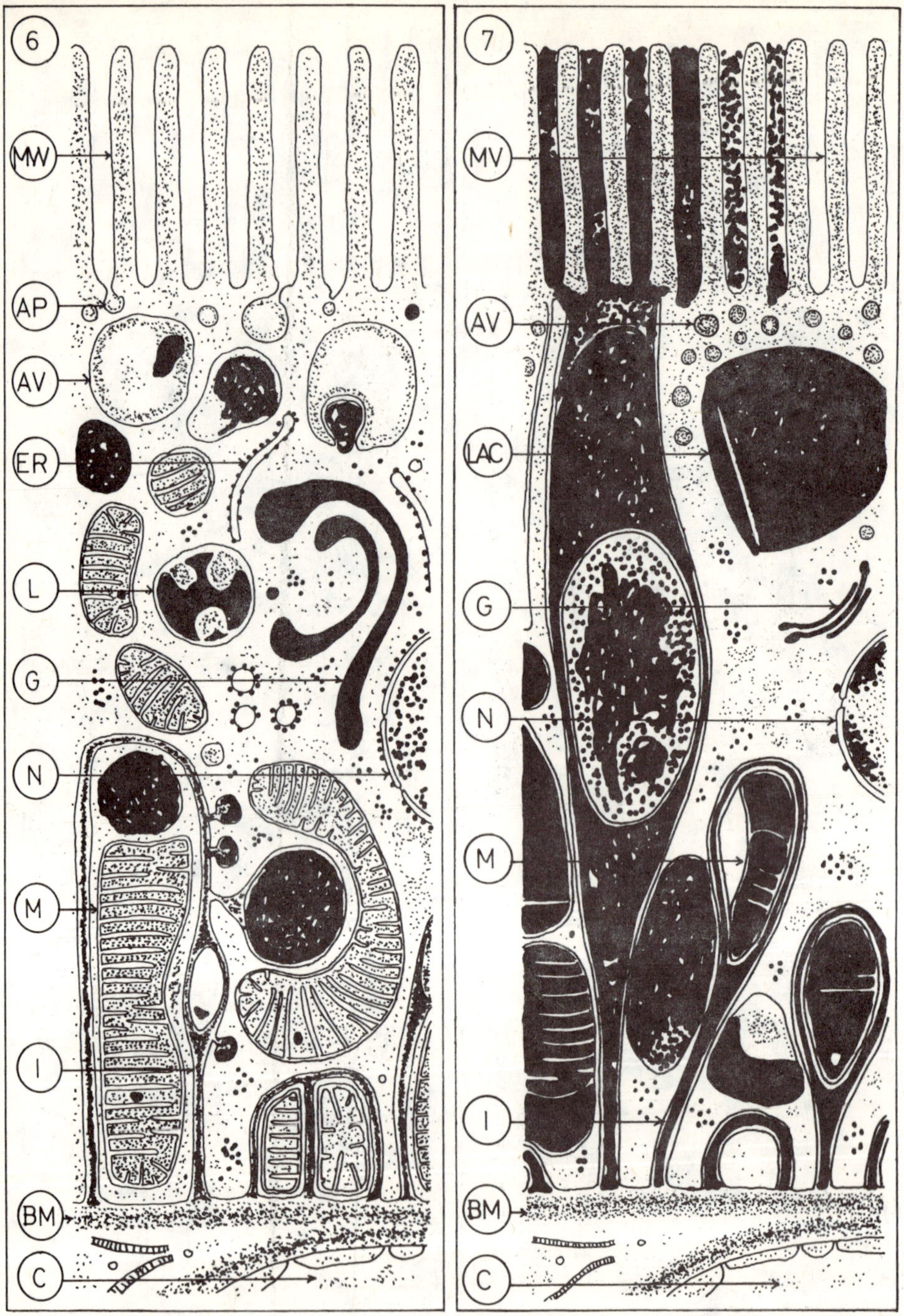

Figs 6-7

THE PURIFICATION AND PROPERTIES OF CYCLODEXTRIN GLYCOSYLTRANSFERASE (CGT-ase) OF BACILLUS MACERANS

G. SERES, [1] E. LÁSZLÓ[2]

[1] CHINOIN Pharmaceutical-Chemical Works, Tó u. 1-5
H-1045 Budapest, Hungary
[2] Institute for Agricultural-Chemical Technology,
Technical University, Gellért tér 4
H-1111 Budapest, Hungary

SUMMARY

The CGT-ase was isolated as an extracellular enzyme from filtrates of three day aged B. macerans cultures. The CGT-ase content of filtrate was purified about 250-fold with affinity chromatography and subsequent gel filtration on SEPHACRYL-S-200 medium. The resulting enzymatically active fractions were pooled and identified as pure CGT-ase, demonstrated by agarose IEF and ion exchange chromatography techniques.

The purified CGT-ase had a molecular weight of 78 K and consisted of a single polypeptide chain established with standard gel filtration procedure on SEPHAROSE CL-6B according to ANSARI and had an isoelectric point of 4,5. The purified enzyme was stable at pH 6,0 and lost its activity at 60 °C after a 30 min incubation. The activity and temperature optima were pH 5,6 and 45 °C respectively.

The main reaction products of CGT-ase action on amylose were α-and β-cyclodextrins, produced in a consecutive reaction pattern.

Studying the IEF patterns of fermentation filtrates it was concluded that CGT-ase activity synthetised by B. macerans is primarily located in the intracellular space and becames extracellular after autolysis of bacterial cells only.

In the fermentation broth another amylolytic factor was found which was synthetised mainly at earlier stage e.g. at the proliferation of cells. This new enzyme was also partially purified.

I. Int. Symp. on Cyclodextrins
Budapest, 1981

2.9.

A RAPID METHOD FOR DETERMINATION OF CGT-ASE ACTIVITY

Zs. PÉTERFI, G. SERES

CHINOIN Pharmaceutical-Chemical Works, Tó u. 1-5
H-1045 Budapest, Hungary

SUMMARY

The known methods for determination of CGT-ase activity are either nonspecific e.g. dextrinizing assay according to KITAHATA et al., or highly subjective e.g. microscopic test reported by TILDEN et al.; moreover they are rather time consuming, as e.g. the glycosyltransferase activity of THOMA et al.

The proposed new method is free of these disadvantages. The CGT-ase action is irreversible under the applied reaction conditions. The assay based on the phenomena that CGT-ase splits the alfa-CD content of a reaction mixture and to the newly formed maltohexaose chain covalently bonds one molecule p-nitrophenyl-beta-D-glucopyranoside (PNFG). Simultaneously the resulting maltoheptaose-PNFG glycoside is hydrolysed to maltose by hog pancreatic alfa-amylase content of the reaction mixture. The amount of maltose - proportional to the CGT-ase activity - was measured as reducing sugar with common DNS method.

The optimised reaction conditions, the CD catalysed hydrolysis of PNFG, the rates of glycosyltransferase coupling reactions with α-CD and PNFG, and K_m values were estimated.

To demonstrate that this method is insensitive against common alfa-amylases of fermentation broth, the assay was accomplished with samples containing amylases of B. subtilis and hog pancreas. No significant effect was observed, i.e. under the applied circumstances the hydrolitic alfa-amylases do not split the α-CD ring.

I.Int.Symp. on Cyclodextrins
Budapest, 1981. 2.10.

ENZYMATIC INVESTIGATIONS WITH CYCLODEXTRINS

I. JODÁL

Institute of Biochemistry, Kossuth Lajos University
Egyetem tér 1
H-4010 Debrecen, Hungary

SUMMARY

It is known that the methylated analogues of β-cyclodextrin dissolve in cold water 10-20 times better than β-cyclodextrin itself and have good complex-forming property.

It has been investigated how these cyclodextrin derivatives influence the activity of some enzymes.

The results of the experiments with lipase indicate that the rate of glyceride-hydrolysis is higher in the aqueous solution of dimethyl-β-cyclodextrin (10 %) than in the presence of the bile as a natural emulgeator.

The activity of the alkaline phosphatase rises both in the presence of β-cyclodextrin and dimethyl-β-cyclodextrin.

In the case of α-amylase (from Aspergillus oryzae) the inhibition of the activity by dimethyl-β-cyclodextrin is similar as that of β-cyclodextrin.

CYCLODEXTRIN-COMPLEXES

I.Int.Symp. on Cyclodextrins
Budapest, 1981 3.1.

STRUCTURE ASPECTS OF CYCLODEXTRIN INCLUSION COMPOUNDS

WOLFRAM SAENGER

Institut für Kristallographie der Freien Universität Berlin
Takustr. 6
1000 Berlin 33, FRG

SUMMARY

α-,ß-, and γ-cyclodextrins have been crystallized as "empty" hydrates and as inclusion complexes and were subjected to X-ray and neutron diffraction analyses. In all cases, the glucoses occur in C1 chair conformation with O_6H hydroxyls directed "away" from the cavity except if hydrogen bonded to included guest molecules. Going from α- to γ-cyclodextrin, the diameter of the macrocycle increases with concomitant reduction of $C_{1'}$-O_4-C_4 angle from 119° to 112.6° and widening of $O_4 \ldots O_{4'}$ distances from 4.23 Å to 4.48 Å. In ß- and γ-cyclodextrins, the intramolecular $O_2 \ldots O_{3'}$ interglucose hydrogen bonds are about 2.83 Å on average and indicative for strong interactions stabilizing the "round" structure of the macrocycle. In the smaller α-cyclodextrin, however, this distance is 3.0 Å, corresponding to weak bonding and considerable flexibility of the macrocycle, leading to an "induced-fit" type complex formation. Host-guest interactions are mainly of the type described by London dispersion forces and dipole-dipole as well as hydrogen bonding forces contribute as well. In crystalline adducts, guests are either trapped in individual cavities closed on both sides by adjacent cyclodextrins to form cage-like structures, or in channel-like cavities produced by stacking of cyclodextrin rings like coins in a roll.

INTRODUCTION

Structural studies on cyclodextrin inclusion compounds and their dynamic behaviour in aqueous solution can be performed using circular dichroism, UV absorption and nuclear magnetic resonance techniques (1, 2). If, however, details concerning host-guest interactions and cyclodextrin conformation are of interest, recourse has to be taken to X-ray or neutron crystallographic methods. In the following, we will concentrate on results derived on their basis.

There have been a great number of crystallographic analyses worked out with α-cyclodextrin as host, some with ß-cyclodextrin and only two, thus far, with γ-cyclodextrin (2). Of the higher known homologs (3), no crystal structure has been carried out. Because individual discussion of all the known crystal structures would lead far beyond the scope of this article, only the salient features of cyclodextrin structure summarized in Tables I, II and host-guest interactions will be discussed.

OVERALL STRUCTURAL FEATURES OF CYCLODEXTRINS

Cyclodextrins are torus-shaped, circular molecules with central cavities in which guest molecules of suitable size can be accomodated. The requirement for this inclusion formation appears to be only spatial because all kinds of guest molecules, ranging from molecular to ionic in character, are accepted. One of the main characteristics of cyclodextrins is that they are hydrophilic outside owing to the secondary and primary hydroxyl groups occupying both rims of the torus and hydrophobic within the cavity because it is lined by C-H groups and ether-like oxygens O_4. O_5. The overall appearance of cyclodextrins is, in fact, not that of a ring but rather of a truncated cone with the wide "open" side formed by secondary O_2H, O_3H hydroxyls

Table I. Some Physical Properties of Cyclodextrins (2).

Cyclodextrin	No. of glucoses	Mol wt.	Water Solubility (a)	Diameter (Å) (b) Cavity	External	Height (Å) (b) of torus
α	6	972	14.5	5.0±0.3	14.6±0.4	7.9±0.1
β	7	1135	1.85	6.2±0.3	15.4±0.4	7.9±0.1
γ	8	1297	23.2	7.9±0.4	17.5±0.4	7.9±0.1

a) g/100 ml solution at room temperature

b) as measured from CPK models

Table II. Some averaged data for α-,β- and γ-cyclodextrin molecules (13). Torsion angles ϕ, Ψ are defined as ϕ, $O_4 \cdots C_1 - O_{4'} - C_{4'}$ and Ψ, $C_1 - O_{4'} - C_{4'} \cdots O_{4''}$; primed and double primed atoms belong to adjacent glucoses.

Cyclodextrin	Angle (°) $C_{1'}-O_4-C_4$	Torsion Angle (°) ϕ	Ψ	Distance (Å) $O_4 \cdots O_{4'}$	Distance (Å) $O_2 \cdots O_{3'}$
α	119.0	166	-169	4.23	3.00
β	117.7	169	-172	4.39	2.86
γ	112.6	165	-169	4.48	2.81

whereas primary O_6H hydroxyls are located on the narrower, more closed side (Figure 1).

In all the crystal structure analyses of cyclodextrin inclusion complexes, the glucoses are in classical $^4C_1$ chair conformation with bonding geometry described in Figure 2. The exact structure of glucoses, expressed by $O_4...O_4$ distances and endocyclic torsion angles (4, 5) varies somewhat depending on cyclodextrin ring size when going from α- to γ-cyclodextrin (Table II), on distortions induced by inclusion of bulky guest molecules and, in more general terms, on crystal packing forces (6). In the present discussion, we can assume glucoses to behave as more or less "rigid" building blocks with two main degrees of freedom, namely rotation about glucosidic links C_4-O_4 and C_1-$O_{4'}$ (where primed atoms denote part of adjacent glucose), and rotations of O_6 primary hydroxyl groups about the C_5-C_6 bond. It is noteworthy that, in a broad sense, in all crystallographically investigated α-,ß- and γ-cyclodextrins, O_4 atoms do not deviate significantly from the common mean plane, i.e. cyclodextrins display truly ring-shaped, toroidal geometry.

Looking at structural parameters more closely, we find that in most cases, except for the "empty" α-cyclodextrin·$6H_2O$ complex with collapsed geometry, Figure 1, torsional angles around glucosidic bonds are fairly restricted in order to enable inter-glucose hydrogen bonding between O_2H and O_3H hydroxyl groups. On the other hand, orientation about the C_5-C_6 bond, defined by torsion angle O_5-C_5-C_6-O_6, is always staggered and indicates a preferred range, -gauche (-60°), directing the primary hydroxyl O_6H "away" from the cyclodextrin cavity. If hydrogen bonding to a guest molecule occurs, O_6 can rotate towards the cavity, changing the range to +gauche (+60°). The trans orientation has not yet been observed.

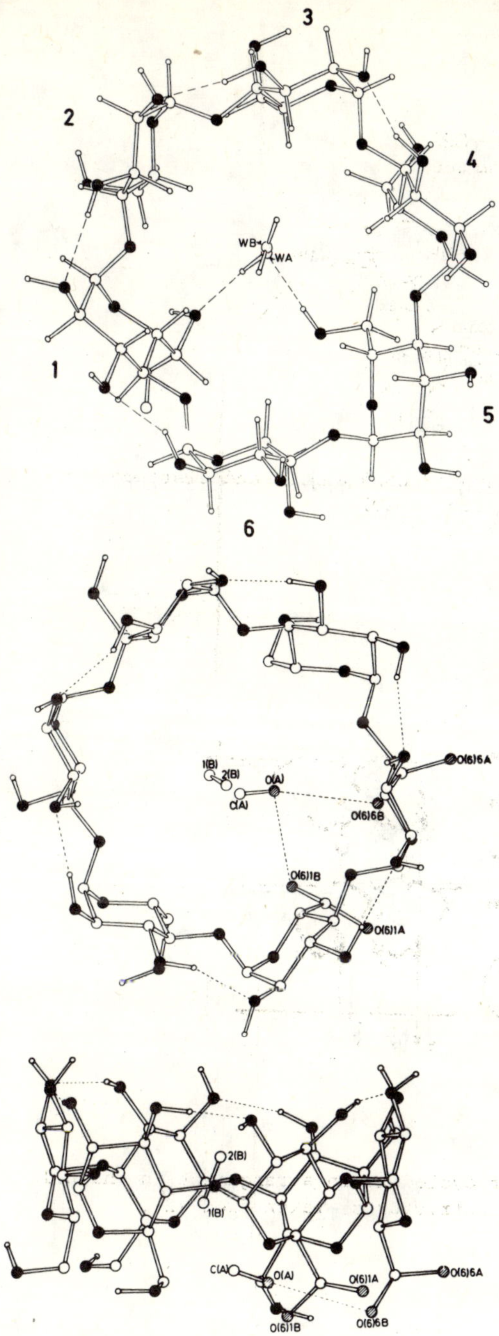

Fig. 1: Results from X-ray structure analyses of α-cyclodextrin complexed with H_2O (top) and with methanol (middle and bottom). o = H, O = C, ● = O. O-H...O hydrogen bonds indicated by dashed lines form a ring between O_2H and O_3H groups of adjacent glucoses in the methanol complex. The ring is disrupted in the water adduct (top) where glucose 5 is rotated "inward" to form a hydrogen bond to included water WA. In the methanol complex, methanol molecules are at only half (statistical) occupation. One of these is hydrogen bonded to two twofold disordered O(6) hydroxyls (shown hatched) and the other one is in van der Waals contact with the interior of the α-cyclodextrin cavity. Taken from (2).

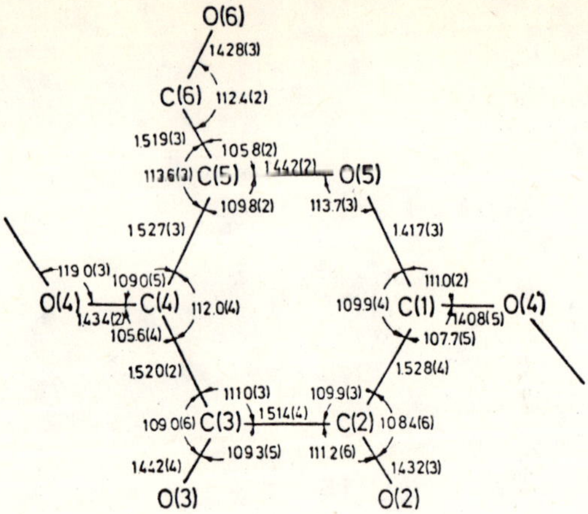

Fig. 2: Nomenclature of atoms and averaged bond angles and distances in an α-cyclodextrin glucose, taken from ref. (20).

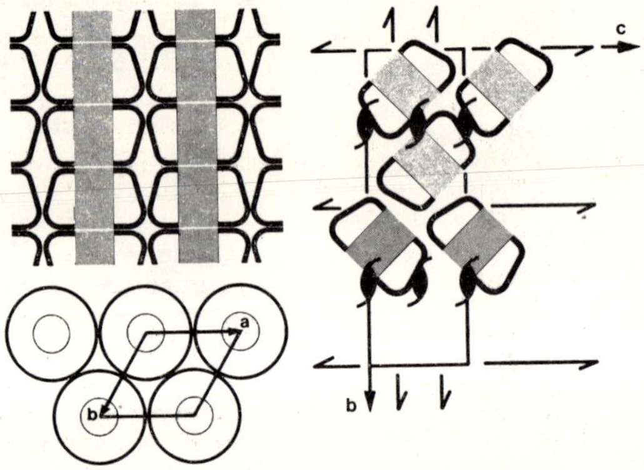

Fig. 3: Schematic representation of cyclodextrin arrangement in channel (left) and cage-type (right) crystal structures. Taken from (2).

INTRAMOLECULAR HYDROGEN BONDS

In all cyclodextrin crystal structures, hydrogen bonding between adjacent glucoses was found, with $O_2 \ldots O_{3'}$ distances in the usually accepted range, 2.8 to 3.1 Å, Table II. Hydrogen bonding between these secondary hydroxyl groups appears to be preferred because their pK, around 12-13, is relatively low, suggesting enhanced polarization and therefore good hydrogen bond donor and acceptor functions (7, 8). The thus formed ring of hydrogen bonds is, in general, fully established in ß- and γ-cyclodextrins but in α-cyclodextrin, where the curvature of the ring produces, on average, longer $O_2 \ldots O_{3'}$ separations (Table II), hydrogen bonding is weaker and can break down, as shown for the "empty" hexahydrate where four $O_2 \ldots O_{3'}$ hydrogen bonds are formed but two are broken, Figure 1 (9, 10). These results are in agreement with spectroscopic data which indicate (11) that ß-cyclodextrin is rigid in aqueous solution whereas α-cyclodextrin is flexible and undergoes conformational change if inclusion of a guest molecule occurs, as proposed in the "induced-fit" concept for proteins (12).

CAGE AND CHANNEL TYPE CRYSTAL STRUCTURES

Depending on size and nature of guest molecules, cyclodextrin inclusion compounds can crystallize in two different forms to produce cage- and channel-type complexes, Figure 3. In the latter, cyclodextrins pile up coaxially like coins in a roll with the cavities aligned into "endless" channels. These are filled with guest molecules which are often disordered in the well-ordered cyclodextrin matrix. In these channel-type complexes, cyclodextrin molecules can be oriented head-to-tail or head-to-head. Thus far, only in one crystal structure, the γ-cyclodextrin·propanol complex, both kinds of stacks are observed simultaneously (13). In ß-cyclodextrin, the head-to-head complex formation is largely preferred, with the wider O_2, O_3 sides hydrogen bonded together to provide a large hydrophobic cavity and in α-cyclodextrin, both head-to-head and head-to-tail orientations are found.

With small molecular guests and especially with water, cage type crystal structures are favoured. In these, cyclodextrin rings are arranged in herringbone pattern such that the cavity of one ring is closed on both ends by contiguous cyclodextrins to create cage-like structures which enclose the guest molecules. In some α-cyclodextrin complexes with guests containing benzene moieties such as p-nitroaniline, cage-type groupings are found where α-cyclodextrin rings are arranged parallel to each other in sheets and consecutive sheets are displaced relative to each other to produce a motif reminiscent of bricks in a wall (14, 15). The reason for this particular packing pattern could be an ellipsoidal distortion of the α-cyclodextrin ring caused by insertion of the flat aromatic guest edge-on into the cavity. These distortions are only observed with α-cyclodextrin and not with higher homologs which also do not crystallize in this pattern.

HOST-GUEST INTERACTIONS

Of special interest are host-guest interactions which are readily derived from crystal structure analyses. These interactions are predominantly of the van der Waals type, as indicated by interatomic distances, and suggest that, in general, strong and specific host-guest contacts do not occur, a finding in overall agreement with thermodynamic data. Forces that stabilize cyclodextrin inclusion compounds are therefore of the type described by London dispersion attraction, by dipole-dipole interactions and, in some cases, by hydrogen bonding.

Even in the near-black crystals formed between α-cyclodextrin and polyiodide, the main host-guest interaction is provided only by hydrogen...iodine van der Waals contacts. The oxygen...iodine charge-transfer forces previously suggested for starch-polyiodide complexes do not play a role because I...O distances are far beyond the expected van der Waals contact of 3.55 Å (16). In these particular crystal structures, iodine atoms are located within channels formed by head-to-head arranged α-cyclodextrin molecules and are so close together that a near-equidistant I...I...I distribution results, indicating charge-transfer along the polyiodide chain which is responsible for the deep blue color.

In most cyclodextrin inclusion complexes, guest molecules located within the cyclodextrin cavity are not engaged in hydrogen bonding with hydroxyl groups of the enclosing cyclodextrin, although they had the potential to do so. This holds especially for larger guests with only little orientational or rotational freedom whereas small guest molecules such as water and alcohols are able to form such hydrogen bonds to primary O_6H hydroxyls directed "inward" in +gauche range whereas secondary O_2H and O_3H hydroxyls do not participate.

Disorder within the cyclodextrin cavity produces considerable problems in many of the investigated crystal structure analyses. It can comprise rather simple statistical disorder resulting in only partially occupied sites as, for instance, in complexes of β- and γ-cyclodextrin with water (17, 18) and of α-cyclodextrin with potassium acetate (19), methanol (20) and krypton (21), which appears to "rattle" in the ~5 Å wide cavity in order to fill it with its only ~3.8 Å diameter sphere. In one especially pathological case, γ-cyclodextrin complexed with n-propanol and water (13), the n-propanol located within the channel provided by γ-cyclodextrin is so badly disordered that it could not be located at all whereas water molecules filling interstices between γ-cyclodextrin stacks are more or less ordered.

Literature Cited

1. Bender, M.L., Komiyama, M.: Cylodextrin Chemistry, Springer Verlag, Berlin-Heidelberg-New York 1978.
2. Saenger. W.: Angew. Chem. Int. Ed. Engl. 19, 344 (1980).
3. Pulley, A.O., French, D.: Biochem. Biophys. Res. Commun. 5, 11 (1961).
4. French, A.D., Murphy, V.G.: Carbohyd. Res. 27, 391 (1973).
5. Pensak, D.A., French, A.D.: Carbohyd. Res. 87, 1 (1980).
6. Harata, K.: Bull. Chem. Soc. Japan 50, 1416 (1977).
7. Van Etten, R.L., Clowes, G.A., Sebastian, J.F., Bender, M.L.: J. Amer. Chem. Soc. 89, 3253 (1967).
8. Chin, T.-F., Chung, P.-H., Lach, J.L.: J. Pharm. Sci. 55, 44 (1968).
9. Manor, P.C., Saenger, W.: J. Amer. Chem. Soc. 96, 3630 (1974).
10. Klar, B., Hingerty, B., Saenger, W.: Acta Crystallogr. B36, 1154 (1980).
11. Rees, D.A.: J. Chem. Soc. B, 877 (1970).
12. Saenger, W., Noltemeyer, M., Manor, P.C., Hingerty, B., Klar, B.: Bioorg. Chem. 5, 187 (1976).
13. Lindner, K., Saenger, W.: Biochem. Biophys. Res. Commun. 92, 933 (1980).
14. Harata, K.: Bull. Chem. Soc. Japan 48, 2409 (1975); Saenger, W., Beyer, K., Manor, P.C.: Acta Cryst. B32, 120 (1976).
15. Harata, K.: Bull. Chem. Soc. Japan 51, 1644 (1978).
16. Noltemeyer, M., Saenger, W.: J. Amer. Chem. Soc. 102, 2710 (1980).
17. Lindner, K., Saenger, W.: Angew. Chem. Int. Ed. Engl. 17, 694 (1978).
18. MacLennan, J.M., Stezowski, J.J.: Biochem. Biophys. Res. Commun. 92, 926 (1980).
19. Hybl, A., Rundle, R.E., Williams, D.E.: J. Am. Chem. Soc. 87, 2779 (1965).
20. Hingerty, B., Saenger, W.: J. Am. Chem. Soc. 98, 3357 (1976).
21. Saenger, W., Noltemeyer, M.: Chem. Ber. 109, 503 (1976).

I.Int.Symp. on Cyclodextrins
Budapest, 1981

STRUCTURAL PROPERTIES OF CHEMICALLY MODIFIED CYCLODEXTRINS: THE CRYSTAL STRUCTURE OF AN INCLUSION COMPLEX BETWEEN ADAMANTANOL AND per-2,6-O-METHYL-β-CYCLODEXTRIN

JOHN J. STEZOWSKI, MÁTYÁS CZUGLER, EMIL ECKLE

Institut für Organische Chemie, Biochemie und Isotopenforschung der Universität, Pfaffenwaldring 55
D-7000 Stuttgart 80, FRG

SUMMARY

Crystal structure determinations for several host-substrate complexes between β-cyclodextrin and organic substrate molecules have revealed the frequent occurence of host β-cyclodextrin dimers formed by hydrogen bond association between secondary hydroxyl groups of monomers. The resultant host has an extended toroidal region that often gives rise to host-substrate complexes with formal stoichiometries 2:2, 2:3 and 2:4. It is probable that this host system is also that responsible for the formation of 2:1 complexes with larger organic molecules such as prostaglandins and vitamins.

A crystal structure determination for a complex with a chemically modified β-cyclodextrin has been carried out. The derivative, per-2,6-O-methyl-β-cyclodextrin is one in which dimerization is very unlikely. A 1:1 complex with adamantanol has been observed in the crystalline state. The host molecule is conformationally similar to its parent but methylation has increased the depth of the hydrophobic torus making it probable that 1:1 complexes with larger substrate molecules can be prepared. This derivative is considerably more water soluble than its parent, consequently, it might be a more potent solubilizer for hydrophobic substrate than β-cyclodextrin.

The chemical structure of the cyclodextrins, displayed in Figure 1 for β-cyclodextrin (β-CD), gives rise to a molecule with many interesting chemical and physical properties. Perhaps the most significant, and certainly the most extensively studied, property of the cyclodextrins is their ability to form host-substrate complexes in which the substrate is either wholely or partly included in the hydrophobic torus of the host.

Interest in cyclodextrin host-substrate complexes arises both from their chemical and physical properties. For example, the cyclodextrins display modest catalytic activity toward the hydrolysis of esters and, to a lesser extent, amides. The catalytic process involves the formation of a host-substrate complex prior to the hydrolysis step and is subject to competitive inhibition. These properties have led to the use of cyclodextrins as models for hydrolytic enzymes. Bender and Komiyama (1) have recently reviewed this aspect of cyclodextrin chemistry.

There is great current interest in the fact that complex formation often increases the stability and/or solubility in aqueous solvent of biologically important molecules. Szejtli (2) has presented numerous examples of possible application of

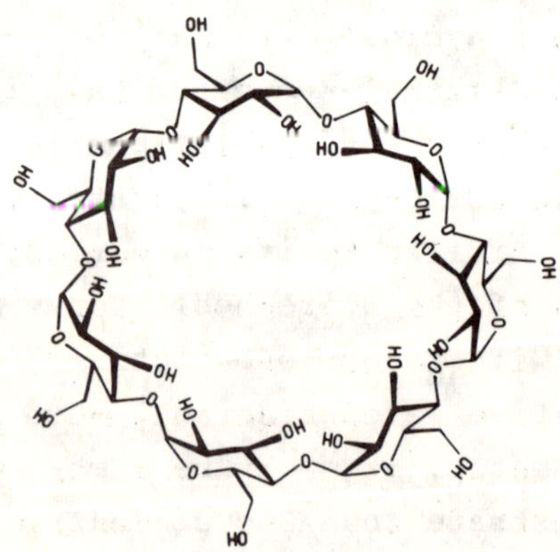

Figure 1. The chemical structure of β-cyclodextrin.

such complexes in the pharmaceutical and food industries.

One of the most effective physical methods for examining the structural properties of host-substrate complexes of the cyclodextrins is by analyzing crystal structure determinations based on single crystal X-ray or neutron diffraction data. Early work in this field concentrated on complexes with α-cyclodextrin (α-CD). Structure determinations from the laboratories of W. Saenger and K. Harata have revealed the geometry of interaction between many different substrate molecules and α-CD.

Our crystallographic studies have been focused on the higher cyclodextrins, particularly on complexes with β-CD and organic substrate. Presently, we are turning our attention to the study of structural properties of chemically modified cyclodextrins and their host substrate complexes. The structure determination for a complex between 2,2',2",2"', 2"",2""', 2""",6,6',6",6"', 6"",6""', 6""" -O-methyl-β-cyclodextrin, henceforth *per*-2,6-O-Me-β-CD, and adamantanol represents the first crystal structure determination in which a chemically modified cyclodextrin serves as the host.

The methylation of β-CD to *per*-2,6-O-Me-β-CD gives rise to a derivative with physical properties that are considerably different from those of its parent. It therefore seems appropriate to review the results of crystal structure determinations for β-CD host-substrate complexes before describing the new complex.

In all cases reported thus far (3-8) crystal structures for complexes of β-CD with organic substrate consist of host molecules associated by hydrogen bonding between secondary hydroxyl groups to give dimers. To date stoichiometries of 2:2, 2:3 and 2:4 (host:substrate) have been observed. Substrate size appears to be important with respect to facilitating formation of the dimer. For example, β-CD crystallized from aqueous methanol or ethanol (9) gives crystals that are apparently isomorphous with those fo the simple hydrate in which the cyclodextrin molecules are not dimerized (10), whereas, crystals obtained from aqueous n-propanol (3,4) or isopropanol (9) contain dimers. Unlike α-CD (11), the conformation of β-CD in the simple hydrate is very similar to that in complexed form, Figure 2.

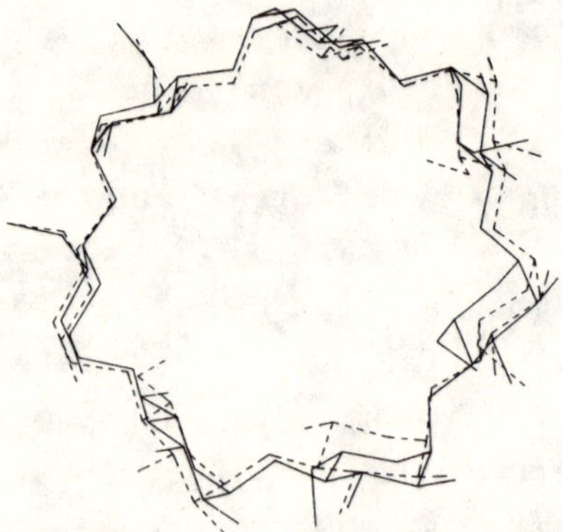

Figure 2. A schematic comparison of the conformations of β-CD in the hydrated and complexed forms. The molecular fit was determined by a least-squares fit of all C and O atoms except the primary hydroxyl oxygen atoms (12).

The crystal structure determinations for β-CD host-substrate complexes have all revealed some degree of disorder in the substrate. This observation remains true even for structures determined with data from crystals cooled to <u>ca</u>. 150 K. Consequently, one is forced to conclude that the "binding forces" are neither very specific nor very strong and that the major driving force for complex formation is most likely entropy. Dimers of β-CD present a considerably extended toroidal region in which substrate can be included. Figure 3 illustrates the structure of the 2:3 complex between β-CD and p-iodophenol. The "third" substrate molecule lies with the plane of the aromatic ring nearly perpendicular to the local 7-fold symmetry axis of the host. There is resolvable disorder in the orientation of this substrate molecule, Figure 4.

The β-CD:(aqueous n-propanol) system has proven to be particularly interesting from the structural point of view. Two polymorphs have been crystallized and their structures

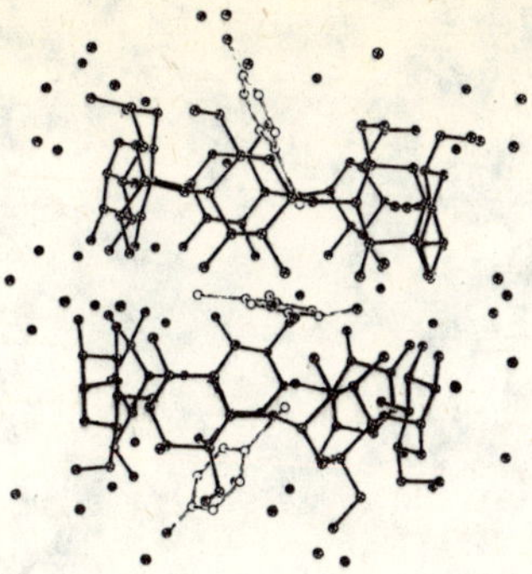

Figure 3. The molecular structure of the 2:3 host-substrate complex between β-CD and p-iodophenol (3).

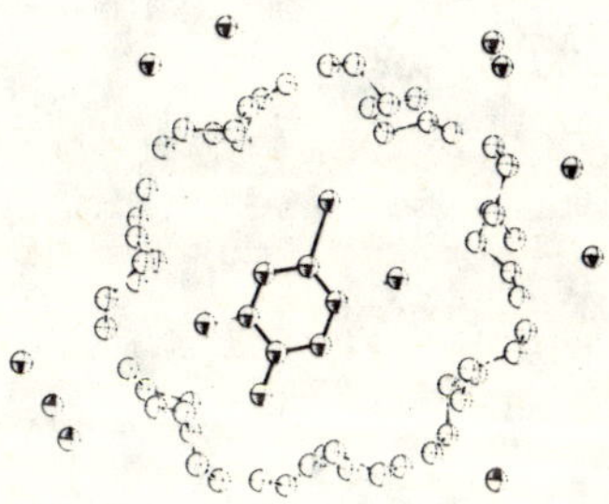

Figure 4. Disorder in the central substrate molecule of the 2:3 β-CD-p-iodophenol complex. The second orientation of the organic substrate is indicated by lighter colored bonds connecting the iodine and oxygen atoms to the ring. Its occupancy factor is less than 50%.

determined (4). One designated C(II) in Figure 5, is obtained by slow cooling of a saturated solution of β-CD in aqueous n-propanol (> 60% n-propanol, v/v). The second, C(I), was prepared by vapor diffusion at room temperature of n-propanol into a similar solution. When crystals of phase C(II) are left in contact with their mother liquor at room temperature, they slowly redissolve with simultaneous appearence of crystals of

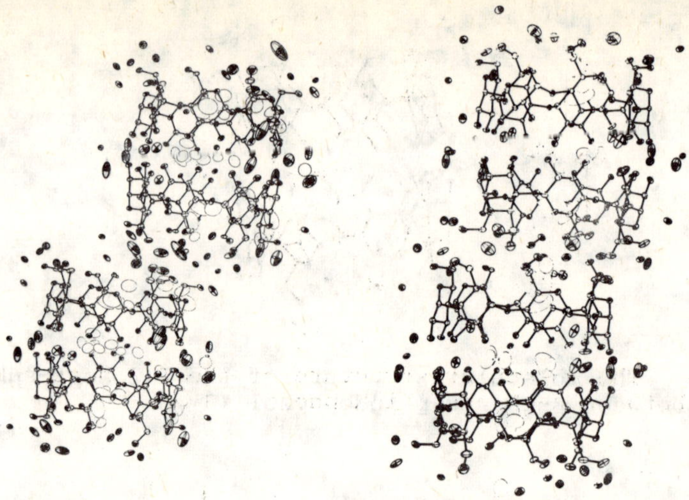

C(I) C(II)

Figure 5. Crystal packing for the thermodynamically stable, C(I), and metastable, C(II), polymorphs of the β-CD:n-propanol host-substrate complexes (4).

phase C(I).

Both crystal forms consist of β-CD dimers with included n-propanol (formal stoichiometry 2:4) but they differ in water content and crystal packing. The dimers display a high degree of conformational integrity as indicated by the superposition (least squares fit) diagram presented in Figure 6. Though it has not been unequivocally demonstrated (13), we have suggested (4,5) that there is considerable reason to believe that the dimer is an important species in concentrated aqueous solutions of β-CD in the presence of organic substrate of suitable molecular size.

As indicated above, we have been turning our attention toward the determination of the structural porperties of chemically modified cyclodextrins. One such derivative, per-2,6-O-Me-β-CD, Figure 7, was thought to be of particular interest because the per-methylation in the 2 position could be expected to eliminate the formation of hydrogen bonded dimers. The pre-

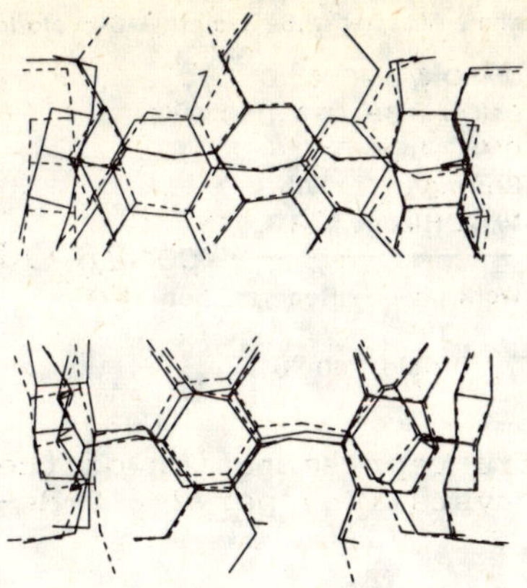

Figure 6. A comparison of the conformations of the β-CD dimers in phases C(I) and C(II).

paration of per-2,6-O-Me-β-CD, carried out by an adaptation of the procedure reported by Boger, et al. (14), is summarized in Figure 8. The crystals of the adamantanol complex (15) were prepared by slow evaporation of an aqueous solution containing a slight excess of substrate.

The complex, Figure 9, displays 1:1 stoichiometry with the crystal packing consisting of columns of complexes stacked "head to tail". As illustrated in Figure 10, there is a modest interpenetration of the host molecules within columns.

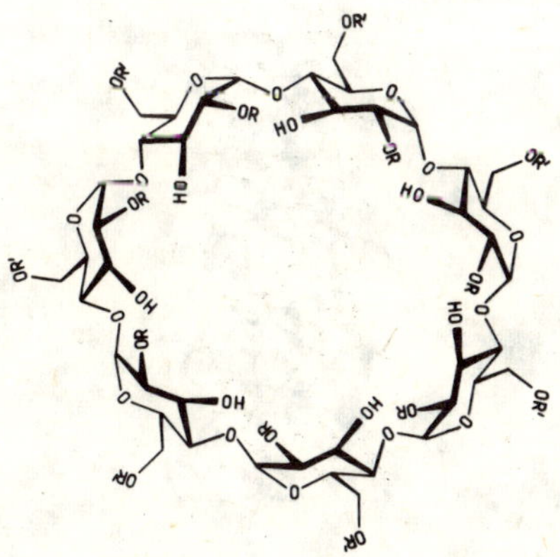

Figure 7. The chemical structure of per-2,6-O-Me-β-CD, R=CH_3 R'=CH_3.

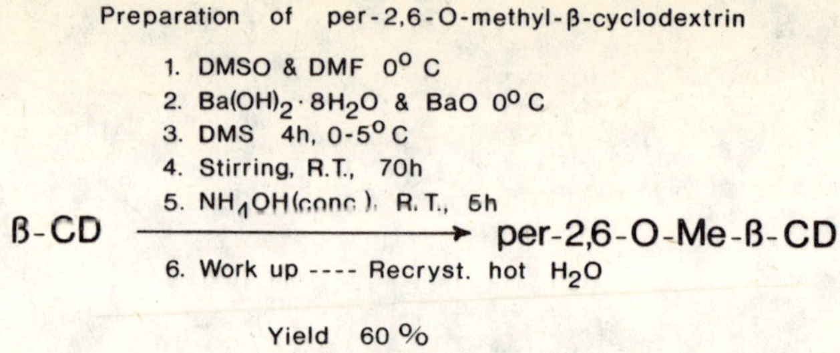

Figure 8. A schematic representation of the preparative method employed to obtain crystalline <u>per</u>-2,6-O-Me-β-CD.

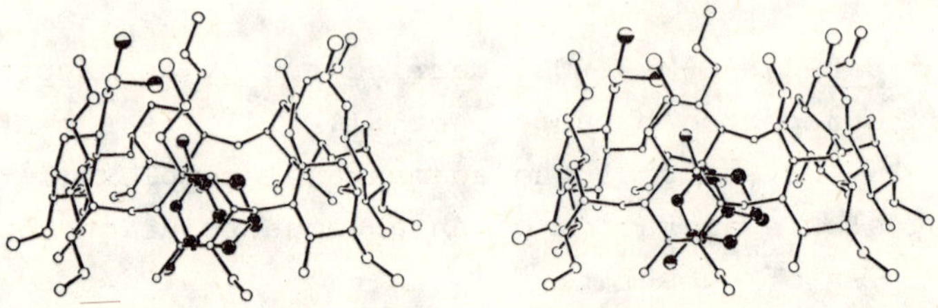

Figure 9. A stereoscopic projection (16) of the molecular structure of the 1:1 <u>per</u>-2,6-O-Me-β-CD:adamantanol complex.

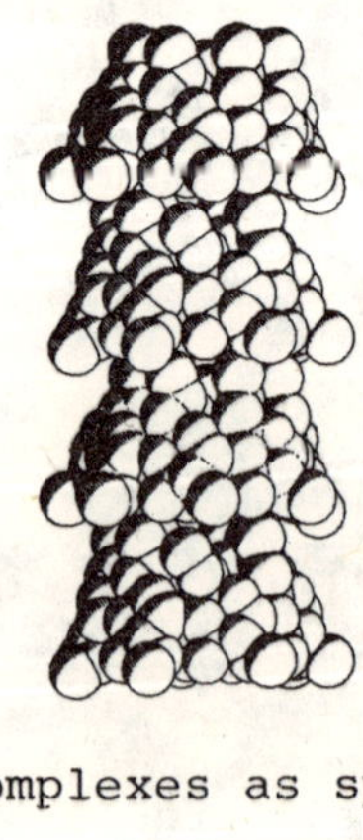

Figure 10. Columns of complexes as space filling models (17).

Once again, there is disorder (resolvable) in the orientation of the included substrate. In one orientation, the adamantanol hydroxyl group is pointed toward the primary (O_6) side of the torus; in the other it is oriented toward the secondary (O_2) side. When the substrate adopts the latter orientation, a methyl group on the primary side is directed toward the center of the torus, thus filling space not occupied by the substrate. These features are illustrated with half atom shading in Figure 9. In each case, one or more water molecules (not shown) form a hydrogen bonded bridge between the host and substrate molecules.

The conformation of per-2,6-O-Me-β-CD in this complex is generally similar to that of β-CD. There is a slight opening of the secondary face as is indicated in Figure 11. The conformations of the C_6, O_6, C_6^{Me} chains are generally such as to extend the depth of the hydrophobic torus. Both of these effects may be due to crystal packing, Figure 10, or may be intrinsic properties of per-6-O-Me-β-CDs. Resolution of that question must await the results of further structural studies.

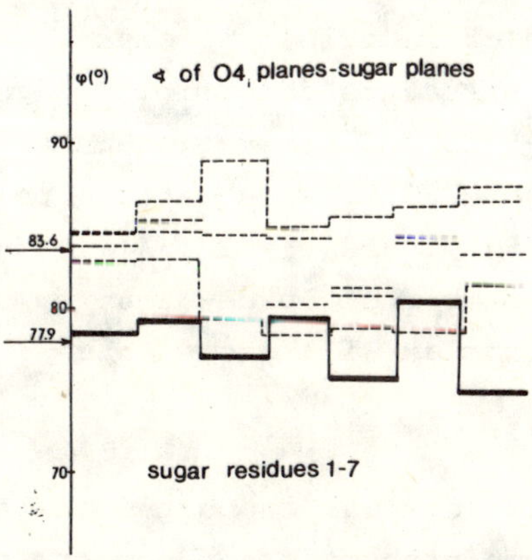

Figure 11. A comparison of the angles between the normal of the mean plane fit to the 7-O_4 atoms and the mean planes fit to atoms C_2, C_3, C_5, & O_5 for per-2,6-O-Me-β-CD (average = 77 deg.) and for β-CD (Average = 83.6 deg). The torus of the modified cyclodextrin is somewhat more open on the secondary side.

The behavior of uncomplexed per-2,6-O-Me-β-CD in aqueous solution deserves some comment here. This derivative displays the opposite solubility-temperature dependence as its parent. In hot water its solubility is very low whereas in cold water it is much more soluble than its parent. Crystals can and have been obtained under both conditions. The space-group is $P2_12_12_1$ in both cases but lattice parameters differ. Those for crystals obtained from hot water are: a = 13.5845(6), b = 17.6530(7) and c = 29.766(1) Å; those from cold water are: a = 13.976(7), b = 20.763(7) and c = 28.807(8) Å. We interpret the similarities and differences in these lattice parameters to be indicative of greater hydration in the crystals obtained from cold water. We expect that crystal structure determinations for these two examples will contribute to our understanding of the hydration of macromolecules and are presently pursuing them. X-ray diffraction data have been measured for both but the structures have not yet been determined.

One more property of per-2,6-O-Me-β-CD merits mention. It is about an order of magnitude more water soluble at room temperature than β-CD. It may therefore be a considerably more effective complexing agent for increasing the solubility of hydrophobic molecules of medicinal interest. We are presently pursuing the crystal structure determination for a complex between per-2,6-O-Me-β-CD and prostaglandin $PGF_{2\alpha}$. The space group symmetry is also $P2_12_12_1$ with lattice parameters: a = 14.1628(7), b = 23.096(1), and c = 27.641(1). Based on the unit cell volume and number of space-group symmetry operators, we suspect that the complex will prove to be of 1:1 stoichiometry.

Acknowledgement: This work was supported in part by the Deutsche Forschungsgemeinschaft grant number Ste 230/3-2. Samples of β-cyclodextrin have been generously supplied by CHINOIN Pharmaceutical and Chemical Works, Budapest.

References:

1) Bender, M.L. and Komiyama, M. "Cyclodextrin Chemistry" Springer-Verlag, Berlin, 1978, 96 pgs.
2) Szejtli, J. Die Stärke-Starch, 29, 26-33 (1977).
3) Stezowski, J.J., Jogun, K.H., Eckle, E., and Bartles, K. Nature, 274, 617-619 (1978).
4) Jogun, K.H. and Stezowski, J.J. Nature, 278, 667-668 (1979).
5) Jogun, K.H., Maclennan, J.M., and Stezowski, J.J., "Fifth European Crystallographic Meeting" Abstracts, 34 (1979).
6) Harding, M.M., Maclennan, J.M. and Paton, R.M., Nature, 274, 621-623 (1978).
7) Hamilton, J.A., Sebesan, M.N., Steinrauf, L.K. and Geddes, A. Biochem. Biophys. Res. Commun., 73, 659-664 (1976).
8) Tokuoka, R., Fujiwara, T., and Tomita, K. Acta Cryst., B37, 1158-1160 (1981).
9) Stezowski, J.J. unpublished observation.
10) Saenger, W. and Lindner, K. Angew. Chem. Int. Ed. Engl., 17, 694-695 (1978).
11) Manor, P.C. and Saenger, W. J. Amer. Chem. Soc., 96, 3630-3639 (1974).
12) Liebman, M.N. "Program HOMLAL" Institute for Cancer Research, Fox Chase-Philadelphia, PA.
13) Harata, K. Bull. Chem. Soc., Jap., 52, 1807-1812 (1979).
14) Boger, J., Corcoran, R.J. and Lehn, J-M. Helv. Chem. Acta, 61, 2190-2218 (1978).
15) Czugler, M., Eckle, E., and Stezowski, J.J. JCS Chem. Commun., in press.
16) Johnson, C.K. "OR TEP II A FORTRAN Thermal Ellipsoid Plot Program for Crystal Structure Illustrations" Rept. No. ORNL-5138 (Oak Ridge National Laboratory), TN., 1971.
17) Motherwell, W.D.S "PLUTO a Program for Plotting Molecular Crystal Structures" University Chemistry Laboratory, Cambridge, England.

APPLICATION OF A STOICHIOMETRIC MODEL OF CYCLODEXTRIN COMPLEX FORMATION

KENNETH A. CONNORS

School of Pharmacy, University of Wisconsin
Madison, WI 53706, U.S.A.

SUMMARY

Let gG represent a substrate S (guest) having 2 binding sites (such as a 1,4-disubstituted benzene) and hH a ligand L (host) having 2 sites, in the case of cyclodextrins the two ends of the cavity. Then there can be four isomeric 1:1 (SL) complexes, four 2:1 (S_2L) complexes, and four 1:2 (SL_2) complexes. If no 2:1 complexes are present, it is inferred that only one end of L can be entered. It follows that there are only two 1:1 complexes (say gH and GH) and one 1:2 complex (gHHG) in the system. Then it can be shown that

$$K_{11} = K_{gH} + K_{GH} \qquad K_{12} = aK_{gH}K_{GH}/K_{11}$$

where K_{gH} and K_{GH} are 1:1 binding constants for sites g and G, K_{11} and K_{12} are experimental stability constants for 1:1 and 1:2 complexing, and $\underline{a}$ is a parameter measuring interaction between the two sites in a 1:2 complex.

K_{11} and K_{12} values are reported for many 1,4-disubstituted benzenes with α-cyclodextrin, and these are interpreted in terms of the model. The dependence of the experimental stability constants on substituent sigma constants is used to establish the principal site of binding. It is concluded that binding site electron density is the primary factor contributing to binding at the site, stability being aided by high site polarizability and opposed by high site polarity. Binding site constants are estimated by making use of chemically reasonable assumptions about the systems. These site constants suggest that substantial fractions of two isomeric 1:1 complexes may co-exist when the two binding sites have similar properties.

Although 1:1 stoichiometry is often assumed in evaluating stability constants for cyclodextrin complex formation, recent observations have revealed the presence of 1:2 complexes in many systems (1-5). (In this paper the substrate S is the guest, the ligand L is the host (cyclodextrin), and stoichiometric relationships are expressed SL, SL_2, etc. The observed stability constants are defined $K_{11} = [SL]/[S][L]$ and $K_{12} = [SL_2]/[SL][L]$.) The existence of such species raises some interesting questions, such as: (i) To what extent does the presence of an unrecognized 1:2 complex compromise the accuracy of a stability constant evaluated without taking it into account? (ii) What are the stabilities, structures, and properties of 1:2 complexes, and how are they related to features of 1:1 complexes of the same substrate? (iii) If some substrates form 1:2 complexes, why do not all substrates form them? A model has been devised that is helpful in designing experiments and interpreting data related to such questions. This paper describes the model, presents stability constants for many systems, and offers interpretations of the data in terms of the model. The stability constants were obtained by combinations of several experimental techniques: solubility, spectrophotometry, kinetics, potentiometry (4,6,7), and a modified competitive spectrophotometric method (8,9,10). Experimental details are available in the original work (4,5,7,10).

THE MODEL

Let h-H represent the host (cyclodextrin) molecule, the lower and upper case letters indicating that the two ends of the cavity are structurally different, and let g-G represent a guest molecule having two binding sites g and G. Then all 1:1 inclusion complexes may be represented as follows, where it is assumed that, within each orientation, the energy distribution of positional isomers is so narrow that only a single complex need be considered:

1:1 complexes	symbol
g-G h-H	gH
g-G H-h	gh
G-g h-H	GH
G-g H-h	Gh

It also is possible, in principle, to form 2:1 and 1:2 inclusion complexes:

2:1 complexes	1:2 complexes
g-G G-g H-h	g-G h-H h-H
g-G g-G H-h	g-G H-h h-H
G-g g-G H-h	G-g h-H H-h
G-g G-g H-h	G-g H-h H-h

Now suppose that no 2:1 complexes exist. Then it is reasonable to infer that the guest molecule can enter only one end of the host. As a consequence, there can be only two 1:1 complexes and only one 1:2 complex. For example, suppose only the H end of the host can be entered. Then the only possible complexes are gH, GH, and GHHg. This case of the general model is applied throughout this paper.

The observed K_{11} value for the system is the sum of the stability constants for all isomeric 1:1 complexes, hence

$$K_{11} = K_{gH} + K_{GH} \tag{1}$$

The 1:2 complex can be formed via the gH or the GH route:

$$gH + h\text{-}H \overset{K'}{\rightleftarrows} GHHg$$

$$GH + h\text{-}H \overset{K''}{\rightleftarrows} GHHg$$

Therefore $$K_{12} = [GHHg]/([gH] + [GH])[h\text{-}H],$$

or $$K_{12} = \frac{K'K''}{K' + K''} \tag{2}$$

It is also readily seen that $$\frac{K'}{K''} = \frac{K_{GH}}{K_{gH}} \tag{3}$$

Writing $K' = aK_{GH}$, where <u>a</u> is a proportionality constant, it follows from Eq. (3) that $K'' = aK_{gH}$. Combination with Eq. (2) gives

$$K_{12} = \frac{aK_{gH}K_{GH}}{K_{11}} \tag{4}$$

The quantity <u>a</u> is a measure of the extent of interaction between the two binding sites in a 1:2 complex. If a = 1, the sites are independent, and with Eqs. (1) and (4) the site binding constants K_{gH} and K_{GH} can be evaluated. If $a < 1$, there is interference between the sites; if $a > 1$ there is a synergistic effect. In the special case that the two binding sites are identical, $K_{gH} = K_{GH}$, and from Eqs. (1) and (4) we get

$$K_{11} = 2K_{gH} \tag{5}$$

$$K_{12} = aK_{11}/4 \tag{6}$$

allowing the site binding constant and the interaction parameter to be estimated.

The model leads to interpretations of experimental stability constants in terms of site binding constants. Since Eqs. (1) and (4) contain three unknown quantities, it is necessary to employ chemically reasonable assumptions about the parameters in order to make use of the equations.

CINNAMIC ACID DERIVATIVES

Table I gives stability constants for α-cyclodextrin complexes of trans-cinnamic acid and some related compounds (5). The two binding sites are the side-chain and the phenyl ring. Because of the length of the side-chain it seems reasonable to set $a = 1$ for these substrates.

Table I. Experimental Stability Constants and Site Binding Constants for Some α-Cyclodextrin/Cinnamate Complexes at 25°C.

Substrate	K_{11}/M^{-1}	K_{12}/M^{-1}	K_X/M^{-1}	K_Y/M^{-1}
Cinnamic acid	2260	60	2199	61
Cinnamate ion	110	15	92	18
3,5-Dimethoxycinnamic acid	1965	0	1965	0
Benzalacetone	105	15	87	18
Methyl cinnamate	1200	60	1148	52

Writing Eqs. (1) and (4) as

$$K_{11} = K_X + K_Y \tag{7}$$

$$K_{12} = \frac{aK_XK_Y}{K_{11}} \tag{8}$$

the site binding constants K_X and K_Y are obtained (see Table I).

Owing to the structural likenesses among these substrates, it is assumed that all larger values can be attributed to one site (K_X) and the smaller constants to the other site (K_Y). Molecular models indicate that the disubstituted phenyl ring of 3,5-dimethoxycinnamic acid cannot penetrate the ligand cavity; hence the model predicts that one of the site binding constants will be zero and that K_{12} will be zero, as observed. This leads to the conclusion that the K_Y values in Table I describe binding to the phenyl ring, and K_X to the side-chain site.

It will be obvious from Table I that the quantitative differences between the sets (K_{11}, K_{12}) and (K_X, K_Y) are practically insignificant. The contribution of the model is in providing a conceptual framework for the interpretation of these numbers in terms of binding site affinities.

4-SUBSTITUTED BENZOIC ACIDS

Table II lists stability constants for α-cyclodextrin complexes of 10 4-substituted benzoic acids. Values were obtained for both the conjugate acid (subscript a) and base (subscript b) forms of the substrate. It was found that $K_{12b} = 0$ for all substrates (7). Note that $K_{11a} > K_{11b}$ for all acids; the conjugate acid complexes more strongly than the base form, which has been observed also with other carboxylic acids (6). Figure 1 is a Hammett plot of K_{11a} and K_{11b}. Although scatter is seen, the trends are clear: K_{11a} has a negative rho value [as shown by Casu and Rava (9)], whereas K_{11b} has a positive rho.

Table II. Stability Constants for α-Cyclodextrin Complexes of 4-Substituted Benzoic Acids, X-C_6H_4-COOH, at 25°C.

X	K_{11a}/M^{-1}	K_{11b}/M^{-1}	K_{12a}/M^{-1}
$NHCH_3$	1301	6.1	0
NH_2	1341	9.0	0
OH	1130	16.6	0
OCH_3	884	3.5	0
CH_3	1091	6.6	0
H	722	11.2	0
F	504	14.2	0
CH_3CO	889	60.3	28.8
CN	471	79.2	25.0
NO_2	350	81.0	20.2

For these substrates it is improbable that the two binding sites are independent, because of their proximity. Nevertheless, the model is useful. We consider that, of the several factors affecting binding affinity within a series, electron density at the binding site is a primary determinant. Figure 1 shows that electron-withdrawing substituents decrease complex stability in the acid series. This is explicable if K_{11a} represents primarily binding at the COOH site of the molecule. This conclusion is consistent with proton NMR data on benzoic acid (11).

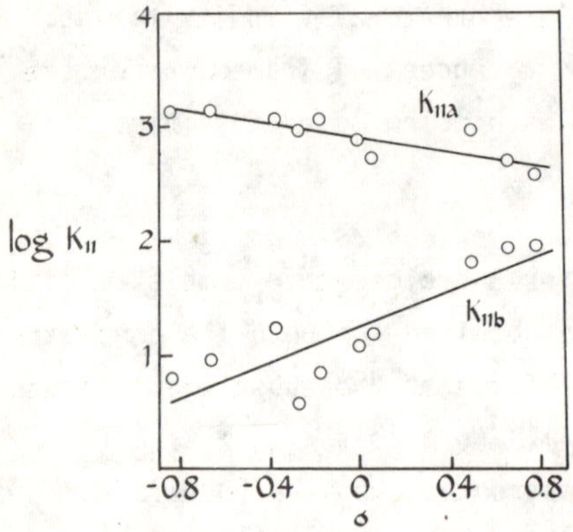

Fig. 1. Hammett plot for K_{11a} (upper) and K_{11b} (lower) for α-cyclodextrin/ benzoic acids.

K_{11b} is smaller than K_{11a} for the same substrate, which is consistent with the view that K_{11a} represents mainly COOH-site binding; upon ionization of this group, its affinity for the cavity will be greatly reduced. Therefore K_{11b} represents binding at the remaining site, namely X.

Now, if K_{11a} describes COOH-site binding (mainly), then K_{12a} must represent mainly binding to the X-site. Therefore this model predicts that K_{12a} should respond to X-substituent effects in the same way that K_{11b} does, since they describe roughly the same process. Table II shows that this is the observed behavior, in a general way; K_{11b} and K_{12a} tend to increase together, and in the opposite sense to K_{11a} along the sigma scale.

Since K_{11b} describes X-site binding, K_{12b} describes binding at the COO^- site. But $K_{12b} = 0$ for all substrates. Moreover the existence of finite K_{12a} values for three substrates means that a in Eq. (8) must be finite, and suggests that it will also be finite for other substrates of similar size and shape. It follows that K_{12b} is zero as a consequence of the site binding constant for COO^- being essentially zero. Letting K_{Yb} represent this quantity, we have $K_{11b} = K_{Xb} + K_{Yb} = K_{Xb}$; that is, K_{11b} can be identified solely with the binding at site X.

Extension to the acid series requires the assumption that $K_{Xb} = K_{Xa}$. If approximately so, then $K_{11a} = K_{Xa} + K_{Ya} = K_{Xb} + K_{Ya} = K_{11b} + K_{Ya}$, so the site binding constant K_{Ya} for the COOH site is given approximately by $K_{11a} - K_{11b}$. With these estimates, Eq. (8) allows $\underline{a}$ to be calculated. For X = CH_3CO, a = 0.51; X = CN, a = 0.38; X = NO_2, a = 0.33. These estimates seem reasonable in magnitude.

A different set of assumptions leads to another estimate of $\underline{a}$. Eqs. (7) and (8) give Eq. (9):

$$K_{11} = \frac{K_{11}K_{12}}{aK_Y} + K_Y \tag{9}$$

If, in a series, $\underline{a}$ and K_Y are non-zero and constant, a plot of K_{11} vs. $K_{11}K_{12}$ will be linear. For the last three substrates in Table II a straight line is obtained, giving a = 0.20, similar to the above estimates.

4-SUBSTITUTED PHENOLS

Table III presents stability constants for α-cyclodextrin/4-substituted phenols. K_{12a} = 0 for all of these systems. K_{11a} and K_{11b} trend in the same way; greater electron-withdrawing ability in the X-substituent is roughly associated with greater complex stability. If site electron density is the general determinant of binding strength, these data suggest that, in both the conjugate acid and base series, the 4-X-site is the site of binding in the 1:1 complexes, with little binding at the phenol or phenolate sites.

Table III. Stability Constants for α-Cyclodextrin Complexes of 4-Substituted Phenols, X-C_6H_4-OH, at 25°C.

X	K_{11a}/M^{-1}	K_{11b}/M^{-1}	K_{12b}/M^{-1}
OCH_3	0	2.7	13.3
CH_3	0	13.9	0
H	0	10.9	0
F	0	15.6	0
I	2316	3955	2.4
Cl	272	488	3.1
Br	704	1221	4.7
CN	158	662	0
NO_2	245	2408	0

Obviously other factors must be having important effects. Complex stability will be determined by binding site electron density, polarizability, and polarity (besides size and shape), the first two factors favoring stability, the polarity opposing it. The data in Table III are described by

$$\log K_{11b} = -1.41 + 5.33\sigma + 0.15R - 0.81\mu$$

$$\log K_{11a} = -0.96 + 0.12R - 0.10\mu$$

where σ is the Hammett substituent constant, R is the molar refraction of $X\text{-}C_6H_5$, and μ is the dipole moment of $X\text{-}C_6H_4\text{-}OH$. These correlations have no quantitative theoretical significance, but they indicate the roles of these factors.

Estimation of a values using the argument from the benzoic acid discussion is not warranted, because of the extensive electron delocalization in these substrates. The only members giving finite K_{12b} values possess substituents capable of tolerating a positive charge by electron release through the resonance effect; this may sufficiently counter the electron release of the phenolate site to allow significant binding at that site.

That $K_{11b} > K_{11a}$ for phenols may seem surprising [the case of 4-nitrophenol has been discussed (11-13)], yet this is explicable in terms of the above findings. Neutral phenols are highly polar species, as revealed by their dipole moments. Upon ionization delocalization of the excess charge occurs, with an increase in electron density at the para- substituent, which is the binding site in both the neutral and anion forms. This factor alone will favor stronger binding in the anion. In addition, the phenol end of the substrate will experience an increase in electron density upon ionization, and because of delocalization this site may actually be less "polar" than the hydroxy group in the neutral form. Highly localized negative charges (as in the benzoates) will result in destabilization, but delocalized negative charge may be stabilizing.

K_{11a} for the phenol $^{-}OOC\text{-}C_6H_4\text{-}OH$ is K_{11b} for the benzoic acid anion $OH\text{-}C_6H_4\text{-}COO^{-}$. Discussing this as a benzoate it was concluded that the COO^{-} binding was negligible. On the other hand it has been concluded that binding at the OH site in phenols is negligible. This inconsistency suggests that it may not be permissible to make a common assignment of binding mode to all members of a series. In this case the charge on the carboxylate may oppose electron release by the hydroxy, leaving it susceptible to binding.

4-SUBSTITUTED ANILINES

Stability data for 9 anilines are given in Table IV; $K_{12a} = 0$ for all of these. The data show a rough trend of K_{11b} with electron-withdrawal by the 4-substituent, indicating that the 4-X-site is the primary binding site in these amines. p-Phenylenediamine is a symmetrical substrate; with Eqs. (5) and (6) we find the site binding constant $K_{NH_2} = 1.1$ and a = 0. However, this estimate of a must be considered in light of the experimental uncertainty in K_{12}. Suppose a = 0.35; then with Eq. (8) and $K_X = K_Y = 1.1$ we calculate $K_{12} = 0.18$. This is experimentally indistinguishable from zero; hence the estimate a = 0 based on $K_{12} = 0$ is not definitive.

Table IV. Stability Constants for α-Cyclodextrin/4-Substituted Anilines at 25°C.

X	K_{11b}/M^{-1}	K_{12b}/M^{-1}	K_{11a}/M^{-1}
NH_2	2.3	0	0
OCH_3	6.7	0	0
CH_3	57.6	3.9	37.1
H	8.8	0	0
Cl	251	0	69
CN	451	--	--
NO_2	635	--	--

CONCLUSION

With the assumptions and the assignments described earlier, site binding constants can be estimated for most of the 1,4-disubstituted benzenes. Table V gives the results, based upon the assumption that a = 0.35 for all systems (see the benzoic acid discussion).

The substituents are arranged in order of sigma values, and the binding constants show the trends with site electron density, modified by polarizability and polarity effects. The calculated K_{11} for any substrate is the sum of the two appropriate entries from the table, allowing X to be each substituent in turn.

The diagonal of the table represents symmetrical substrates, whose study will assess the validity of the model and of the site assignments. The estimates in Table V imply the presence of appreciable fractions of two isomeric 1:1 complexes from substrates lying close to the diagonal, and this phenomenon should be most readily detected for substrates lying in the lower right corner of the table. Substrates whose entries lie farthest from the diagonal are most likely to generate simple systems of one 1:1 complex.

Table V. Site Binding Constant Estimates for the X-Site in $Y\text{-}C_6H_4\text{-}X$.

Y \ X	O^-	NH_2	OH	Me	H	COO^-	I	Cl	COOH	CN	NO_2
O^-				13.9	10.9		3948	479		662	2408
NH_2		1.1		43	8.8	0		251	1341	451	635
OH						0	2316	272	1130	158	245
Me	0	15	0			0			1091		
H	0		0			0			722		
COO^-		9.0	16.6	6.6	11.2					79	81
I	7		0								
Cl	9		0								
COOH		0	0	0	0					88	73
CN	0		0			0			383		
NO_2	0		0						277		

LITERATURE CITED

1. Gelb, R. I., Schwartz, L. M., Murray, C. T., and Laufer, D. A., J. Am. Chem. Soc., 100, 3553 (1978).
2. Gelb, R. I., Schwartz, L. M., and Laufer, D. A., J. Am. Chem. Soc., 100, 5875 (1978).
3. Gelb, R. I., Schwartz, L. M., Johnson, R. F., and Laufer, D. A., J. Am. Chem. Soc., 101, 1869 (1979).
4. Connors, K. A. and Rosanske, T. W., J. Pharm. Sci., 69, 173 (1980).
5. Rosanske, T. W. and Connors, K. A., J. Pharm. Sci., 69, 564 (1980).
6. Connors, K. A. and Lipari, J. M., J. Pharm. Sci., 65, 379 (1976).
7. Connors, K. A., Lin, S.-F., and Wong, A. B., J. Pharm. Sci., in press.
8. Lautsch, V. W., Bandel, W., and Broser, W., Z. Naturforsch., 11b, 282 (1956).
9. Casu, B. and Rava, L., Ric. Sci., 36, 733 (1966).
10. Lin, S.-F., Ph.D. Thesis, University of Wisconsin-Madison, 1981.
11. Bergeron, R. J., Channing, M. A., and McGovern, K. A., J. Am. Chem. Soc., 100, 2878 (1978).
12. Bergeron, R. J. , Channing, M. A., J. Am. Chem. Soc., 101, 2511 (1979).
13. Dunn, M. F. and Bernhard, S. A., Chap. XI in "Investigation of Rates and Mechanisms of Reactions", Part I, E. S. Lewis, editor, p. 624 (Vol. VI of "Techniques of Chemistry"), Wiley-Interscience, N.Y., 1974.

I.Int.Symp.on Cyclodextrins
Budapest, 1981 3.4.

THE COMPLEX FORMATION OF CYCLODEXTRINS AND RELATED (1→4)-α-D-GLUCANS AS REVEALED BY THE SPIN LABELING TECHNIQUES

MASAAKI OHARA

Faculty of Pharmaceutical Sciences, Josai University
Keyakidai, Sakado, Saitama 350-02, Japan

SUMMARY

The complex formation and inclusion phenomena observed in aqueous solutions of cyclodextrins and some related linear (1→4)-α-D-glucans were investigated f''rthermore by means of Spin-Label-Induced NMR Relaxation techniques.

A solution of the carbohydrates was incubated with spin labels such as stable nitroxide radicals and paramagnetic metal ions of the lanthanide series. ^{1}H-NMR spectra of these systems indicated that spin labels enhanced the relaxation of the carbohydrate proton nuclei in the vicinity of the binding sites. Remarkable spectral perturbations for cyclodextrins, which may be attributed to the inclusion complex formation, were brought about only when some nitroxide radicals as TEMPO were introduced. This findings were in harmony with the ESR evidence on the matter.

The complex formation phenomena of cyclodextrins and amylose in aqueous solution have drawn the attention of many investigators(1, 2, 3, 4). Various physico-chemical studies, the ^{1}H NMR spectroscopic ones in especial demonstrated that the helical nature and the central cavity of these glucans with α-D-(1→4) linkages are responsible for the binding modes and sites in this process, respectively(5-10).

For the aqueous solution of cyclodextrins large ^{1}H NMR spectral perturbations are observed with regard to chemical shifts and/or the nuclear magnetic relaxation of resonances by the addition of various diamagnetic aromatic guest molecules to the systems. The addition of various paramagnetic species as a guest are expected to induce extremely large NMR spectral perturbations(11, 12).

This paper describes an approach to the problem of employing spin-label-induced nuclear magnetic resonance relaxation method(13). This method is a sort of the spin labeling techniques which have been extensively applied in recent years to various biological systems(14-19). Here we have made use of some stable nitroxide radicals and paramagnetic metal ions of the lanthanide series as the spin-label guests. The carbohydrate systems examined were those of α-, β-, γ-cyclodextrin, some maltooligosaccharides and amyloses of different degree of polymerization.

Materials and Methods

Preparation, identification and physical data of the employed cyclodextrins and nitroxide radicals (TEMPO, TEMPONE, TEMPOL; their structures are as follows) were described in our previous literature(19).

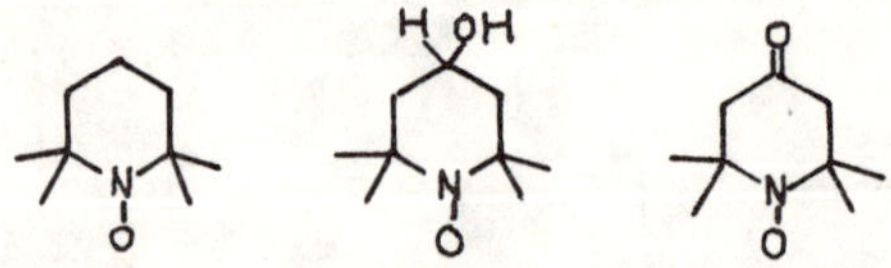

TEMPO TEMPOL TEMPONE

Maltopentaose, maltohexaose, maltoheptaose and maltooctaose with high purity were obtained from Seishin-Seiyaku Co., Inc. Amylose of $\overline{DP}$ ca. 17 were a potato product purchased from Sigma Chemical Co., Inc. Amylose of $\overline{DP}$ ca. 1500 were obtained from Hayashibara Biochemical Lab., Inc. The source of each lanthanide ion, Gd(III), Eu(III), La(III) was $Gd(NO_3)_3 \cdot 6H_2O$, $Eu(NO_3)_3 \cdot 6D_2O$ and $La(NO_3)_3 \cdot 6H_2O$, respectively (from Merck, Darmstadt, Germany).

The OH protons of the carbohydrate samples and protons in the water of crystallization of lanthanide salts were deuterated with D_2O (Merck, Darmstadt, Germany; 99.75 % grade) prior to the NMR measurements.

General Procedures for the ^{1}H-NMR Measurenents

5 mg (4.41×10^{-6} mol) of β-cyclodextrin was added to a D_2O solution (1 ml) of various amount of TEMPO (ca. 1×10^{-9} to ca. 3×10^{-5} mol) and then the mixture was transferred into glass tubes (diameter 5 mm) and sealed.

The ^{1}H-NMR spectra were measured with a JNM-FX 200 type (200 MHz) and a JNM-PFT-100 type (100 MHz) spectrometer, each at 23 ± 1 °C referenced to DSS.

Nitroxide Radical Induced Perturbations in the ^{1}H NMR Spectra of Cyclodextrins and Related (1→4)-α-D-Glucans

The addition of the nitroxide radicals to cyclodextrins in aqueous solution caused broadenings of the cyclodextrin proton resonance lines. This is ascribed to the paramagnetic relaxation effect of the nitroxide radical. Figure 1 shows the effect of increasing concentrations of TEMPO on the ^{1}H NMR spectrum of β-cyclodextrin. As the concentration of TEMPO was increased, some proton signals assigned to H(3) and H(5) (5, 6, 7) were broadened significantly. The H(2) and H(4) signals were affected to a lesser extent. The anomeric proton peaks were virtually unaffected even at higher concentrations

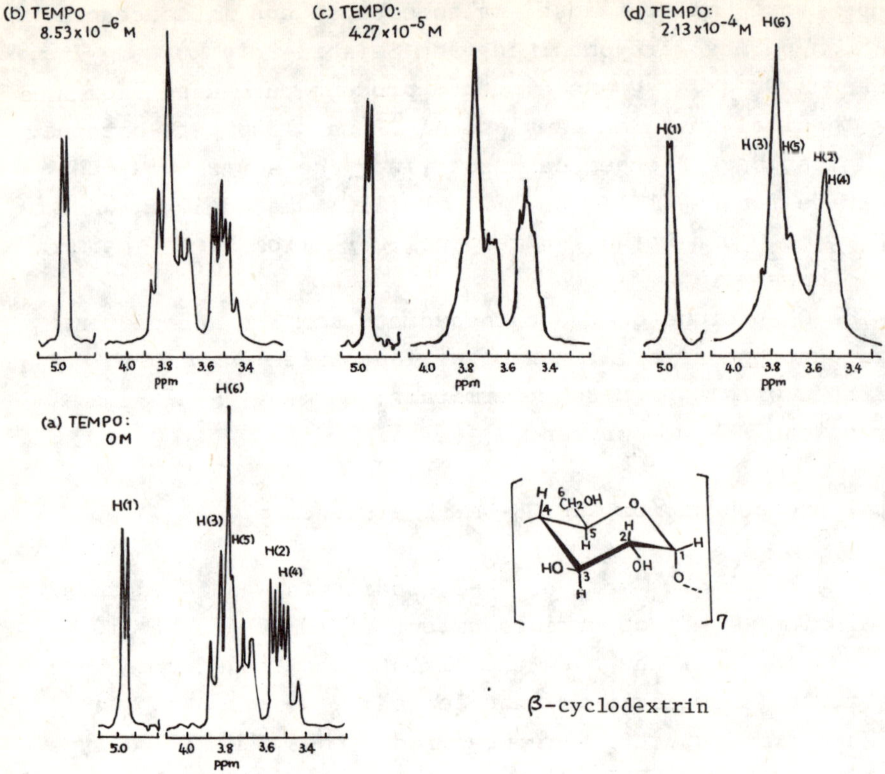

Fig. 1. TEMPO-induced ^{1}H-NMR spectral perturbations (at 200 MHz) for β-cyclodextrin (4.41 mM in D_2O). The spectra are observed by the pulse-FT method with the WEFT mode at 23±1°C referenced to DSS.

of TEMPO.

Figure 2 shows the effect of increasing concentrations of TEMPO on the T_1 values of β-cyclodextrin proton resonance lines determined. Generally speaking, the T_1 values of the signals assingned to H(3) and H(5) (c, d, f) were decreased markedly, while those of H(1), H(2) and H(4) (a, b, k, l, n) were almost constant at moderate concentrations of TEMPO.

The nitroxide radicals were also applied to some malto-oligosaccharide and amylose systems. Any significant effects were not detectable in the ^{1}H NMR spectrum at moderate concentrations of the nitroxide radicals except to the system of amylose of $\overline{DP}$ ca. 1500. A slight yet dintinct effect of the differential broadenings was observed in the CH resonances of the polysaccharide sample. This effect was similar to that of cyclodextrins.

Paramagnetic Ion Induced Perturbations in the ^{1}H NMR Spectra of Cyclodextrins and Related (1→4)-α-D-Glucans

Gd(III) produced rather pronounced changes in the ^{1}H NMR spectrum for every glucan system. Any differential line-broadening was not detectable in the CH resonances. On the addition of Eu(III) neither paramagnetic shift nor line-broadening were observed in the ^{1}H NMR spectrum of cyclodextrins and malto-oligosaccharides. La(III) exerted no appreciable effect on the ^{1}H NMR spectrum of cyclodextrins and maltooligosaccharides.

Binding Modes and Sites of Various Spin Labels to Cyclodextrins and Related (1→4)-α-D-Glucans

The present NMR investigations (line-broadening observations and T_1-measurements) show the existence of the strong mutual interactions between particular proton nuclei as H(3) and H(5) of cyclodextrins and an unpaired electron of the nitroxide spin labels. It has been known that the protons H(3) and H(5) are directed to the interior of the cyclodextrin to-

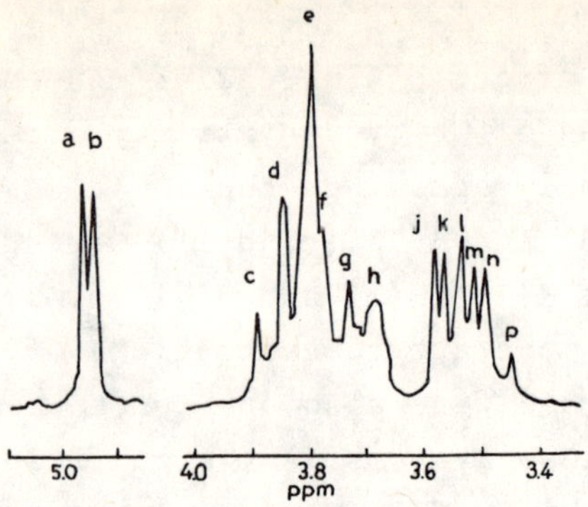

^{1}H NMR spectrum, 200 MHz, of β-cyclodextrin referenced to DSS.

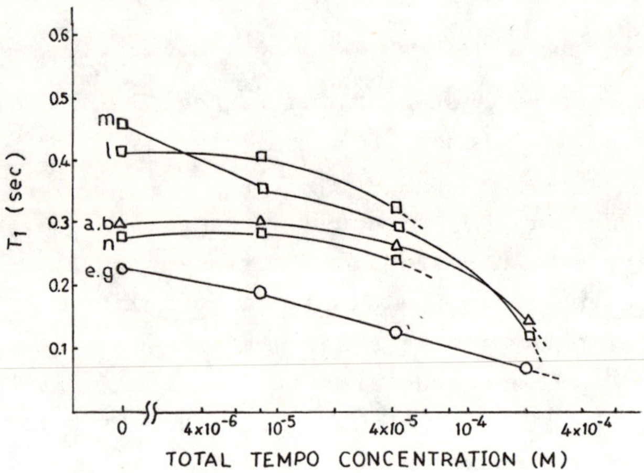

Fig. 2. Changes in the proton T_1-values of β-cyclodextrin (4.41 mM in D_2O) in the presence of TEMPO. The numbered resonances correspond to the graphical changes. In other signals of c, d, f and j a rapid decrease in the T_1 values is observed. In the signals of h, k and p a slow decrease in the T_1 values is observed. T_1 values are measured by the Hahn's 180°-90°-pulse method at 23 ± 1 °C.

rus (2). Therefore, it is clear that the interior molecular surface is responsible for the binding of nitroxide spin labels to cyclodextrins. These results are in good accordance with our previous ESR measurements (19).

The ^{1}H NMR spectra (Fig. 1.) consist of the superposition of the signals of free cyclodextrins and TEMPO bound cyclodextrins. Hence, exchange rate of nitroxide spin labels between these these two sites of β-cyclodextrin is expected to be greater than the NMR time scale. From the dissociation constant of TEMPO — β-cyclodextrin complex, K_d, 3.9×10^{-3}M (measured at 30 °C), the amount of TEMPO bound β-cyclodextrin is roughly estimated at ~ 5 %.

As a consequence of the foregoing, it follows that remarkable line-broadenings may be attributed to fast exchange rate of the complex formation process.

Symbols used

TEMPO, 2,2,6,6-Tetramethyl-4-piperidine-1-oxyl; TEMPONE, 2,2,6,6-Tetramethyl-4-piperidone-1-oxyl; TEMPOL, 2,2,6,6-Tetramethyl-4-piperidinol-1-oxyl; $\overline{DP}$, Average degree of polymerization.

Literature Cited

1. Cramer, F.: Einschlussverbindungen, Springer Verlag, Berlin-Göttingen-Heidelberg 1954; p.61, p.86-109.
2. Bender, M.L., Komiyama, M.: Cyclodextrin Chemistry, Springer Verlag, Berlin-Heidelberg-New York 1978.
3. Pfannemüller, B., Mayerhöfer, H., Schulz.: Biopolymers 10 243-261 1971
4. Moulik, S. P., Gupta, S.: Carbohydr.Res. 81 131-143 1980
5. Demarco, P. V., Thakkar, A. L.: J.Chem.Soc.Chem.Commun. 2-4 1970
6. Bergeron, R., Rowan, R. III.: Bioorg.Chem. 5 425-436 1976

7. Wood, D. J., Hruska, F. E., Saenger, W.: J.Am.Chem.Soc. 99 1735-1740 1977
8. Behr, J. P., Lehn, J. M.: J.Am.Chem.Soc. 98 1743-1747 1976
9. Suzuki, M., Sasaki, Y.: Chem.Pharm.Bull. 27 1343-1351 1979
10. Uekama, K., Hirayama, F., Koinuma, H.: Chem.Lett. 1393-1396 1977
11. Mildvan. A. S., Weiner, H.: Biochemistry 8 552-562 1969
12. Roberts, G. C. K., Hannah, J., Jardetzky, O.: Science 165 5o4-506 1969
13. Krush, T. R.: Spin-Label-Induced Nuclear Magnetic Resonance Relaxation Studies of Enzymes, in Spin Labeling, Ed. by Berliner, L. J., Academic Press, Inc., New York-San Francisco-London 1976; p.339-372.
14. Wien, R. W., Morrisett, I. D., McConnell, H. M.: Biochemistry 11 3707-3716 1972
15. Berliner, L. J.: Spin Labeling, Academic Press, Inc., New York-San Francisco-London 1976.
16. Morishima, I., Ishihara, K., Tomishima, K., Inubushi, T., Yonezawa, T.: J.Am.Chem.Soc. 97 2749-2756 1975
17. Lord, S. T., Breslow, E.: Biochem.Biophys.Res.Commun. 80 63-70 1978
18. Lenkinski, R. E., Dallas, J. L., Glickson, J. D.: J.Am.Chem.Soc. 101 3071-3077 1979
19. Ohara, M., Hettler, H., Gauss, D., Cramer, F.: Bioorg.Chem. 8 211-217 1979

I.Int.Symp. on Cyclodextrins
Budapest, 1981 3.5.

A SIMPLE RULE FOR PREDICTING CIRCULAR DICHROISM INDUCED IN AROMATIC GUESTS BY CYCLODEXTRIN HOSTS IN INCLUSION COMPLEXES

MÁRTON KAJTÁR,[1] CSILLA HORVÁTH-TORÓ,[1] ÉVA KUTHI,[1]
JÓZSEF SZEJTLI[2]

[1] Institute of Organic Chemistry, Eötvös Loránd University
Múzeum krt. 4/B
H-1088 Budapest, Hungary
[2] CHINOIN Pharmaceutical-Chemical Works, Tó u. 1-5
H-1045 Budapest, Hungary

SUMMARY

Circular dichroism (CD) spectra of inclusion complexes of aromatic hydrocarbons, of substituted benzoic acid derivatives and of 1- and 2-naphthols with water-soluble β-cyclodextrin polymer, 2,6-dimethyl-β-cyclodextrin and β-cyclodextrin, respectively, were recorded. For interpreting the sign and intensity of induced optical activity in these complexes, a descriptive model based on the theory of coupled oscillators is presented. The correlation between structure and chiroptical properties of cyclodextrin complexes of aromatic molecules is expressed in the form of a simple sector rule by which either the relative arrangement of guest and host in a cyclodextrin complex or the polarization of the electronic transitions of the guest can be estimated from the CD spectrum.

INTRODUCTION

The appearance of circular dichroism in the absorption bands of achiral molecules as a result of their complexing with cyclodextrins has been observed by a number of authors in the last decade. A deeper analysis of the experimental material shows that the sign and intensity of the induced CD bands are dependent on the structure of the complex and on the polarization and dipole strength of the electronic transitions in the guest molecule. For interpreting the phenomenon of induced optical activity and for calculating the Cotton effects found in the CD spectra of the cyclodextrin complexes of aromatic molecules, Harata and Uedaira [1] and later Shimizu et al. [2,3] applied the Tinoco [4] formulation of the coupled oscillator model of Kirkwood [5]. Their results are in qualitative and, in most of the instances, in fairly good quantitative agreement with the experimental data. However, the application of their method requires rather laborious and still only approximate computations for each individual case. Therefore we are going to give here a qualitative description of the coupled oscillator model as applied to the inclusion complexes of aromatic molecules with cyclodextrins and, on the basis of this model and of experimental data, to formulate a simple rule for correlating the chiroptical properties of the complexes with their structures and with the nature of their electronic transitions. In our opinion, this simple rule can be more easily and generally applied in practice than the calculations and, at the same time, with almost the same reliability.

MATERIALS AND METHODS

For the preparation of the complexes β-cyclodextrin, 2,6-dimethyl-β-cyclodextrin [6], and a water-soluble polymer built up of β-cyclodextrin residues connected by glycerol epichlorohydrine (β-cyclodextrin content of about 50%) [7] were used. The concentrations and (guest)/(host) ratios of the solutions were chosen to assure practically complete complexation of the guest. The UV and CD spectra were measured at ambient temperature in quartz cells using a SPECORD spectropolarimeter and a Roussel-Jouan Dichrographe N° III (Jobin-Yvon), respectively.

RESULTS

Examples of our experimental results are summarized in the following tables. The intensity of induced optical activity is expressed in terms of the anisotropy number ($g = \Delta\varepsilon/\varepsilon$).

Table I. Optical activity induced in the ^1L_b band of aromatic hydrocarbons in their complexes with β-cyclodextrin polymer

Guest molecule	λ,nm ($g.10^4$)	Guest molecule	λ,nm ($g.10^4$)
benzene	255 (-0.33)	1,2,3,5-tetramethylbenzene	278 (-1.10)
toluene	263 (-0.92)	1,2,4,5-tetramethylbenzene	279 (+0.53)
tert-butylbenzene	261 (-1.16)	naphthalene	312 (-1.12) 278 (-0.82)*
p-xylene	268 (-1.14)		

*^1L_a band (polarized perpendicular to the long axis)

Table II. Optical activity induced in the inclusion complexes of substituted benzoic acids with 2,6-dimethyl-β-cyclodextrin

Guest molecule	Band	λ,nm ($g.10^4$)	Guest molecule	Band	λ,nm ($g.10^4$)
benzoic acid	^1L_b ^1L_a	280 (-0.98) 233 (+2.04)	4-hydroxy-benzoic acid	^1L_a	256 (+2.40)
2-hydroxy-benzoic acid	^1L_b ^1L_a	304 (-0.50) 240 (+1.64)	2-amino-benzoic acid	^1L_b ^1L_a	340 (+0.20) 247 (+0.43)
3-hydroxy-benzoic acid	^1L_b ^1L_a	297 (-0.94) 242 (+2.23)	4-amino-benzoic acid	^1L_a	287 (+1.40)

Table III. Induced optical activity of the β-cyclodextrin complexes of 1-naphthol and 2-naphthol

Guest molecule	Band*	Angle of polarization**	λ,nm ($g.10^4$)
1-naphthol	^1L_b	-24°	324 (+1.07)
	^1L_a	$+82^o$	295 (-0.49)
	^1A_g	-3°	234 (+2.23)
	^1B_b	-1°	215 (+2.17)
2-naphthol	^1L_b	-62°	329 (-1.16)
	^1L_a	$+78^o$	287 (-1.26)
	^1B_b	$+9^o$	225 (+1.64)

*cf. lit.[8] ** Measured from the long axis of the molecule; positive values toward the hydroxyl group.

DESCRIPTIVE MODEL OF COUPLED OSCILLATORS AS APPLIED TO CYCLODEXTRIN COMPLEXES

According to the model of coupled oscillators [5,9], an optically active system is composed of (at least) two oscillating electric dipoles which must not be either coplanar or perpendicular simultaneously to each other and to the line connecting their centers. The system formed by the coupling of these oscillating dipoles will possess not only a resultant oscillating electric moment but an oscillating magnetic moment, too, which being either parallel or antiparallel to the electric moment will give rise in the system to a positive or a negative rotational strength, respectively. The relative directions of the resultant electric and magnetic moments, i.e. the sign of the optical activity, depend on the spatial arrangement, the helicity, of the two oscillators.

If the two oscillators are of different energy and in such a disposition to each other that, when in phase (i.e. the torsion angle of the two vectors being less than 90°), they determine a right handed helix (Fig.1a), then the electric and magnetic moments resulting from their coupling will be parallel, and the rotational strength belonging to the oscillator of lower energy will be of positive sign.

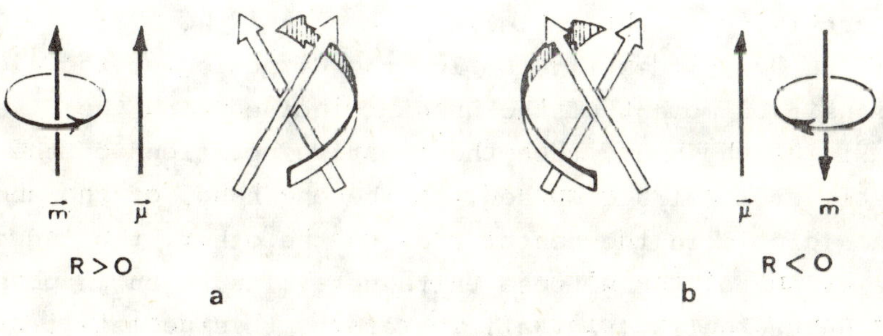

Fig.1. Coupling of two oscillators determining a right handed (a) and a left handed (b) helix.

In the inverse case, i.e. if the relative position of the pair of oscillators is of left handed type, the resultant electric and magnetic moments are antiparallel, and the rotational strength belonging to the lower energy oscillator will be negative (Fig.1b).

This qualitative model of the coupled oscillators can be applied to the cyclodextrin complexes of aromatic molecules. The electric moment belonging to the selected electronic transition of the aromatic molecule in the cavity of the cyclodextrin host can be regarded as the oscillator of lower energy. The oscillating dipoles induced by the electric field of light in the σ-bonds of the cyclodextrin can be summed up to moments attributable to the individual glucose units (Fig.2). These

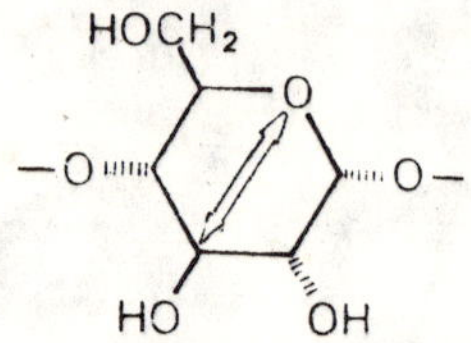

Fig.2. Oscillating electric moment induced by light in a glucose unit.

dipoles will be regarded in the simple model as the oscillators of higher energy which, by coupling with the electric transition moment of the aromatic guest, give rise to the induced optical activity of the latter.

If in a cyclodextrin complex the direction of the electric transition moment of the guest coincides with the C_7 symmetry axis of the host, then the relative positions of the coupled oscillator pairs composed, on the one hand, of the above mentioned dipole in the center and, on the other, the induced dipoles of any of the glucose units determine, when in phase, a right handed helix (Fig.3a). Therefore, this geometric arrangement of the "oscillators" results in a positive induced Cotton effect in the absorption band of the guest.

The helicity of the oscillator pairs remains unchanged even in cases when the direction of the transition moment of the guest somewhat deviates from that of the symmetry axis of the cyclodextrin. At an angle of about 30°, however, the transition moment becomes parallel with one of the glucose dipoles, and by further increase of the deviation the relative arrangement of some of the oscillator pairs will change from the right handed to the left handed type. At an angle of about 30-40° of the deviation, the contributions of the different oscillator pairs compensate each other, and the optical activity vanishes. By further increasing the angle between the transition moment and the symmetry axis, we arrive at a geometry at which most of the oscillator pairs become of left handed type and the resulting rotational strength of the system changes to negative sign. The optimal situation for a negative induced optical activity is the one with the transition moment of the guest perpendicular to the symmetry axis of the host (Fig.3b).

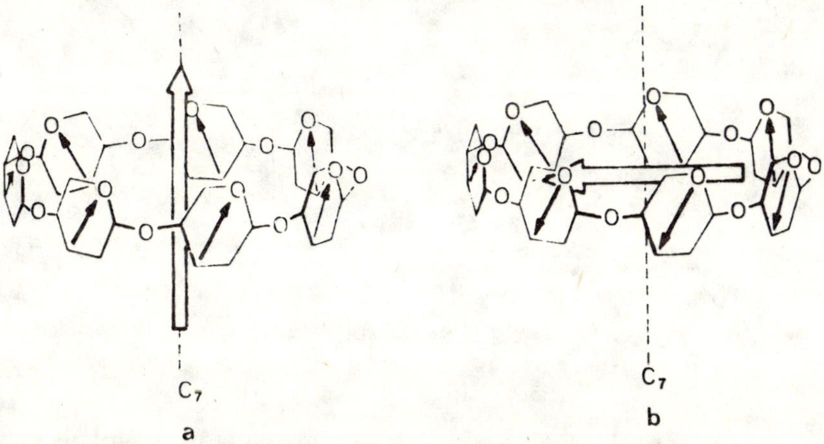

Fig.3. Coupled oscillator systems composed, on the one hand, of the induced dipoles of the individual glucose units (black arrows) and, on the other, of the electric transition moment of the guest (empty arrow) polarised parallel (a) with and perpendicular (b) to the C_7 symmetry axis of the host.

SECTOR RULE

The conclusions from the simple model can be summarized in the following rule. If the structure of the inclusion complex is such that the electric moment belonging to the investigated transition of the guest molecule falls within an imaginary double cone whose center coincides with that of the cyclodextrin and whose constructor forms an angle of about 30° with the C_7 symmetry axis (Fig.4), then the induced CD maximum will be positive. However, the CD maximum induced in a transition whose electric moment is outside of this double cone, is of negative sign. The positive or the negative CD is of maximal intensity if the transition dipole moment is parallel or perpendicular, respectively, to the symmetry axis of the cyclodextrin. This rule can be best represented in a cross sectional sector diagram (Fig.5).

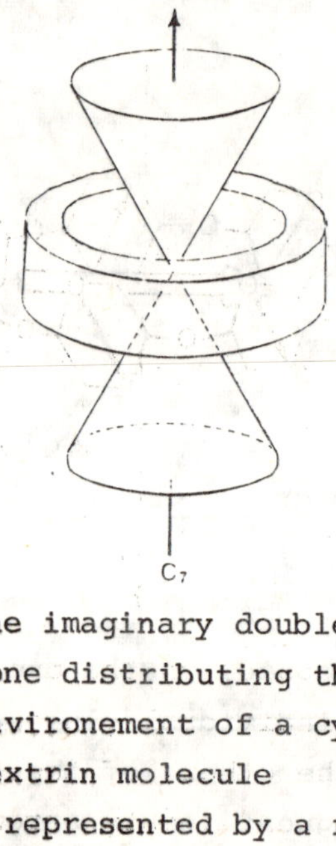

Fig.4. The imaginary double cone distributing the environement of a cyclodextrin molecule (represented by a ring) into sectors of different signs.

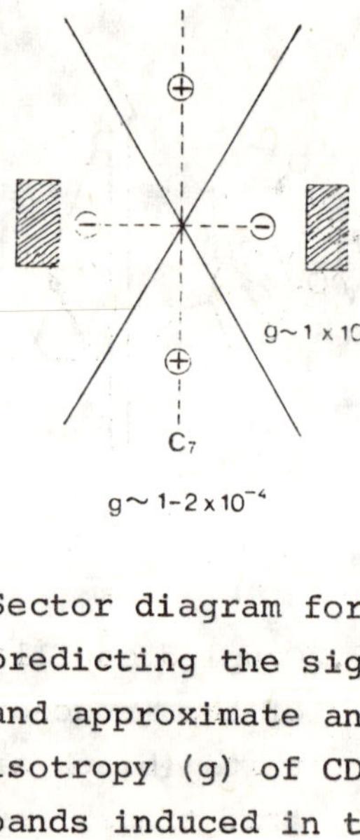

Fig.5. Sector diagram for predicting the sign and approximate anisotropy (g) of CD bands induced in the cyclodextrin complexes of aromatic molecules.

For characterizing the intensity of induced CD bands we used values of the anisotropy number ($g = \Delta\varepsilon/\varepsilon$) instead of those of experimentally measured $\Delta\varepsilon$. The dipole moments induced by light in the glucose units of the cyclodextrin molecule are of about the same magnitude within the spectral range of the aromatic bands investigated. The magnitude of the "perturbing" dipoles being constant, the value of the induced rotational strength due to the coupled oscillator mechanism must be proportional to the square of the transition electric moment of the guest, and, therefore, the value of the anisotropy ($\Delta\varepsilon/\varepsilon$) must be constant, irrespective of the intensity (ε) of the absorption band in question. The values of g are indeed of the same order of magnitude (10^{-4}) for all the induced CD bands of cyclodextrin complexes. The variations falling within one order of magnitude of the g values reflect the geometric arrangement of guest and host in the inclusion complex. Based on the experimental data, approximate values relating to the anisotropy number of the induced CD maxima are also assigned to the sector pairs in the diagram of Fig.5.

DISCUSSION

Most of our experimental results and of those found in the literature [1-3] can well be rationalized in the light of the simple sector rule.

The cyclodextrin complexes of simple aromatic hydrocarbons have not been studied yet, therefore no X-ray data relating their crystal structure are available. Being relatively small and highly symmetric, the benzene molecule can be accommodated by the β-cyclodextrin host (inner diameter 0.75 nm) in any of the possible orientations. If the plane of the benzene molecule were perpendicular to the C_7 symmetry axis of the cyclodextrin, all the possible directions of the electric moment belonging to the ^1L_b transition (polarized in the plane of the ring) would be concomitant, according to the sector rule, with an induced CD band of negative sign. Of the other possible positions of the benzene ring, those having the longer axis of the molecule (crossing the opposite C atoms) along the symmetry axis of the cyclodextrin seem to allow more favourable van der Waals interactions with the host than those in which the benzene axis is perpendicular to the cyclodextrin axis. In arrangements of the former type, one of the possible directions of the ^1L_b electric moment is lying along the line of the maximal negative rotational strength as shown in the sector diagram (Fig.6a), while the other possible directions coincide with the section lines of zero optical activity. Thus the negative sign as well as the low value of the anisotropy of the CD band induced in the ^1L_b transition of benzene, the most symmetric aromatic molecule, in its β-cyclodextrin complex (Table I) can be explained by applying the sector rule.

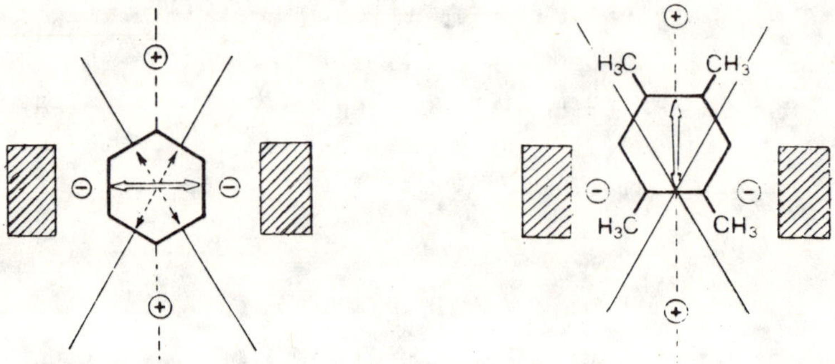

(a) (b)

Fig.6. The sector diagram as applied to the cyclodextrin complex of benzene (a) and 1,2,4,5-tetramethylbenzene (b).

In the complexes of alkylbenzenes listed in Table I, the relative orientation of guest and host can be taken as unambiguous from simple geometric considerations. The C_2 symmetry axes running through the one or the two <u>para</u> positioned alkyl groups are to be lying parallel with the cyclodextrin axis. The directions of the electric moments of the ^1L_b transition (obtained easily from Platt's spectroscopic moments [10]) get in negative sector pairs, in accordance with the sign of the experimentally found CD maxima. In the complex of durene (1,2,4,5-tetramethylbenzene), however, the transition moment is parallel with the cyclodextrin axis (Fig.6b), and the induced Cotton effect is indeed positive. The relatively small value of anisotropy due to the CD band of this complex may probably be ascribed to an asymmetric position of the benzene ring inside the cyclodextrin. The pair of ortho methyl groups entering the cavity first do not let the benzene ring itself reach the center of the cyclodextrin (cf. Fig.6b).

All the monosubstituted benzene derivatives studied by us and by other workers [2] give a negative induced CD band in the ^1L_b and a positive one in the ^1L_a transition in accordance with the rule and with the direction of their transition moments. Neither the structure of the complexes nor the directions of the transition moments are, however, straightforward in the case of ortho and meta disubstituted benzene derivatives. The experimental CD data obtained with o- and m-hydroxybenzoic acids and with o-aminobenzoic acid (Table II) suggest, according to the sector rule, the structures shown in Fig.7. (The polarisation of the electronic transitions are taken from PPP--calculations of Kiss and Szőke [11,12].)

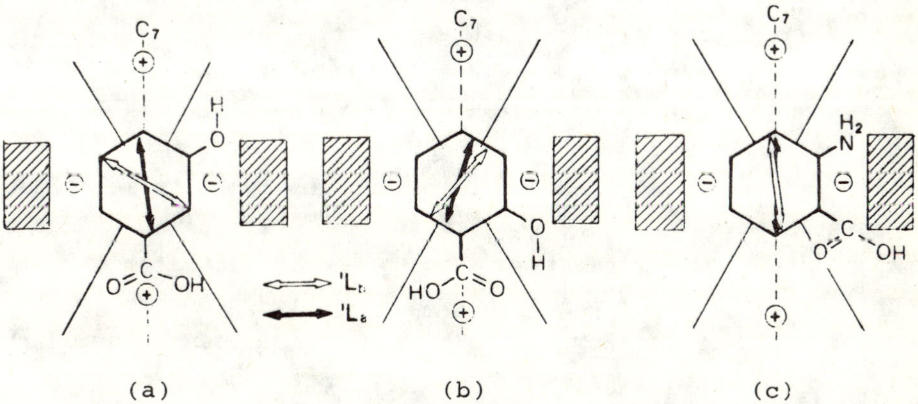

Fig.7. Structures for the cyclodextrin complexes of m-hydroxybenzoic (a), o-hydroxybenzoic (b) and o-aminobenzoic (c) acids as suggested by the sector rule.

It can be expected that the structure of the complex of m-hydroxybenzoic acid is analogous to that of the m-nitrophenol complex known from X-ray diffraction studies [13], and this is in agreement with our supposition. At present, no data relating to the structure of the cyclodextrin complex of an ortho disubstituted derivative can be found in the literature. The structure suggested for the salicylic acid complex (Fig.7b) seems to be fairly plausible. This cannot be told for the complex of anthranylic acid (Fig.7c). However, the low values of anisotropy of the Cotton effects induced in this complex indicate that it may probably assume more than one structure the CD effects of which partly compensate each other.

By comparing the CD spectra of the two naphthol complexes with the data on the polarization of the individual bands (Table III), a full agreement between experimental data and predictions of the sector rule can be established if we assume that the naphthol molecules accommodate in the cavity of the host with their long axes parallel to the symmetry axis of β-cyclodextrin. The same arrangement of the naphthalene ring can been deduced from the calculations of Harata and Uedaira [1,14] as well as from the negative sign of the ^1L_a Cotton effect found by us in the CD spectrum of the complex of naphthalene (Table I). The most remarkable feature of the CD spectra of the naphthol complexes is the nice separation of the ^1L_b and ^1L_a bands of 1-naphthol, though the corresponding bands in the UV spectrum are strongly overlapping. The separation of overlapping bands of inverse signs is often mentioned as an advantage of the CD spectrum over the UV spectrum in the assignment of electronic transitions. The possibility of recording the CD spectra of achiral aromatic compounds in the form of their cyclodextrin complexes offers a good opportunity for the extension of this method of spectral analysis. The CD study of the naphthol complexes show that, if the structure of the complex is known, the sector rule can also be applied for estimating the approximate polarizations of the transitions in aromatic molecules.

LITERATURE CITED

1. Harata, K., Uedaira, H.: Bull. Chem. Soc. Japan, 48, 375 (1975).
2. Shimizu, H., Kaito, A., Hatano, M.: Bull. Chem. Soc. Japan, 52, 2678 (1979).
3. Shimizu, H., Kaito, A., Hatano, M.: Bull. Chem. Soc. Japan, 54, 513 (1981).
4. Tinoco, I. Jr.: Advan. Chem. Phys., 4, 113 (1962).
5. Kirkwood, J.P.: J. Chem. Phys., 5, 479 (1937).
6. Szejtli, J., Dósa, É., Lipták, A., Nánási, P., Jodál, I., Jánossy, L.: Hung. Patent Appl. 1140/80, 1980.
7. Szejtli, J., Zsadon, B., Fenyvesi, É., Szilas, M., Tüdős, F.: Hung. Patent Appl. 1960/80, 1980.
8. Suzuki, S., Fujii, T., Baba, H.: J. Mol. Spectrosc., 47, 243 (1973).
9. Urry, D.W. (Ed): "Spectroscopic Approaches to Biomolecular Conformation", American Medical Association, Chicago, 1970, p. 65.
10. Platt, J.R.: J. Chem. Phys., 17, 484 (1949); 19, 263 (1951).
11. Kiss, Á.I., Szőke, J.: Acta Chim. Acad. Sci. Hung., 90, 133 (1976).
12. Kiss, Á.I., Szőke, J.: unpublished results
13. Harata, K., Uedaira, H., Tanaka, J.: Bull. Chem. Soc. Japan, 51, 1627 (1978).
14. Harata, K.: Bull. Chem. Soc. Japan, 52, 1807 (1979).

I.Int.Symp. on Cyclodextrins
Budapest, 1981 3.6.

TRANSFER OF ORGANIC MOLECULES THROUGH AQUEOUS LAYERS MEDIATED BY CD'S AND CD-DERIVATIVES

A. SCHUBERT, Z. NOSZTICZIUS, G. PATONAY, T. HASKÓ

Institute of Physics, Technical University, Gellért tér 4
H-111 Budapest, Hungary

SUMMARY

Organic (e.g. hydrocarbon) molecules were transferred from one hydrogen bubble stream to another through an aqueous barrier. The transferred molecules were detected by flame ionization detector. Dissolving CD's or CD-derivatives in the water, the transfer could be facilitated by about one or two orders of magnitude. Non-volatile guest molecules in the solution acted as competitive inhibitors of this "facilitated transport". Kinetic analysis of the transport and the inhibition provided simple means for estimating the stability constants of CD-complexes for both volatile and non-volatile guest molecules in dilute aqueous solution.

Introduction

In the classical "carrier mediated transport" experiments the carrier facilitates the permeation of hydrophylic substances (e.g. ions) through lipid membranes of biological or artificial origin. The ability of CD's of forming water soluble complexes with highly hydrophobic guest molecules suggests an inversion of this situation: the CD's may act as carriers facilitating the transfer of these guest molecules through aqueous barriers. The higher is the equilibrium concentration of the complex in aqueous solution, the more efficient facilitation can be attained. Moreover, the transfer of a given substance can be hindered by adding another type of guest molecule to the solution. Both the facilitation itself and the inhibition can be used for estimating the stability constants of CD-complexes in dilute aqueous solutions.

Experimental procedure

All the applied chemicals were of reagent grade. CD's and CD-derivatives were prepared and supplied by the Biochemical Laboratory of the Chinoin Pharmaceutical Works, Budapest.

A schematic diagram of the complete experimental setup is depicted in Fig. 1. In the experiments transfer was induced from a H_2 stream (upstream gas) containing volatile organic substance (permeant) to a pure H_2 stream (downstream gas) through an aqueous solution. Electronically controlled constant flows of H_2 were produced by electrolytic H_2 generators. The upstream and the downstream gas flows were bubbling through the solution simultaneously without direct contact between them; this could be attained using a

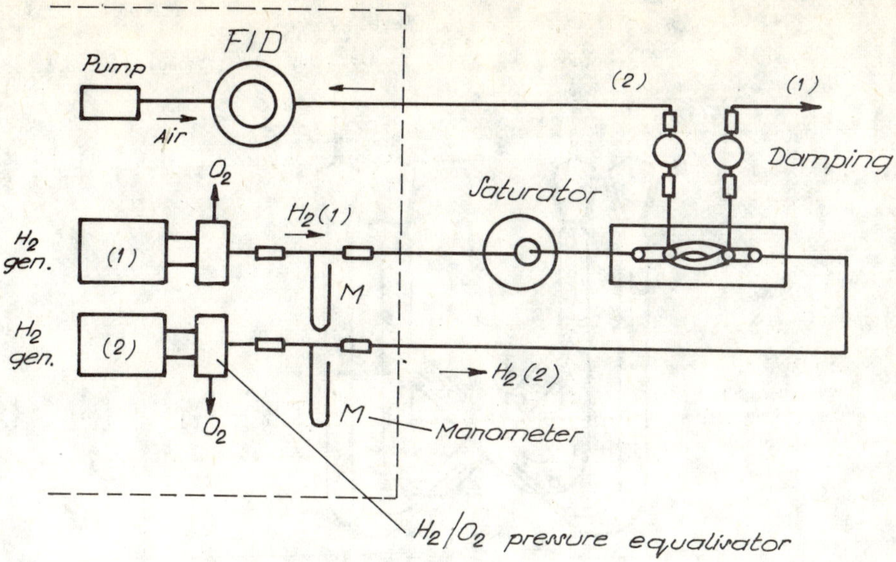

Fig.1. Schematic diagram of the experimental setup

specially designed cell (Fig. 2.). The volume of the cell was about 15 cm^3. The permeant content of the upstream inlet gas was controlled by the temperature of the saturator. The transferred amount of the permeant was monitored in the downstream outlet gas by flame ionization detector (FID). Fig. 3. illustrates the appearance of the apparatus.

In the course of a series of experiments the total CD-concentration, c_o, in the solution was fixed. After the partial pressure, p, of the permeant in the upstream inlet gas had been adjusted a stationary transfer rate was reached within 1 - 2 minutes. The electric current, I, of the FID

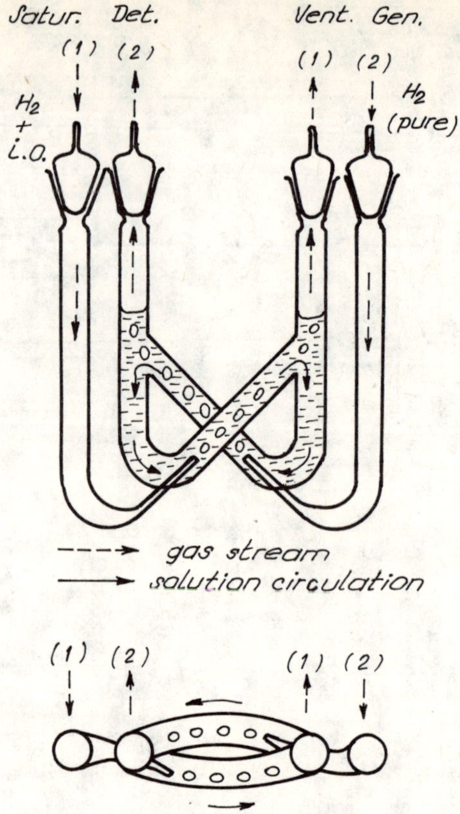

Fig.2. The transfer cell

was then recorded and the experiment was repeated with a new *p* value.

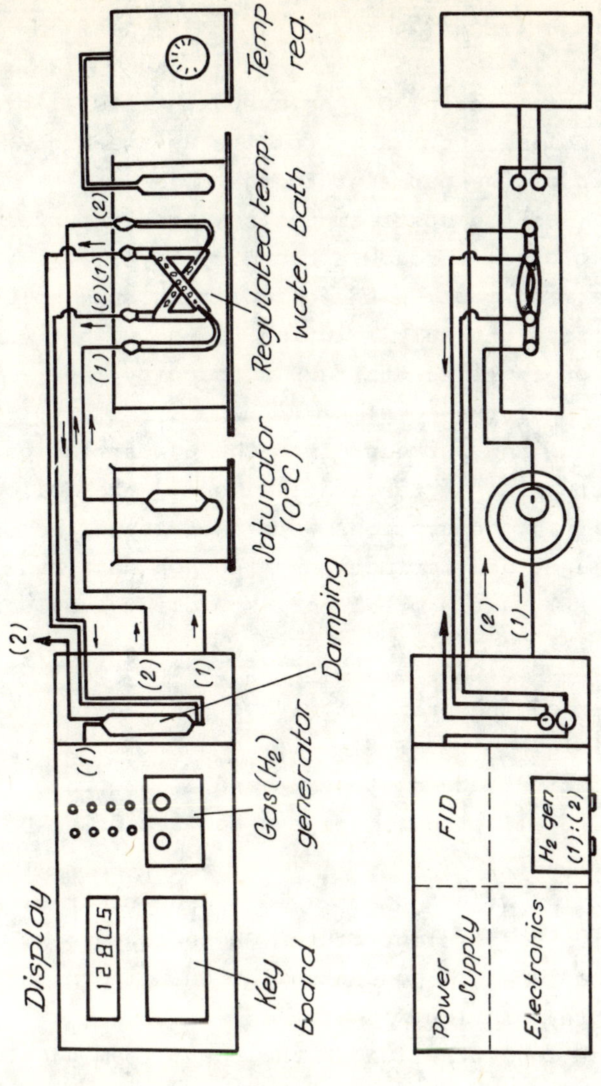

Fig. 3. Schematic drawing of the apparatus

Theoretical interpretation

The elementary steps of the overall transfer process are:

1. absorption of the permeant from the upstream gas;
2. association of the permeant to the CD;
3. diffusion from the upstream-gas/solution interface into the bulk solution through a boundary layer;
4. convective transport in the circulating bulk solution;
5. diffusion from the bulk solution to the downstream gas/solution interface through a boundary layer;
6. dissociation of the complex;
7. desorption of the permeant into the downstream gas.

In constructing a quantitative model, the following set of simplifying assumptions were used supported partly by order-of-magnitude estimations based on data available from literature, partly by direct experimental evidences.

/i/ Only 1:1 complexes are present in the solution.

/ii/ Transfer of the permeant without CD-facilitation is negligible.

/iii/ In the above scheme steps 3 and 5 are rate limiting, therefore all other steps can be treated as being in equilibrium.

/iv/ Association with a guest molecule does not alter the diffusion coefficient of the CD in aqueous solution, thus the total CD-concentration is uniform throughout the whole solution.

/v/ Permeant concentration in the downstream gas is negligible as compared to that in the upstream gas.

/vi/ The signal of the FID is directly proportional to the transfer rate.

Let us consider first the transfer of a single permeant through CD-solution. Following the assumptions, we have the equilibrium conditions

$$c = s.p \quad \text{(Henry's Law)} \qquad /1/$$

$$Kc\ (c_o - c_x) = c_x \quad \text{(complex equilibrium)} \qquad /2/$$

on the upstream gas/solution surface. Here c is the surface concentration of the permeant in the solution, s is its solubility, p is its partial pressure in the upstream gas; c_o is the total CD-concentration, c_x is the surface concentration of the complex, K is the equilibrium (stability) constant. The stationary transfer rate can be given as

$$r = P\ (c_x - c_x\ (bulk)) = Pc_x\ (bulk) \qquad /3/$$

(P is the permeability of the boundary layer), since on the downstream gas/solution surface together with the permeant concentration the complex concentration vanishes too. From Eqs. /1/, /2/ and /3/

$$r = \frac{Pc_o}{2} \frac{p}{1/Ks + p} . \qquad /4/$$

Finally the stationary current of the FID is

$$I = kr = \frac{kPc_o}{2} \frac{p}{1/Ks + p} \qquad /5/$$

(k is the calibration constant of the FID). On a double reciprocal scale we have

$$\frac{1}{I} = \frac{2}{kPc_o} \left(1 + \frac{1}{Ks} \frac{1}{p}\right), \qquad /6/$$

that is, the intercept with the $1/p$ axis of the $1/I$ vs. $1/p$ plot is $-Ks$. Provided the solubility, s, of the permeant is known, the stability constant, K, can be calculated.

Let us consider now the case, when the CD-solution contains also a non-volatile substance forming complex with the CD. This substance acts as a competitive inhibitor of the permeant transfer. Instead of Eq. /2/ we have now two equilibrium equations:

$$Kc\,(c_o - c_x - c_y) = c_x\,, \qquad /7/$$

$$K'(c' - c_y)\,(c_o - c_x - c_y) = c_y \qquad /8/$$

where, in addition to the former notations, c' is the total concentration of the inhibitor, c_y is the surface concentration of the inhibitor-CD complex and K' is its stability constant. From Eqs. /1/, /3/, /5/, /7/ and /8/ we may obtain

$$K' = Ksp\,\frac{I_{max}}{I}\,\frac{1 - I/I_o}{c'-c_o\,(1 - I/I_o)}\,, \qquad /9/$$

where I is the actual FID current, I_o is that under the same conditions but without inhibitor and I_{max} is the maximum current (the $1/I$ intercept of the $1/I$ vs. $1/p$ plot).

Results and discussion

Only very preliminary results can be reported at the moment. The facilitation of the transfer of hydrocarbons by α-CD, β-CD and dimethyl-β-CD (DIMEB) is illustrated in Table I. The FID currents presented were measured using 1% solutions of the CD's and an upstream inlet H_2 gas saturated by the hydrocarbon at 0^oC.

Table I.

	pure water (nA)	1% α-CD (nA)	1% β-CD (nA)	1% DIMEB (nA)
n-pentane	6.6	11.3	25.8	12.6
n-hexane	1.1	3.7	12.3	5.7
n-heptane	0.2	3.4	15.6	3.7
n-octane	0.03	1.0	8.1	1.7
i-octane	0.15	*	0.8	21.7

*the complex formed white precipitate

The presence of transfer facilitation is quite obvious, and the i-octane/DIMEB system proved to be a specially good model for further studies. The dependence of transfer rate on partial pressure is shown in Table II.

Table II.

p (torr)	13	18	22	32	38
I (nA)	23.5	37.1	38.1	47.3	52.9

The double reciprocal plot is fairly linear, and the calculated value of Ks is 0.0145 $torr^{-1}$.

It is only mentioned here that definite transfer inhibition in the i-octane/DIMEB system could be brought about using various guest molecules, such as prednisolone, dexamethasone and lidocaine; the detailed results of these investigations will be published in the near future.

I.Int.Symp. on Cyclodextrins
Budapest, 1981 3.7.

RESOLUTION OF CHIRAL SULFINYL COMPOUNDS VIA β-CYCLODEXTRIN INCLUSION COMPLEXES

J. DRABOWICZ, M. MIKOŁAJCZYK

Center of Molecular and Macromolecular Studies, Polish Academy of Sciences, Department of Organic Sulfur Compounds
Boczna 5
90-362 Łodz, Poland

SUMMARY

The optical resolution of racemic aralkyl sulfoxides via their inclusion complexes with β-cyclodextrin was recently reported from our laboratory. To obtain more information concerning the relationship between the chirality and optical purity of the included sulfoxides and the structure of their inclusion complexes, the resolution of a series of o- and m-tolyl alkyl sulfoxides was carried out. The influence of other factors such as pH and the presence of inorganic salts on the stereoselectivity of the inclusion of sulfoxides was also investigated. The results on the resolution of sulfinamides via β-CD complexes are also presented.

Cyclodextrin inclusion complexes with racemic compounds (guests) are mixture of diastereomers which can be formed in unequal amounts. Therefore, stereospecific inclusion into cyclodextrins (hosts) can be applied for optical resolution of racemic compounds. By this method, first reported by Cramer and Dietsche [1], we were able to resolve a series of chiral sulfoxides 1, sulfinates 2, and thiosulfinates 3 [2].

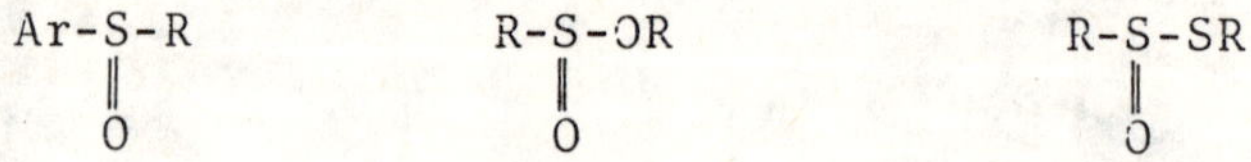

1 2 3

The relationship observed between the stereospecificity of the inclusion of sulfoxides 1 into β-cyclodextrin and the chirality of the preferably included sulfoxide led us to propose that the two concurrent inclusion complexes 4A and 4B are formed the ratio of which is dependent on the nature of alkyl substituent at sulfur. In the case of t-butyl aryl sulfoxides the formation of the inclusion complex 4B is favoured due to steric reasons.

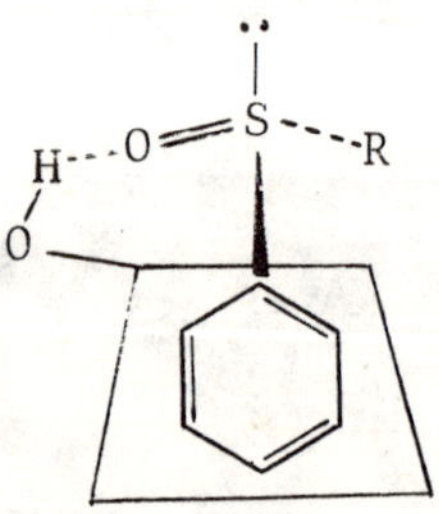

4A

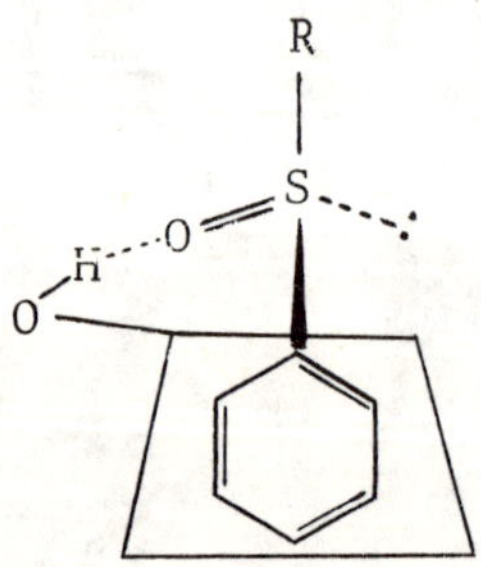

4B

To provide further experimental evidence supporting the proposed nature of the β-cyclodextrin-sulfoxide binding the optical resolution of a series of ortho- and meta-tolyl alkyl sulfoxides 5 and 6 was carried out.

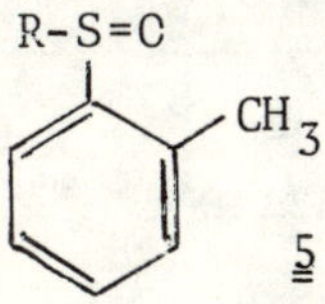

5

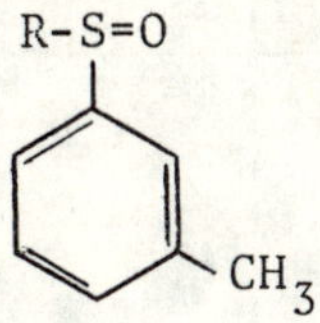

6

Resolution of 5 and 6 was carried out in water solution according to the standard procedure described in our previous paper [2]. The optical rotations, optical purities and absolute configurations of the included sulfoxides are summarized in Table 1. For the comparison purposes it contains also the pertinent data for para-analogues taken from our earlier work.

An inspection of the data in Table 1 shows that the inclusion of sulfoxides 5 and 6 into β.CD was stereospecific in all cases. As expected, however, the optical purities of the included sulfoxides are strongly dependent on the position of the methyl group in the aromatic ring. Thus, the highest degree of resolution was observed for ortho-tolyl alkyl sulfoxides 5 and the lowest for meta- analogues 6. Such relationship is in a full agreement with the asumption that the inclusion of aryl alkyl sulfoxides into β-cyclodextrin consits in the incorporation of the aromatic ring into β-cyclodextrin cavity because the lowest stereospecificity of the inclusion of meta-tolyl sulfoxides 6 resulted from much weaker penetration of the aromatic ring of 6 into the cavity of β·CD.

TABLE I. Resolution of Racemic Alkyl Tolyl Sulfoxides R-S(O)Tol, via β-Cyclodextrin Inclusion Compounds

No	Tol	R	$[\alpha]_{589}$(c,solvent)	Opt.[a] [%] purity	Abs.[a] conf.
5a	o	Me	+34.90(0.51,acetone)	12.3[a,b]	R
5b	o	Pr^i	-50.00(1.06,acetone)	21.8[a,c]	S
5c	o	Bu^n	-25.00(0.40,acetone)	22.3[a,d]	S
6a	m	Me	+ 5.80(1.47,acetone)	3.0[a,e]	R
6b	m	Pr^i	+ 0.50(0.62,acetone)	0.5[a,f]	R
6c	m	Bu^n	+ 0.30(1.57,acetone)	0.5[a,g]	R
	p	Me	+11.50(2.77,ethanol)	8.0[h]	R
	p	Pr^i	+ 3.45(1.39,ethanol)	1.9[h]	R
	p	Bu^n	+ 6.70(1.18,acetone)	3.6[h]	R

[a] The chirality at sulfur and optical purity of alkyl tolyl sulfoxides have been estimated chemically by means of the conversion of optically active R-S(O)OR[1] into the corresponding alkyl tolyl sulfoxides, R-S(O)Tol, using appropriate Grignard reagents, TolMgBr,assuming that the Grignard reaction proceeds with full inversion of configuration at the sulfinyl sulfur atom.

[b] based on the conversion of (+)-(R)-Me-S(O)OiPr, $[\alpha]_{589}=+62^o$ (ee,25.6%) into (+)-(R)-methyl tolyl sulfoxide, $[\alpha]_{589}=+72.5^o$

[c] based on the conversion of (+)-(R)-Pr^i-S(O)OMe, $[\alpha]_{589}=+1.44^o$ (ee,1.28%) into (+)-(R)-isopropyl o-tolyl sulfoxide, $[\alpha]_{589}=+2.0^o$

[d] based on the conversion of (+)-(R)-Bu^n-S(O)OPr^n, $[\alpha]_{589}=+22.2^o$ (ee,20.2%) into (+)-(R)-n-butyl p-tolyl sulfoxide, $[\alpha]_{589}=+22.6^o$

[e] based on the conversion of (+)-(R)-Me-S(O)OPr^{i}, $[\alpha]_{589}$=+62° (ee,25.6%) into (+)-(R)-methyl m-tolyl sulfoxide,$[\alpha]_{589}$= +42.50^{o}

[f] based on the conversion of (+)-(R)-Pr^{i}-S(O)OMe, $[\alpha]_{589}$=+1.44 (ee,1.28%) into (+)-R-isopropyl m-tolyl sulfoxide $[\alpha]_{589}$= +1.07.

[g] based on the conversion of (+)-(R)-Bu^{n}-S(O)OPr^{n},$[\alpha]_{589}$=+22.2 (ee,20.2%) into (+)-(R)-n-butyl m-tolyl sulfoxide, $[\alpha]_{589}$= =+10.60^{o}

[h] taken from reference[2].

TABLE II. Resolution of Benzyl Alkyl Sulfoxides Bz-S(O)R, in the presence of Trifluoroacetic Acid[a]

No	R	CF_3COOH[g]	$[\alpha]_{589}$(solvent)[b]	Opt.[%] purity	Abs. conf.
9a	Me	0	- 8.5(Et)	8.0	R
		1.2	- 3.3(Et)	3.0	R
9b	Et	0	+ 5.0(Chl)	4.7	R
		1.2	- 7.0(Chl)	6.8	S
9c	Pr^{n}	0	+ 2.0(Et)	3.6	R
		1.2	+ 2.0(Et)	3.6	R
		6.0	+ 1.0(Et)	1.8	R
9d	Bu^{t}	0	+45.0(Et)	1.5	R
		1.2	c		

[a] Resolution was carried out in water solution (50 ml) containing 0.75 g of β-CD

[b] Et = ethanol, Chl = chloroform

[c] Inclusion complex was not formed.

It is interesting to note that (-) isopropyl o-tolyl sulfoxide 5b and (-)-n-butyl o-tolyl sulfoxide 5c recovered from their inclusion complexes with β·CD have the S-configuration whereas the chirality at sulfur in (+)-methyl o-tolyl sulfoxide 5a is R. Most probably, in the latter case the inclusion complex of the structure similar to 4A is preferentially formed. However, in the case of 5b and 5c the combination of non-bonding steric repulsions between bulky isopropyl and n-butyl groups, o-methyl group and the cyclodextrin ring favours the formation of the complex having the structure 4B in which the included sulfoxides have the opposite S-chirality at sulfur.

In the next step of the present study the optical resolution of ortho- and para-methoxyphenyl methyl sulfoxides 7 and 8 was attempted in the hope to get some information concerning the effect of the methoxy group on the stereospecificity of the inclusion. In this context, one should note that methoxy group is able to form a hydrogen bond with the cyclodextrin hydroxy groups.

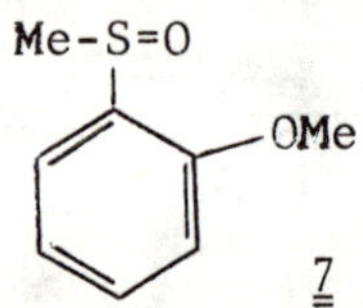

7

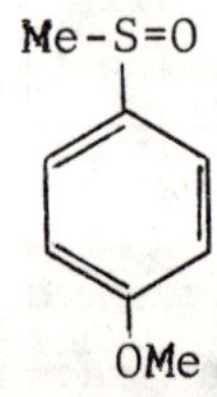

8

It was found that sulfoxide 8 does not practically precipitate from the water solution of β·CD in the form of its inclusion complex. On the other hand, the racemic sulfoxide 7 forms very easily with β·CD the solid inclusion compound from which (-)-$\underline{7}$, $[\alpha]_D$ -25.8° (o.p.8%) was recovered. The optical purity of (-)-$\underline{7}$ obtained and its absolute configuration S was determined by chemical correlation shown below.

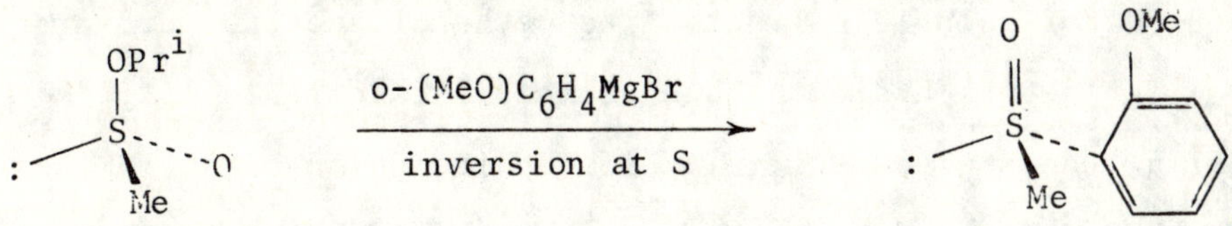

(+)-(R), $[\alpha]_D$ +62°(o.p.25.6%) (+)-(R)-$\underline{7}$, $[\alpha]_D$ +81.9°(o.p.25.6%)

The preferential inclusion of (-)-(S)-$\underline{7}$ into β·CD contrasts with the behaviour of ortho-tolyl methyl sulfoxide $\underline{\underline{5a}}$ the R-enantiomer of which is preferably included into β·CD. This change in the chirality of the included sulfoxides $\underline{7}$ and $\underline{\underline{5a}}$ is not induced by the steric interaction since the effective steric bulk of the ortho- methoxy group is smaller than that of the ortho- methyl group[3,4]. In our opinion this change may be due to the formation of the hydrogen bond between ortho-methoxy group as shown below.

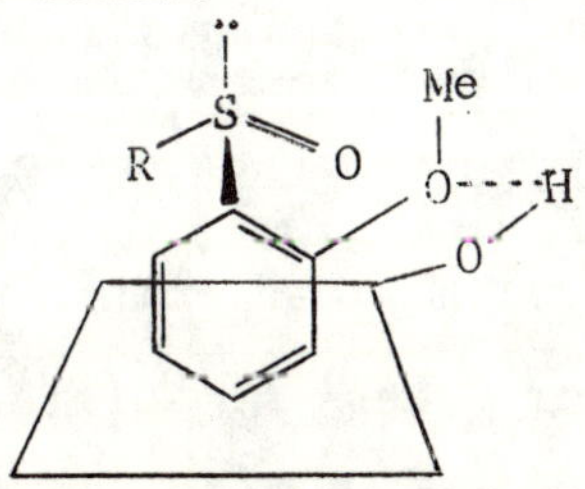

If hydrogen bonding plays some role in the formation of the inclusion complexes between β·CD and sulfoxides, one may expect that the stereospecificity of the inclusion will be dependent on the pH of a solution in which the formation of such complexes takes place. The results collected in Table II concerning the resolution of benzyl alkyl sulfoxides 9 carried out in the presence of trifluoroacetic acid clearly indicate that this is the case. Similar effect was also observed by adding water-soluble solvents like methanol,acetone, dioxane which can act as hydrogen bond acceptors (see Table III). Finally, the stereospecificity of the inclusion of sulfoxides into β·CD was found to be affected by the addition of inorganic salts to the water solution (see Table IV). For example, resolution of isopropyl benzyl sulfoxide in the presence of small amount of KSCN is quite effective in contrast to the non-stereospecific inclusion of this sulfoxide taking place in water solution.

As part of this study the optical resolution of sulfinamides 10 was also investigated.

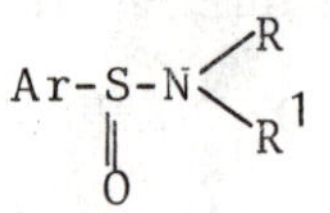

R=H or alkyl

10

Since sulfinamides are insoluble in water, the β-cyclodextrin inclusion complexes were obtained either by grinding β·CD hydrate with a 4-5 molar excess of sulfinamide or by shaking a water solution of β·CD with a solution of sulfinamide in organic solvent. Optical rotations, optical purities and chirality at sulfur in 10 recovered from the inclusion complexes are given in Table V. These preliminary data indicate that the steric bulk of alkyl substituents bonded to the nitrogen atom

TABLE III. Resolution of Sulfoxides in Water Solution Containing a Water-Soluble Solvent[a]

Sulfoxide	Solvent[b] (ml)	$[\alpha]_{589}$ (solvent)[b]	Opt. [%] purity	Abs. conf.
Bz-S(O)Me	Me(5)	c		
	Ac(5)	c		
	DMF(1.5)	- 5.05(Et)	4.5	R
	without	- 8.50(Et)	8.0	R
Bz-S(O)Et	Ac(5)	+ 9.0(Chl)	8.5	R
	DMF(1.5)	+ 4.7(Chl)	4.5	R
	without	+ 5.0(Chl)	4.7	R
Bz-S(O)Pr^{i}	Ac(5)	+30.1(Et)	23.2	S
	Me(5)	0 (Et)		
	Di(5)	c		
	DMF(12)	- 6.7(Et)	5.1	R
	without	- 0.2(Et)	0.15	R
pTol-S(O)Et	Me(15)	+ 8.56(Ac)	6.5	R
	Ac(10)	+24.0(Ac)	17.5	R
	Is(10)	c		
	Is(2)	+11 (Ac)	7.9	R
	Et(10)	+ 6.07(Ac)	4.5	R
	without	+11.5(Ac)	8.0	R

[a] Resolution was carried out in water solution (50 ml) containing 0.75 g of β-CD

[b] Ac = acetone, Me = methanol, Et = ethanol, Is = isopropanol, Chl = chloroform, Di = dioxane, DMF = dimethylformamide

[c] Inclusion complex was not formed

TABLE IV. Resolution of Benzyl Akyl Sulfoxides, Bz-S(O)R, in Water Solution Containing Inorganic Salts[a]

Sulfoxide	Salt[g]	$[\alpha]_{589}$(solvent)[b]	Opt.[%] purity	Abs. conf.
Bz-S(O)Me	KSCN(1.8)	c		
Bz-S(O)Et	KSCN(1.8)	- 3.0(Et)	5.9	R
	KCl(1.3)	- 5.0(Et)	9.5	R
	NaCl(1.1)	+ 2.4(Chl)	2.0	R
	without	+ 5.0(Chl)	4.7	R
Bz-S(O)Pr^i	KSCN(1.8)	-15.4(Et)	11.8	R
	KJ (2.7)	- 9.3(Et)	7.0	R
	without	- 0.2(Et)	0.15	R
pTol-S(O)Et	KSCN(1.8)	+ 7.02(Ac)	4.7	R
	without	+11.50(Et)	8.0	R

[a] Resolution was carried out in water solution (50 ml) containing 0.75 g of β-CD

[b] Et = ethanol, Ac = acetone, Chl = chloroform.

[c] Inclusion complex was not formed.

TABLE V. Resolution of Arylsulfinamides, Ar-S(O)NRR[1], (10) via β-Cyclodextrin Inclusion Complexes

No	Ar	R	R^1	$[\alpha]_{589}$(c,solvent)	Opt.[a] [%] purity	Abs. conf.
a	Ph	H	Me	+12.5(1.82 Ac)	7.4	S
b	p-Tol	H	Cy	+ 3.6(1.10 Ac)	-	S
c	p-Tol	Me	Me	+16.7(3.2 Ac)	9.0	S
d	p-Tol	Et	Et	0		
e	p-Tol	Pr^i	Pr^i	0		

[a] Optical purity values were calculated based on the following data: $[\alpha]_D = +170.0^o$ (acetone) for R-10a from C.W.Schroeck, C.R.Johnson, J.Am.Chem.Soc. 93, 5385(1971); $[\alpha]_D = +159.3^o$ (acetone) for R-10c from S.Colonna, R.Giovini, F.Montanari, Chem.Commun., 1968, 865.

in 10 has considerable effect on the stereospecificity of the inclusion.

LITERATURE CITED

1. Cramer, F., Dietsche, W.: Chem.Ber. 92 378 (1959).

2. Mikolajczyk, M., Drabowicz, J.,: J.Am.Chem.Soc. 100, 2510 (1978).

3. Taft, R.W.: J.Am.Chem.Soc., 74 3120 (1952).

4. Gordon, A.J., Ford, R.A.: Chemist Companion, Wiley, New York, N.Y. 1972; p.109.

I.Int.Symp. on Cyclodextrins
Budapest, 1981 3.8.

INTERACTIONS OF β-CYCLODEXTRIN WITH IONIC DETERGENTS

K. KRAĽOVA, Ľ. MITTERHAUSZEROVÁ

Chemical Institute of UK, Kalinčiakova 8
880 34 Bratislava, Czechoslovakia

SUMMARY

Association constants of β-cyclodextrin with 3 derivatives of cationic surfactants N,N'-bis/4-/aryloxy/butyldimethyl/-1,6-hexanediammonium dibromid, namely 1-naphtoxy-, 2-naphtoxy and p-tert.butoxyderivative, estimated at 20°C are 601 ±178 lM^{-1}, 629 ±93 lM^{-1}, 1743 ±200 lM^{-1} respectively. The surface activity of aqueous solutions of these derivatives in the presence of CD is lowered and critical micelle concentration /CMC/ is shifted to the higher concentrations. Such a shift of CMC for these surfactants was observed also using solubilization study.

Interactions of CD with 8 cationic surfactants of the type N,N'-bis/alkyldimethyl/-1,6-hexandiammonium dibromid /alkyl = nonyl bis hexadecyl/ were studied by spectrophotometry in the presence of methylorange /MO/. Absorption spectrum of MO is influenced by the interaction of CD with surfactant molecule and the degree of spectral changes depends on the association of surfactant molecules. This effect can be used for investigation of association process in the system MO-surfactant as well as in the system MO-surfactant-CD /C_{CD} = const./. In the presence of CD of aqueous surfactant solutions is shifted to the higher concentrations, and the measure of this shift is proportional to the interaction of components. Ratio of CD bound to one surfactant molecule,evaluated from the CMC shift is lowered with lengthening of alkyl part of molecules. Analogous results were obtained using conductometric method.

INTRODUCTION

The interaction of surface-active compounds with β-cyclodextrin and the formation of inclusion complexes present an interesting field of study with respect to their influence on the physico-chemical properties of these compounds. The cationic surfactant - β-cyclodextrine complexes were also studied by japanese chemists /1,2/.

The present study is aimed at obtaining experimental information on the interactions of β-CD with quaternary bis-ammonium salts having long alkyl- and aryloxy-chains, with special orientation to the problem of the surface-active and solubilizing properties of these cationic surfactants, and their electric conductivity.

EXPERIMENTALS

Materials used:

The homologous series of cationic surfactants of the following types: N,N'-bis-/alkyldimethyl/-1,6-hexanediammonium dibromide (alkyl: nonyl- /NoDiBr/; decyl- /DeDiBr/; undecyl- /UnDiBr/; dodecyl- /DoDiBr/; tridecyl - /TriDiBr/; tetradecyl- /TeDiBr/; pentadecyl- /PeDiBr/; hexadecyl- /HeDiBr/); as well as N,N'-bis-[4-/aryloxy/butyldimethyl]-1,6-hexanediammonium dibromide (aryloxy: 1-naphthoxy- /1-NaphDiBr/; 2-naphthoxy- /2-NaphDiBr/; 4-tert.butylphenoxy- /TebDiBr/), were synthesized at the Pharmaceutical faculty of Komenský University, Bratislava /3/.

β-cyclodextrin /CD/ /90.14% dry solids content/ produced by Chinoin Budapest.

Methyl orange /MO/ as well as pyrene, supplied by Lachema, Brno.

Methods:

Spectrophotometric determination of association constants of the surfactant-CD complex: Absorption spectra of the pure components as well as of the complexes were registered on a double-ray SPECORD spectrophotometer /type UV VIS, Zeiss, Jena/ at 20°C. Ketelaar's equation /4/ has been used to calculate the complex' association constants. This equation is a modification of the Benesi-Hildebrand equation /5/ which is valid if $C_D > C_A$:

$$C_A/(A-A^o) = [K_C \cdot C_D(\epsilon_C-\epsilon_A-\epsilon_D)]^{-1} + (\epsilon_C-\epsilon_A-\epsilon_D)^{-1}$$

where C_A,C_D are molar concentrations of the acceptor and donor, respectively; ϵ_C,ϵ_A,ϵ_D are molar concentrations of the complex,acceptor and donor; A is the system absorbancy whereas A^o the absorbancy sum of the free components /both absorbancy values determined with a 1 cm thick layer/; K_C = association constant.

Determination of the surface activity: Stalagmometry, with droplet weighing at 20°C has been used to determine surface tension of aqueous surfactant, and surfactant-CD-water systems /C_{CD} = const./; redistilled water has been used to prepare the solutions.

Determination of the solubilizing capacity: In aqueous solutions of 1-NaphDiBr, and in 1-NaphDiBr-CD-water systems /C_{CD} = const./,solid crystalline pyrene was shaken until equilibrium was reached.The pyrene portion that has not been dissolved was then filtrated, and the concentration of solubilized pyrene in the filtrate was determined by spectrophotometry.

Determination of surfactant-CD interaction in aqueous solutions in the presence of methyl orange by spectrophotometry: MO absorption spectra in the visible range /SPECORD UV-VIS, Zeiss, Jena/ were determined in systems of MO-surfactant-CD-redistilled water /C_{MO} = const.; C_{CD} = const./,depending upon increasing surfactant concentration,at 20°C. MO concentration used was approx. 3.10^{-5}M.

Determination of electric conductivity: These determinations in aqueous surfactant solutions, and surfactant-CD-water systems /C_{CD} = const./ have been carried out using a Radelkisz, Budapest, type OK 102 conductometer with a platinum electrode. Temperature at the determinations was 20°C.

RESULTS AND DISCUSSION

The formation of surfactants-CD complexes has an influence upon the absorption spectra of 3 aryloxyderivatives of the surfactants examined /1-NaphDiBr, 2-NaphDiBr, TebDiBr/.

The extremely close values of CD-1-NaphDiBr /601±178 lM^{-1}/ and CD-2-NaphDiBr /629±93 lM^{-1}/ complexes, association constants suggest that, with respect to the overall molecular conformation of the surfactant, incorporation of the naphthalene core into the CD-cavity does not occur in a pure equatorial,or pure axial position as it had been found at complexes of CD with 1-naphthol and 2-naphthol /6,7/. The association constant of the TebDiBr - CD complex, in comparison with naphthoxy derivatives of surfactants, was approx. 3x higher /1743±200 lM^{-1}/, which suggested that, with this derivative, the non-polar p-tert.butyl group enters significantly the CD interaction.

The surface activity /γ/ vs.log C_t dependence /C_t = surfactant concentration/ for aqueous solutions of 1-NaphDiBr, and 1-NaphDiBr-CD-water systems is shown in Fig.1. Based upon the γvs.logC_t behaviour in aqueous 1-NaphDiBr solutions it can be assumed that in the concentration range studied, i.e. 5.5 - 15 mM, a certain reconstruction of the micellary associations previously formed took place, since the respective values decreased with an increasing surfactant concentration, slowly but evidently. The presence of CD pronouncedly affec-

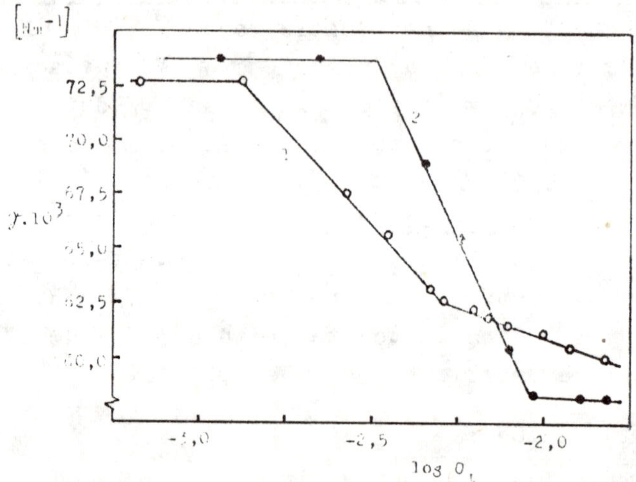

Fig.1. Dependence of surface activity on logarithm of 1-NaphDiBr concentration in aqueous solutions(1)/ and in the presence of CD;C_{CD} = 7,949 mM (2)

ted the surface activity of aqueous 1-NaphDiBr solutions expressed not only by a shift in the concentration threshold from which on a decrease is observed with an increasing surfactant concentration, but by a CMC shift towards higher values as well. Similar results were obtained in the case of TebDiBr.

Fig.2 shows the influence of 1-NaphDiBr in aqueous solutions, and that of 1-NaphDiBr-CD-water systems upon pyrene solubilization. Due to the CD interaction, CMC of aqueous solutions of 1-NaphDiBr shifted towards higher concentrations proportionally with complex formation. The solubilizing capacity of micellary surfactant solutions was not affected by CD.

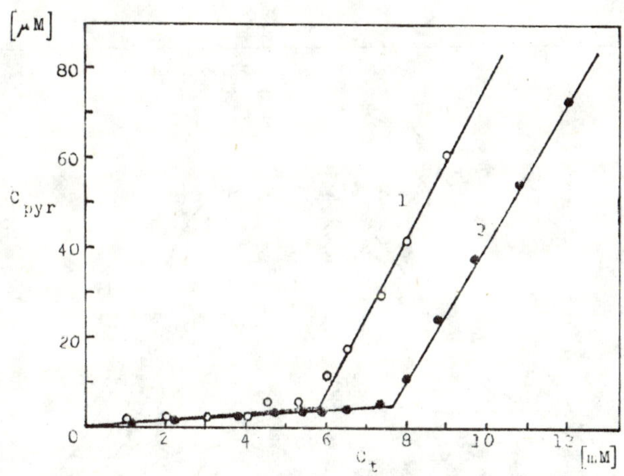

Fig.2. Solubilization of pyrene in 1-NaphDiBr aqueous solutions(1) and in the presence of CD (2); C_{CD}=4 mM

Due to an intense interaction between bis-ammonium type quaternary salts and methyl orange in aqueous solutions, a pronounced change of the MO absorption spectrum in the visible wavelength range can be observed. In extremely low surfactant concentrations /corresponding to monomeric concentration ranges/, the absorbancy value of λ = 465 nm / i.e. the MO absorption band maximum in redistilled water/ drops sharply while the whole MO-absorption band maximum is shifted towards lower wavelengths. With a further increase of surfactant concentration, the maximum again shifts back towards higher wavelengths, and after having reached the critical micelle concentration, it remains constant, with practically no further change occuring with the absorbancy value of the absorption band in this concentration range /or showing only minimum change/. Prior to reaching of the CMC in a relatively narrow concentration range, surfactant presence increase yields in a sharp rise of absorbancy of the MO-absorption band at the wavelength evaluated /Fig.3/. β-cyclodextrin with MO gives in aqueous solutions inclusion complexes /8/, however in distilled water, the shift of the maximum MO-absorption band towards lower wavelengths, and also its intensity drop due to CD-complex formation, is not very pronounced /Fig.3/.

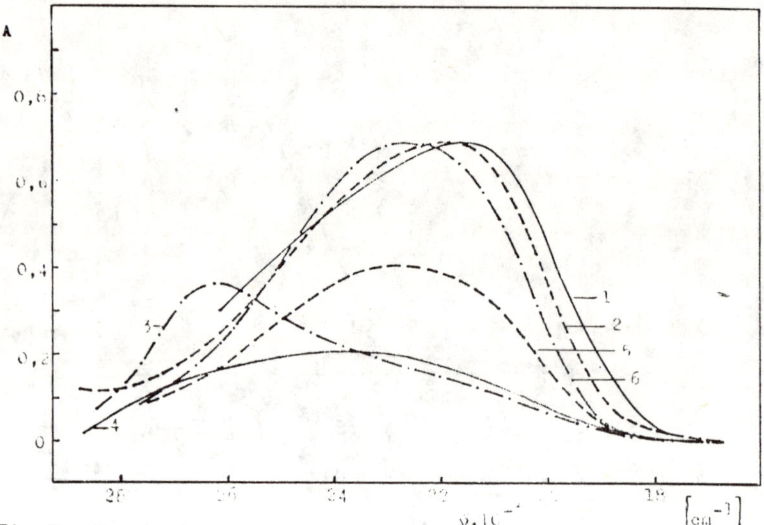

Fig.3. Absorption spectrum of methylorange in aqueous solution (1), in aqueous solution of β-cyclodextrin (C_{CD}=4.37 mM (2), and in systems of MO-DoDiBr/C_{DiDiBr} = = 0,58 mM (3); 0,884 mM (4); 0,983 mM (5) and 1,769 mM (6).

The dependence of MC-absorption band /λ = 465 nm/ absorbancy, in the presence of surfactants, and with respect to the MO-absorption band having the same concentration in water, /% A/ upon surfactant concentration, is shown in Fig.4 with aqueous DoDiBr solutions, and DoDiBr-CD-water systems.

With all of the other surfactants studied /i.e. members of the N,N'-bis-/alkyldimethyl/-1,6-hexanediammonium dibromide homologous series/, the % A vs.C_t - dependence in surfactants' aqueous solutions as well as in CD-systems was similar to the DoDiBr behaviour.

In MO-surfactant-water systems having extremely low surfactant concentrations /with respect to the CMC-value/ at first a significant drop of % A is observed in a relatively narrow concentration range. Subsequently - in the surfactant's monomeric concentration range, the % A value shows none /or only small/ change with an increasing surfactant concentration, until the concentration is reached when associate i.e. fractional micelles start forming which is accompanied by a sharp change in the dependence line. The % A value, in the $C_t >$ CMC concentration range, practically does not change any more with increasing surfactant concentrations. The CD complex formation intensively affects the process of association in surfactants' aqueous solution. In the presence of CD, each phase of this process /i.e. monomer occurence, fractional micelles formation range, micellary zone/ is shifted towards higher surfactant concentrations. Based upon the % A vs. C_t

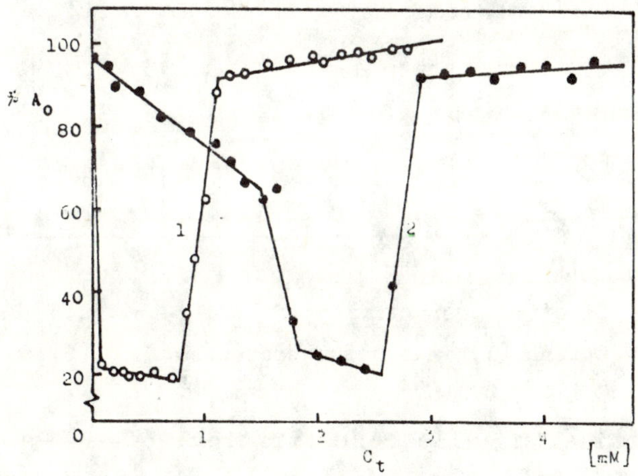

Fig.4. Dependence of % A_o in methylorange aqueous solutions on DoDiBr concentration (1) and in the presence of CD (2); C_{CD}= 3,972 mM

dependence, the CMC in aqueous solutions of the surfactant, and in CD-containing systems, can be evaluated.

The effect of surfactant concentration upon specific conductivity /ϰ/ of aqueous solutions of DoDiBr, and DoDiBr-CD-water systems, is shown in Fig.5. The dependence with each surfactant studied, i.e. members of the series N,N'-bis/alkyldimethyl/-1,6-hexane diammonium dibromide,was similar to the DoDiBr. With CD present,a decreased molar conductivity of surfactant monomeric solutions occured,together with a CMC-shift towards higher surfactant concentrations. The decrease in molar conductivity of surfactant monomeric solutions in the presence of CD could be explained by the weight increase of the monomers due to CD linked to the lateral chains in the surfactant molecules. The molar conductivity of micellary surfactant solutions remains practically unaffected by the presence of CD.

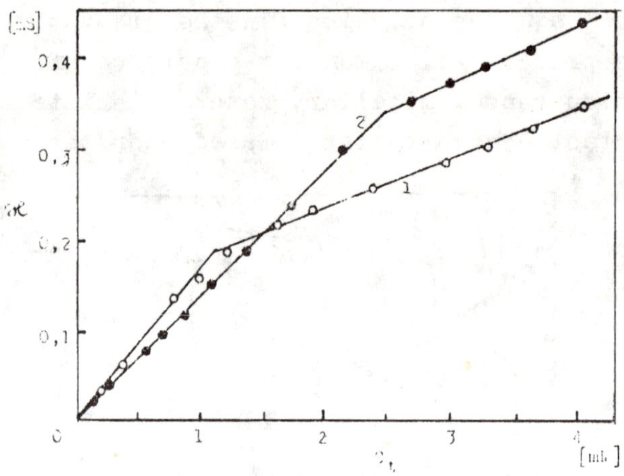

Fig.5. Electric conductivity of DoDiBr in aqueous solutions (1) and in the presence of CD (2); C_{CD} = 2,072 mM

The critical micelle concentrations of aqueous solutions of the type N,N'-bis-/alkyldimethyl/-1,6-hexanediammonium dibromide, and that of systems containing CD with C_{CD} = const., were determined conductometrically as well as using spectrophotometric method in the presence of MO.Based upon the known

CD concentration, and of the CMC shift, the rate of $\Delta CMC/C_{CD}$ can be calculated which shows the CMC shift due to specific CD concentration /Tab.1/. The $\Delta CMC/C_{CD}$ rate dependence upon the surfactant's alkyl chain length is linear in the case of the concentration CD 2 mM studied; the rate decreases with a lengthening alkyl chain of the surfactant /Fig.6/. It can be assumed from the $\Delta CMC/C_{CD}$ rate dependence upon the alkyl chain length of surfactants, members of the N,N'-bis/alkyldimethyl/-1,6-hexanediammonium dibromide homologous series,that the interaction with CD is taking place on the lateral chains of the surfactant's molecules.

Tab.1. Dependence of the $\Delta CMC/C_{CD}$ on the surfactant's alkyl chain lenght

Alkyl	C_{CD}, mM	$\Delta CMC/C_{CD}$	Method
nonyl	3,994	0,569	MO
decyl	1,987	0,867	el.con.
	3,847	0,463	MO
	6,005	0,914	el.con.
undecyl	1,829	0,760	MO
	3,971	0,611	el.con.
dodecyl	2,072	0,626	el.con.
	3,972	0,445	MO
	4,007	0,543	el.con.
	5,994	0,537	el.con.
tridecyl	1,986	0,715	el.con.
	2,031	0,511	MO
tetradecyl	2,006	0,491	el.con.
	3,973	0,510	MO
	4,012	0,484	el.con.
	6,017	0,523	el.con.
	8,023	0,458	el.con.
pentadecyl	1,986	0,521	el.con.
	2,030	0,472	MO
hexadecyl	1,986	0,267	el.con.
	2,018	0,262	MO

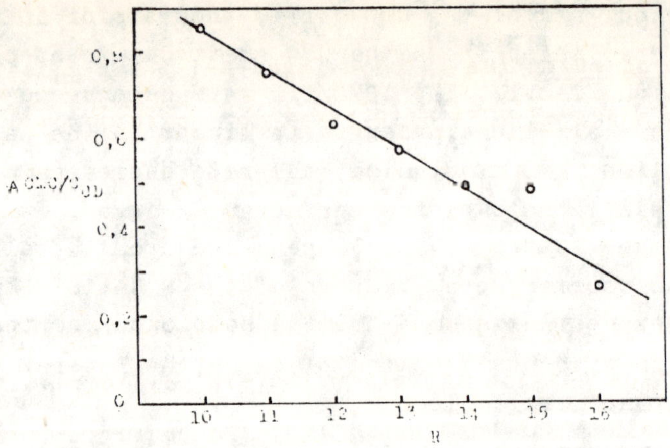

Fig.6. Dependence of the ratio of CMC shift and concentration of CD on the alkyl chain length

REFERENCES

1. Okubo, T., Kitano, H., Ise, N.: J.Phys.Chem. 80, 2661 /1976/
2. Sunamoto, J., Okamoto, H., Taira, K., Murakami, Y.: Chem. Lett. 371 /1975/
3. Imam, T.: Preparation of some organic bis-quaternary ammonium bromides and a study of their physico-chemical and biological properties, Diss., Scientific-Research Institute, Pharm.School KU, Bratislava, Czechoslovakia /1979/
4. Ketelaar, J.A.A., van de Stolpe, C., Goudsmit, A., Dzcubas, W.: Rec.Trav.Chim. 71, 1104 /1952/
5. Benessi, A.A., Hildebrand, J.H.: J.Amer.Chem.Soc. 71, 2703 /1949/
6. Harata, K.,Uedaira, H.:Bull.Chem.Soc.Japan 48, 375 /1975/
7. Mitterhauszerová, Ľ., Kráľová, K., Unpublished results
8. Szejtli, J., Budai, Z., Kajtár, M.: Magyar Kém.Folyóirat 84, 68 /1978/

I.Int.Symp. on Cyclodextrins
Budapest, 1981 3.9.

THERMOFRACTOGRAPHY OF CYCLODEXTRIN-COMPLEXES BY CLASSICAL AND OVERPRESSURED TLC TECHNIQUES

L. KERNÓCZI,[1] P. TÉTÉNYI,[1] E. MINCSOVICS,[1] I. ZÁMBÓ,[1] E. TYIHÁK,[1] J. SZEJTLI[2]

[1] Research Institute for Medicinal Plants
H-2011 Budakalász, Hungary
[2] CHINOIN Pharmaceutical-Chemical Works, Tó u. 1-5
H-1045 Budapest, Hungary

SUMMARY

Cyclodextrin complexes and simple mechanical mixtures can be rapidly differentiated from each other /e.g. in case of essential oils and pharmacons/ by thermofractography and by classical or overpressured thin-layer chromatographic techniques. In this way the fact of inclusion complex formation, stability againts heat and diminution of volatility can be well demonstrated.

INTRODUCTION

A number of methods are used to verify the fact of inclusion complex formation in solid phase: X-ray diffraction, mass spectroscopy, Warburg's method, differential thermogravimetry /DTG/, differential scanning calorimetry /DSC/, evolved gas analysis /EGA/. These methods are generally not suitable for routine assays. The aim of our work is the study of the combination of the temperature-gradient sampling method with classical TLC and overpressured thin-layer chromatography /OPTLC/ for qualitative and semiquantitative analysis of cyclodextrin complexes.

RESULTS AND CONCLUSION

When thermo-fractionating of essential oil-β-cyclodextrin complexes /1,2,3/ is carried out in nitrogen stream, that is to say increasing the material transport, it was found that components of the essential oil appear already from 40°C on the layer chromatogram. The components occuring in a temperature interval below 80°C derive from superficially bound essential oil. During raising temperature release of additional essential oil quantity is caused by the leave of water content of the complex /steam distillation/, then from about 130°C essential oil liberation from the complex increases abruptly and continues until 280 to 300°C. The strong decomposition starting at 130°C is due to the fact that complexes are usually non-stoichiometric composition /4/, the guest-molecules may not only occur in holes of cyclodextrin complexes, but in holes of crystal structure, too. In this way these molecules escape from loose structures. Above 200°C the β-cyclodextrin begins to decompose and the oil enclosed continues to release and between 280 and 300°C the rest complex carbonizes. It is obvious by the thermofractograms /Figures 1,3,5/ of the complexes of thymus oil, dill oil and sage oil that the loss of essential oil bound superficially and that

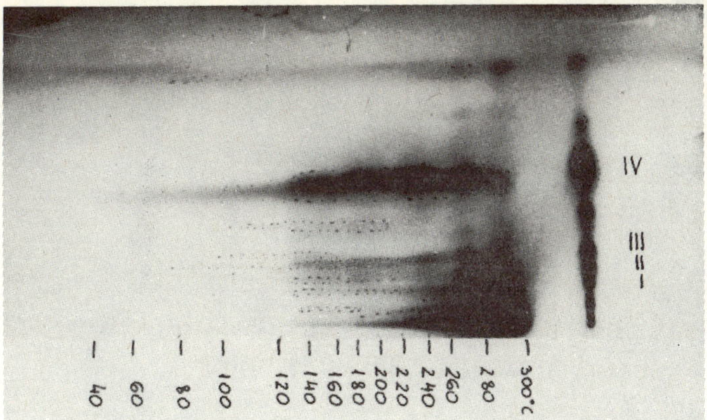

Fig. 1. Thermofractogram of thymus oil-β-cyclodextrin complex

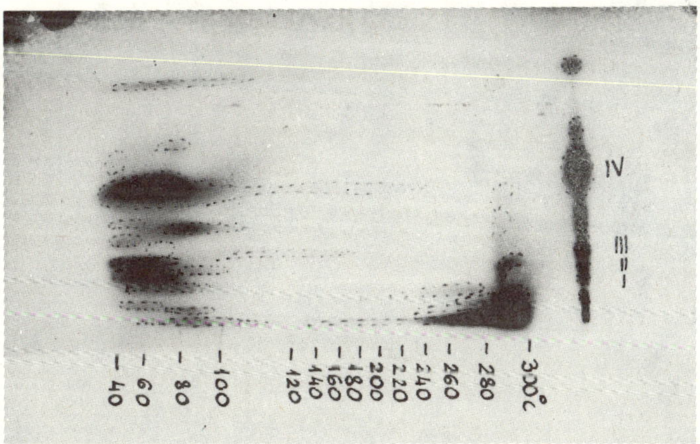

Fig. 2. Thermofractogram of thymus oil-β-cyclodextrin mechanical mixture
I geraniol II borneol III linalool IV thymol

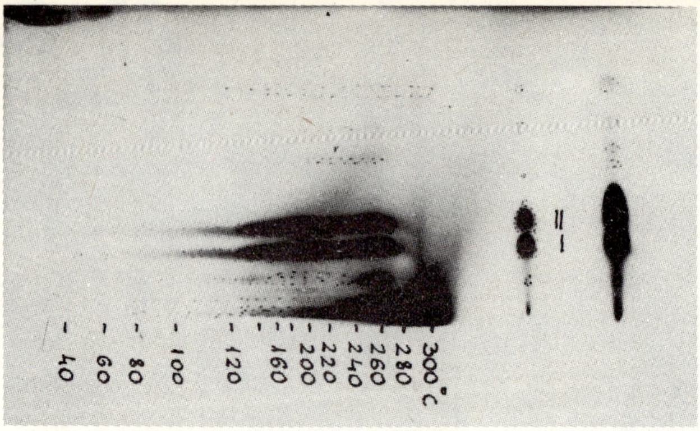

Fig. 3. Thermofractogram of dill oil-β-cyclodextrin complex

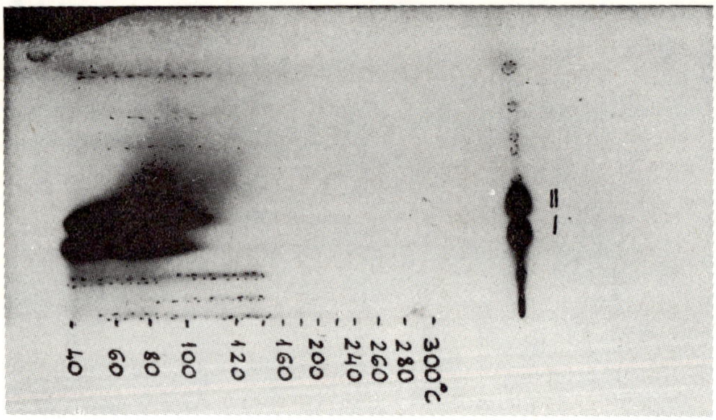

Fig. 4. Thermofractogram of dill oil
I carvone II 3,7-dimethyl-4,5,6,9-tetrahydrocoumarane

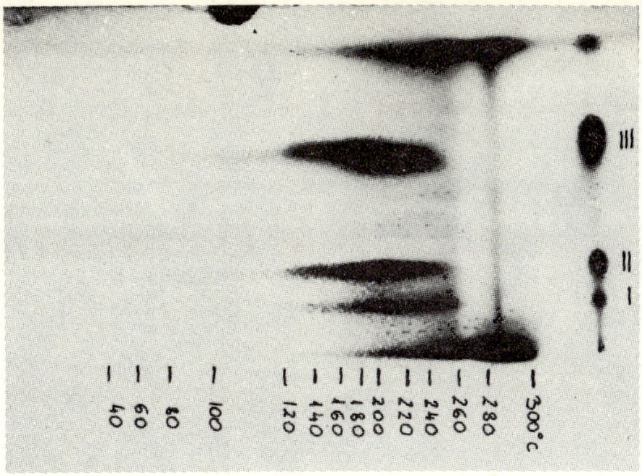

Fig. 5. Thermofractogram of sage oil- β-cyclodextrin complex

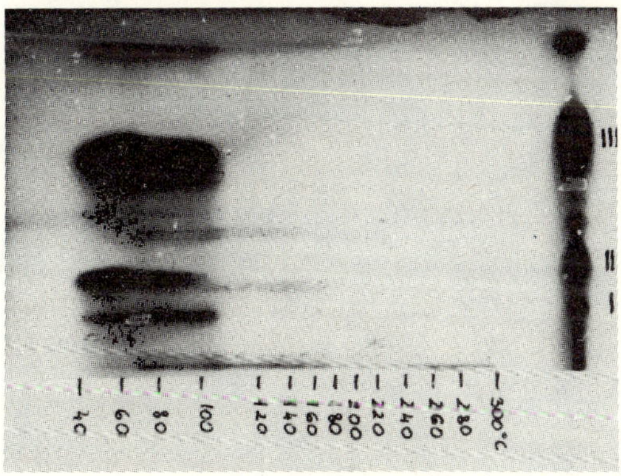

Fig. 6. Thermofractogram of sage oil
I geraniol II linalool III linalylacetate

released by the leave of the own water content of β-cyclodextrin /6 per cent/ amounts to 10-20 per cent, which is in accordance with data of GC pyrolysis /5/. Parallel with the decomposition initiating above 120°C the complete spectrum of essential oil appears on the chromatogram. The optimum temperature for qualitative analysis of complexes is 190 to 200 °C, since at this temperature the decomposition products of β-cyclodextrin do not result in any error during the evaluation of chromatogram.

The chromatograms of Figure 2,4 and 6 - which prove the fact of complex formation - have been obtained at the thermofractionating of physical mixture consisting of thymus oil, dill oil and sage oil, resp. and β-cyclodextrin. The thermofractograms show that, in contrast with the decomposition of β-cyclodextrin complexes, the components of free essential oil are evaporated almost completely below 120°C. The same was proved by heat accumulation experiments at 40°C /for 15 minutes/, 60°C and 80°C /each for 10 minutes/. As an example: about 95 per cent of the components of lavander oil, muscat sage oil volatilized at 40°C, those of dill oil and tarragon oil at 60°C, of pine oil and thymus oil at 80°C and finally those of nutmeg oil, clove oil and camomile oil at 100°C. The thermofractogram of the physical mixture of essential oil and β-cyclodextrin is similar to that of the free oil, except the spectrum of decomposition product occuring from 200°C, namely the majority of oil components volatilize prior to the evaporation of water content of β-cyclodextrin, thus complex formation is undetectable even in traces.

With full knowledge of the temperature limits described, the thermofractograms supply information on the amount of enclosed and superficially adsorbed guest-molecule in the complex produced by crystallization.

For the study of terpenoid components of essential oil-β-cyclodextrin complex new planar liquid chromatographic technique, so called overpressured thin-layer chromatography /OPTLC/ was used.

In the OPTLC /6,7/ using a pressurised ultramicro chamber /PUM chamber/ the sorbent layer is completely covered with a flexible cover membrane under an external pressure so that in the closed chamber a layer of water /a water-cushion/forms between the Plexiglass cover-plate and the flexible and fixed cover membrane. Therefore, in this chamber the vapour space above the sorbent layer is virtually eliminated. The eluent feeder micropump delivers the solvent into the sorbent layer through an sorbent inlet opening by an adjustable overpressured flow. Main advantages of this technique are the following: rapid and efficient separation of a large number of samples, applicability of longer migration distance even on fine-particle layers, with relatively short development time; excellent resolution with large separation number and theoretical plate number; low solvent consumption and the advantages of classical TLC as application of specific reagents; visual evaluation etc.

Figure 7a and 8a show the separation of thermofractionated essential oil from Thymus vulgaris and its complex with β-cyclodextrin using normal silica gel chromatoplate and saturated chamber.

The thermofractograms obtained with linear OPTLC technique illustrate the more efficient separation of the components of the thymus oil in free and complex form than in classical TLC. It follows from our preliminary results that the combined application of the temperature-gradient sampling method with OPTLC technique offers a good possibility for a more selective separation of a large number of components in essential oil than earlier planar liquid chromatographic techniques.

The thermofractography can be used as well to the investigation of those pharmacons, at wich the temperature interval of fractions obtained by a method of temperature-gradient TLC is lower, than its complexes.

Menadione melts about 60°C, totally volatilizes at 130-140°C, because the boat of TAS is empty, but it can be driven out only at 180°C from the end of the capillary. Greater part

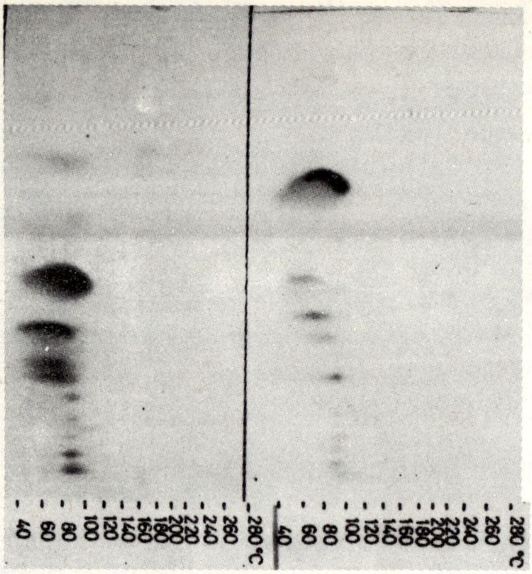

Fig. 7. Thermofractogram of thymus oil with classical TLC and linear OPTLC technique

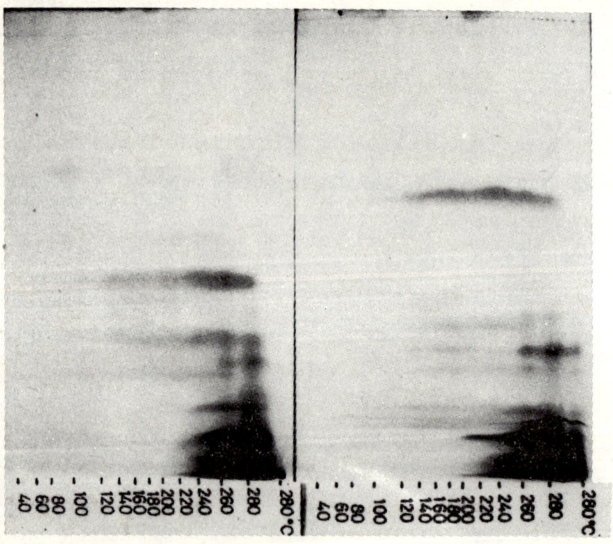

Fig. 8. Thermofractogram of thymus oil-β-cyclodextrin complex with classical TLC and linear OPTLC technique

of menadione volatilizes at 130°C / 2 minutes this is shown by the thermofractograme of menadione-β-cyclodextrin mechanical mixture. The figure demonstrates that the microsteam distillation is caused by crystal water of β-cyclodextrin. Menadione was not adsorbed on the surface of β-cyclodextrin (with 4.5% cont.), decomposition of complex has began and continued after 120°C to 280°C. Guest molecules adsorbed on surface are contained in β-cyclodextrin complexes with cont of 8.0 to 12+5%. These molecules are well separable from those which are closed in complex with holding at 120°C to 10 minutes or 130°C to 2 minutes.

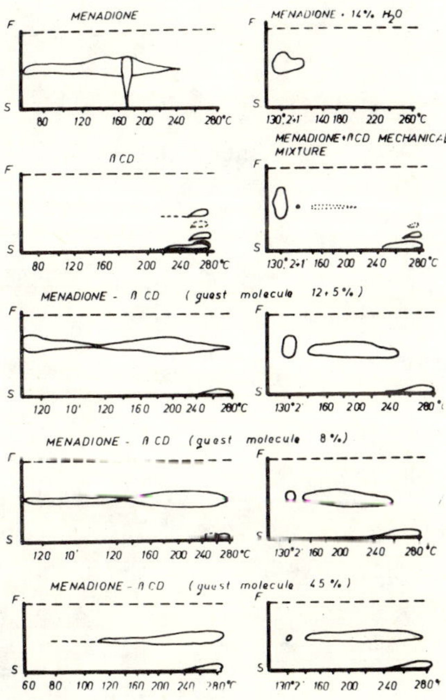

Fig. 9. Thermofractograms of β-cyclodextrin,menadione,menadione-β-cyclodextrin mechanical mixture and complexes

In this case thermofractography is particulary used as well for the differentiation of menadione closed in complex and adsorbed on surface of the β-cyclodextrin complex.

MATERIAL AND METHODS

β -cyclodextrin /CHINOIN/
Thymus oil /Thymus vulgaris L./ and its β-CD complex
Dill oil /Anethum graveolens L./ and its β-CD complex
Sage oil /Salvia sclarea L./ and its β-CD complex
TASOMAT /DESAGA/, without temperature programme-controller
Sample, 0.5 to 2 mg; complex, 2 to 20 mg
Temperature programming from 40° to 300°C
Rate of movement, 0.275 cm/min Carrier gas, N_2 16 ml/min
CHROMPRES 10 /Labor MIM/ Overpressured Thin-Layer Chromatograph
External pressure on membrane, 1.2 MPa
Solvent velocity, 1.4 cm/min Temperature, 25°C
Chromatoplate: normal Silica gel $60F_{254}$/Merck/
Solvent system: Benzene-ethylacetate /98:2 v/v, N_{us} chamber/
methylene chloride /N_s and PUM chamber/
Detection: 1% vanilline and 10% sulfuric acid in ethanol,
10% phosphorus molibdenic acid in ethanol

LITERATURE CITED

1. Stahl,E., Fuchs, J.: Dtsch.Apoth.Ztg. 108 1227, /1968/
2. Kernóczi, L., Tétényi, P., Mincsovics, E., Szejtli, J.: Quart.J.Crude Drug Res. 16 153. /1978/
3. Kernóczi, L.,Tétényi, P., Szejtli,J.:Herba Hung.19.63/1980/
4. Lammers, J.N.J.J.: Stärke 19 70, /1967/
5. Pobozsny, K., Kernóczi, L.,Tétényi, P., Héthelyi, É., Szejtli, J.: Planta Med. 42. 255. /1981/
6. Tyihák, E., Mincsovics, E.: J.Chromatogr. 174 75. /1979/
7. Mincsovics, E., Tyihák, E., Kalász, H.: J. Chromatogr. 186 293, /1980/

I.Int.Symp. on Cyclodextrins 3.10.
Budapest, 1981

THERMOANALYTICAL INVESTIGATIONS ON CYCLODEXTRIN INCLUSION COMPOUNDS

J. SZTATISZ,[1] S. GÁL,[1] J. KŐMIVES,[1] Á. STADLER-SZŐKE[2]
J. SZEJTLI[2]

[1] Institute for General and Analytical Chemistry, Technical University, Gellért tér 4
H-1111 Budapest, Hungary
[2] CHINOIN Pharmaceutical-Chemical Works, Tó utca 1-5
H-1045 Budapest, Hungary

SUMMARY

Inclusion compounds /complexes/ of different substances with β-cyclodextrin were studied with the TG, DSC, and EGA methods. In most cases, thermal analysis was found to be suitable to determine the complex character of the samples. The complex nature can be verified by [illegible] of the three methods in different cases, e.g. EGA is the most suitable for the detection of ethereal oil - β-cyclodextrin complexes, DSC is for paracetamol and TG for HBr complexes.

INTRODUCTION

A crystalline product obtained in a process of cyclodextrin complex preparation is not necessarily a real inclusion compound /complex/, moreover, the crystalline product may be a mixture, containing the complex as a component. In order to prove the complex nature of products, different analysis methods have been applied: x-ray diffraction, Warburg oxidation for unsaturated compounds, IR spectroscopy, TLC etc., however, thermoanalytical methods - being quite fast and reliable - have been used very seldom in this field /1, 2/. The goal of the present work was to show the applicability of thermal analysis methods to answer the questions.

a. Is a particular product-obtained from β-cyclodextrin

/β-CD/ and another substance - an inclusion compound /complex/ ?

b. How much of the guest material present is bound in the complex and what is the percentage of the free fraction ?

Of the several well-known methods of thermal analysis, the present authors found EGA, TG /DTG/ and DSC in the order of importance to be the most useful to study cyclodextrin complexes.

Evolved gas analysis /EGA/ means specific measurement of the rate of evolution of gases or vapours leaving the sample, during a temperature program.

Differential Scanning Calorimetry /DSC/ is the measurement of the rate of heat evolved of or absorbed by the sample, during a temperature program.

Thermogravimetry /TG/ covers measurement of the mass change of the sample /TG curve/ and the rate of the mass change /DTG curve/, during a temperature program.

EXPERIMENTAL

The present investigations were carried out on the DuPont 990 Thermal Analysis System. $5^{O}C$/min heating rate and air flow were applied in the 910 DSC cell and the 951 thermobalance. In the 916 TEA apparatus, following the evolution of organic gases and vapours by means of a hydrogen flame ionization detector, samples were heated in flowing nitrogen with a rate of $8^{O}C$/min.
Pure β-cyclodextrin, the guest substances of interest, their simple /physical/ mixtures and assumed complexes with CD were measured. Within one series of experiments the sample mass of the pure guest substance was equal to its mass in the mixture and the complex.

RESULTS AND DISCUSSION

Thermal behaviour of β-cyclodextrin is described in Fig.1.

The sample lost its water content up to 100°C /about 14%/. An endothermic peak was found on the DSC curve at 220°C representing a reversible transformation of CD. The thermal decomposition - accompanied by oxidation in air - started at 250°C, melting occurred near 300°C /DSC curve/ influencing the shape of the DTG and EGA curves as well: the decomposition rate decreased temporarily. Ignition took place above 300°C. EGA curves of allyl mustard oil samples are shown by Fig.2.

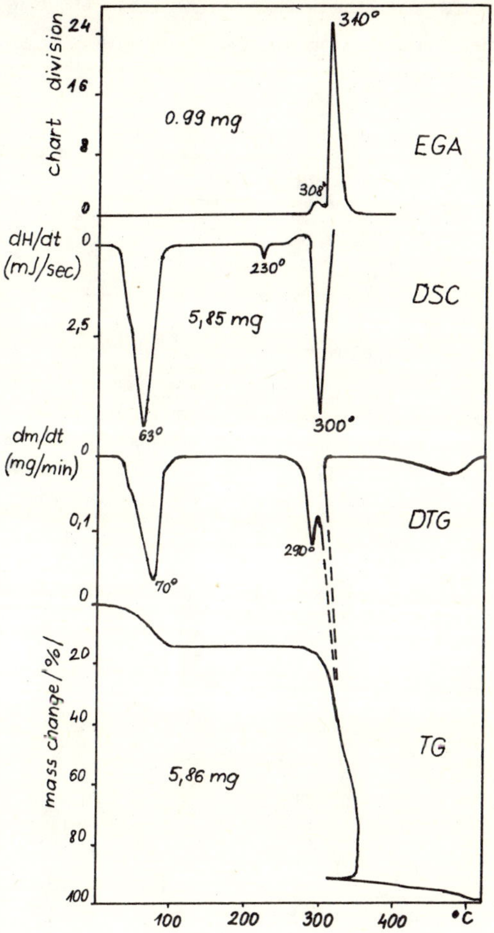

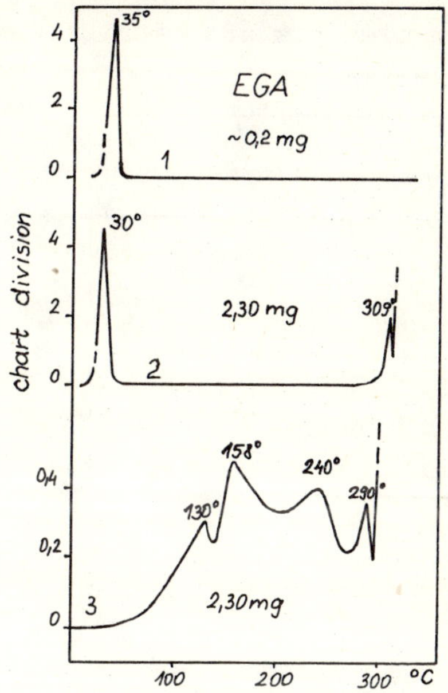

Fig.2. EGA curves of allyl mustard oil /1/, mixture with CD /2/, and CD complex /3/

← Fig.1. Thermoanalytical curves of β-cyclodextrin

Both the mixture and the complex contained 10% mustard oil,

which evaporated from the pure sample and the simple mixture at about 35°C quantitatively. However, the complex did not start to loose mass up to 50°C, and the evolution of the decomposition products stretched to the decomposition of CD itself. Complexes and mixtures of other ethereal oils can be characterized in the same way.

A similar difference was found between the physical mixture and the inclusion compound of vitamin D_3 /Fig. 3/. Free vitamin D_3 decomposed in a range near 200°C, while vitamin in the complex was stable in this range and decomposed simultaneously with the CD.

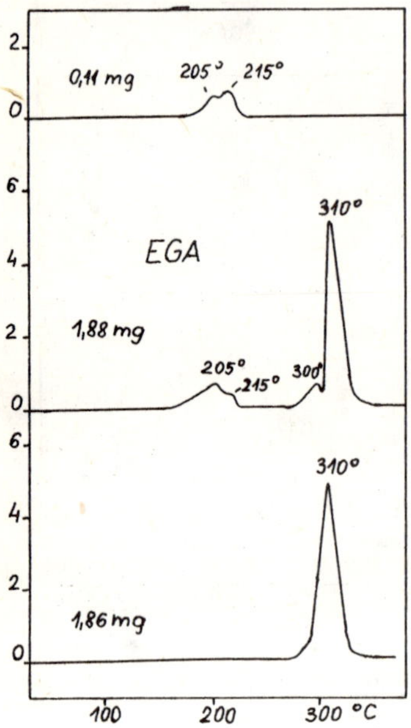

Fig.3. EGA curves of vitamin D_3 /1/, mixture /2/, and complex /3/ with CD

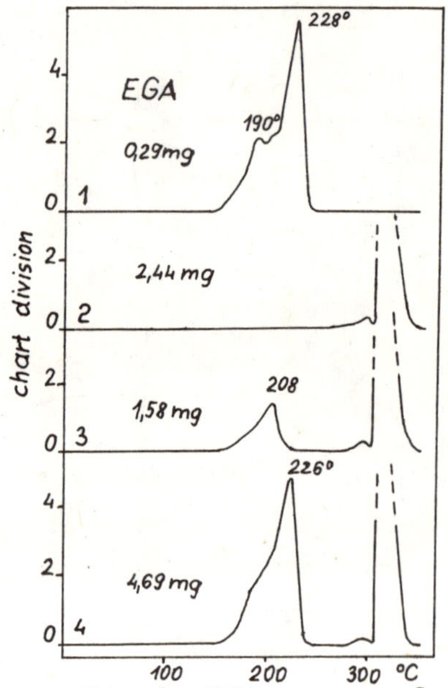

Fig.4. EGA curves of vitamin E acetate /1/ and CD complexes /2,3,4/

For some materials EGA can tell how much of the active agent is in the complex form, and what is the amount of the free agent. This determination is not so straightforward with other analytical methods. As an example, curves of vitamin E acetate samples are presented in Fig. 4. The pure inclusion compound did not show the decomposition peak of the free vitamin E acetate; the proportion of the free active agent in the other samples could be estimated from the peak area. The total vitamin content /which was measured in solution with a spectrophotometric method/ and its free fraction /related to the total of the vitamin/ were:

Number of sample /see Fig. 4./	1	2	3	4
Total vitamin E acetate /%/	100	12.3	19.3	6.4
Free fraction /%/	100	0.0	10.0	80

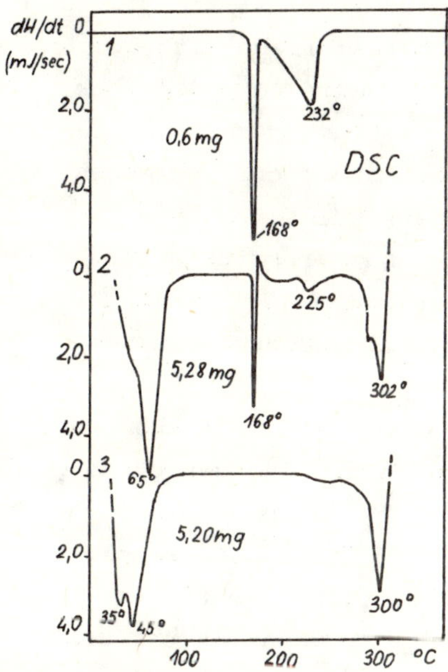

Fig. 5. DSC curves of paracetamol /1/, mixture /2/, and complex /3/ with CD

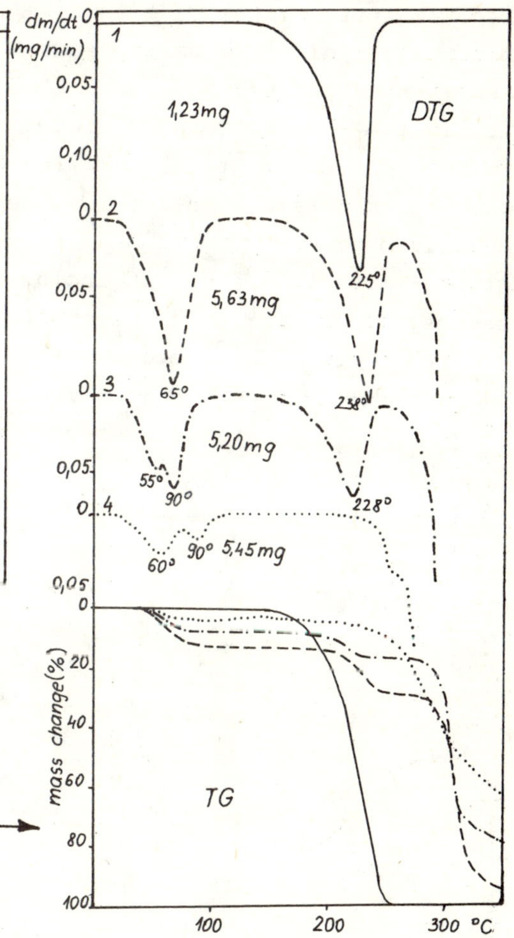

Fig. 6. TG and DTG curves of permetrin /1/, mixture /2/, and complexes /3,4/

The use of the DSC method can be demonstrated on paracetamol /p-hydroxyacetanilide/. This substance melted at 168°C, decomposition started in the same region /Fig. 5./ The DSC curve of the simple mixture resembles the sum of the curves of the two pure substances, however, after melting, a small exotherm peak was recorded suggesting complex formation. The fraction of the complex in the product obtained in this way was found 80% /related to the total paracetamol content/ on the basis of EGA. The complex /lower curve/ did not show the melting peak of the guest substance and the decomposition of paracetamol started above 220°C.
Behaviour of permetrin samples were characterized with TG measurements /Fig.6/. Sublimation of the pure compound took place in the 150-250°C range. Permetrin evaporated similarly from the physical mixture, too. The two other curves represent cyclodextrin - permetrin samples prepared differently. The lyophilized product /curve 3/ gave the same results as the mixture - i.e. it was not an inclusion compound - while the mass loss of permetrin in the crystallized product became significant above 220°C only.

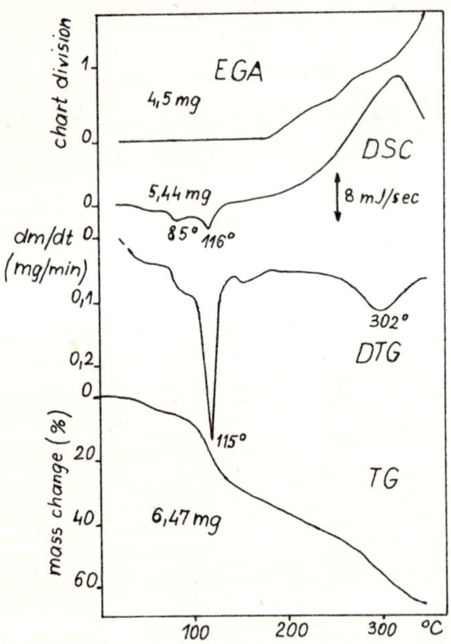

Fig.7. Thermoanalytical curves of HBr-CD complex

The curves in Fig. 7 show a CD complex of an inorganic substance: hydrogen bromide. The amount of HBr in the sample was 10%, that of water 6%. The sample lost water in the vicinity of 50°C, then the decomposition of the complex started, yielding inorganic matter and carbon compounds of higher oxidation states /not detected by the EGA method used/ up to about 170°C. Thus, the evolution of HBr shifted to much higher temperatures in comparison to the pure substance, which is a vapour at room temperature, and, at the same time, the bonds of the inclusion compound were strong enough to change the decomposition processes of β-cyclodextrin as well, causing a shift to lower temperatures.

CONCLUSIONS

Thermoanalytical methods, firstly EGA and TG/DTG are suitable to determine whether a particular product is really a complex if the guest substance undergoes some change before the start of β-CD decomposition /250-300°C/. This change may be evaporation, decomposition, oxidation, melting or polymorphic transition.

In several cases, the applied methods were found suitable for quantitative determination, too. This offers a possibility

a. to control the quality of complex products /i.e. to estimate the percentage of "free" guest substance in a product,

b. to establish the relative amounts of cyclodextrin and guest compound in the complex during experimental work on a new inclusion compound.

LITERATURE CITED

1. Szejtli, J., Szente, L., Bánky-Előd, E.: Acta Chim. Acad. Sci. Hung. 101, 27 /1979/
2. Kurozumi, M., Nambu, N., Nagai, T.: Chem. Pharm. Bull. 23, 3062 /1975/

I.Int.Symp. on Cyclodextrins
Budapest, 1981

CHIRAL CONFORMATION OF BILIRUBIN, BILIVERDIN AND BENZIL IN ASSOCIATION WITH CYCLODEXTRIN

G. LEBAS, C.DE RANGO, G. TSOUCARIS

Laboratoire de Physique, Centre Pharmaceutique
92290 Chatenay-Malabry, France

It often happens for molecules with very low barrier of internal rotation, that chiral conformers are not stable enough, even at the lowest accessible temperature, to exhibit a detectable optical rotation or circular dichroism spectrum. A well known example, important in biology, are the molecules of bilirubin (br) and biliverdin (bv). However, these molecules are normally associated in the blood with albumin or other proteins and they exhibit, in this association, very intense circular dichroism (cd) bands. We have now obtained a very similar cd spectrum in association with β-cyclodextrin (cyd), which here acts fully as a biomimetic molecule. This establishes that the chiral conformation, responsible of the cd spectrum, is characteristic of bv and br themselves, and must correspond to a well defined minimum energy chiral conformer. It is probable that the structure of this chiral conformer is very similar to that observed in the crystal state of pure br.

The molecular dimensions of bv and br are too large to allow inclusion into the β-cyd cavity, as this happens with smaller molecules. The association may involve only a small part of the bv or br molecule, or alternatively two cyd molecules form a dimer forming a large cavity. This "chiralising" aptitude of cyd is also exhibited towards smaller molecules, like benzil.

Crystal structure analysis is being carried out on the benzil complex : Space group : $C222_1$; a = 19.58 Å, b = 24.00 Å, c = 32.84 Å, Z - 8 .

INTRODUCTION -

The inclusion phenomenon in cyclodextrin (cyd) enables the guest molecule to exhibit different or even new properties relatively to these of the free molecule. Several such properties (physical, chemical biological) have been extensively studied during the recent years (1) The occurence of an induced circular dichroism (cd) is of particular interest, as it provides a quick and easy way of testing the effective inclusion of a guest (2)

The induced cd can arise from a chiral electronic perturbation or from occurence of a chiral conformer, or both. We report here experimental evidence that the cd spectra of several guest molecules in cyd complexes are to be ascribed mostly if not entirely, to a conformational isomerism. Indeed, it often happens for molecules with very low barrier of internal rotation, that chiral conformers are not stable enough, even at the lowest accessible temperature, to exhibit a detectable optical rotation or circular dichroism spectrum. But embedding in a chiral host milieu, may be a sufficient "impetus" to induce a chiral stable conformation of the guest. This milieu can be an inclusion complex in solution, for instance like cyd, as well as a crystalline clathrate ; cyd can be used again, but also other chiral clathrates (Tri-ortho-thymotide, urea, etc...)

A new application is then foreseen : determination of the absolute configuration, in relation with the cd spectrum, of conformationally labile (in solution) molecules. This can be achieved by determining the structure of the cyd complex.

I.BILIRUBIN AND BILIVERDIN -

Bilirubin (br,I) and biliverdin (bv,II) respectively are normal catabolites of heme (Fig. 1.)

In case of insufficient or improper catabolism of br, its accumulation in the blood leads to jaundice and associated severe troubles. These molecules are carried normally in association with blood proteins, mostly albumin and, also, ligandin.

From the displayed chemical formulas, the actual molecular conformation is not obvious. But, the discovery of a very intense cd spectrum in the complexes with the above proteins (3) clearly indicates that br and bv exhibit a chiral stable conformation. This has been confirmed by crystallographic analysis of several crystal form of br (4) All crystals studied are racemic, being formed by equal number of left and right hand molecules associated in the crystal by an inversion center. We note also that the crystal structure of the Albumin-br complex has not been determined, owing to the imperfect crystalline quality of the crystals.

We report now that br and bv can be associated in solution with β-cyclodextrin, and the complex exhibits a cd spectrum. Moreover, the form of the spectrum is practically identical with that obtained in the protein complexes.

Fig. 2 and 3 show the cd spectra of β-cyd•br and β-cyd•bv

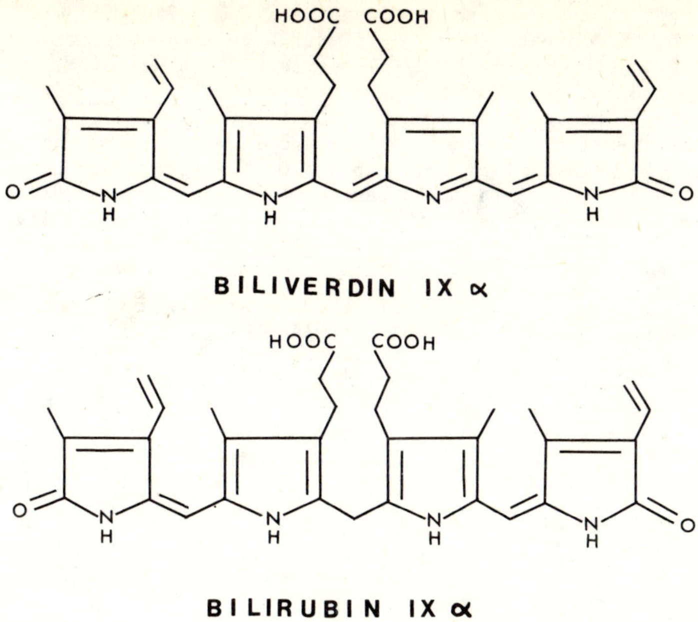

Fig. 1.

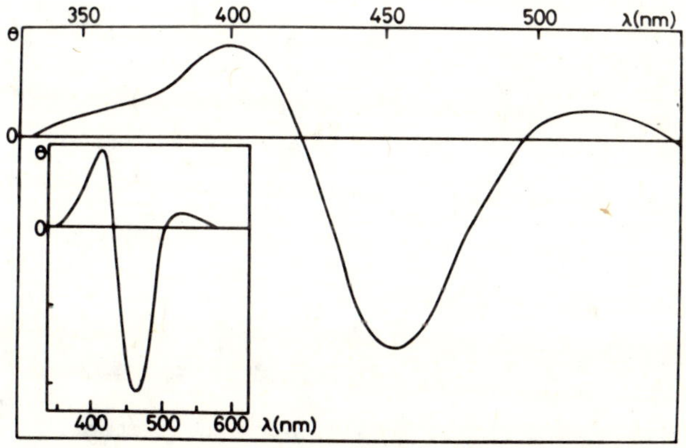

Fig. 2. Cd spectrum of br•βcyd, pH = 10.2, at a molar ratio of βcyd to br of 100 mol/mol, θ → 1cm = 4000 deg.cm^{2}dmole^{-1}. Inset shows the cd spectrum of 1/1 complex of br-ligandin, pH = 7.4, (5)

compared respectively to the spectra of ligandin•br (pH = 7.4) (5) and serum albumin•bv (6). It appears that β-cyd acts here as a biomimetic molecule. These facts show clearly that there exists a stable chiral conformation of br and bv, and that the same conformer , of same chirality, is formed as well in association with cyd, as with proteins.

The molecular dimensions of bv and br are too large to allow inclusion into the β cd cavity, as this happens with smaller molecules The association may involve only a small part of the bv or br molecule, or alternatively two cyd molecules form a dimer allowing a large cavity.

II BENZIL AND DIACETYL

These molecules exhibit conformational freedom, arising from the rotation around the central C-C bond. In benzil, moreover, there is an additional rotation parameter around the phenyl-carbonyl bond .

The cd spectrum of benzil is given in fig. 4. The table below summarizes the important features of all observed spectra.

β-cyd complex	λ max (nm)	θ max (deg.$cm^2 dmole^{-1}$)	pH
benzil	265 320 380	3 000 200 200	7
diacetyl	275	-100	7
biliverdin	375 665	35 000 -17 000	10.2
bilirubin	400 455 510	12 000 -27 000 4 000	10.2

The crystal structure of pure benzil has been determined (10). The molecule, in the crystal state, is chiral ; the two-fold molecular axis coincides with the crystallographic two-fold axis of the chiral space group $P3_121$ or $P3_221$. The dihedral angle O=C-C=O is of 68°. The phenyl group is almost coplanar with the keto group :

$$C_{phenyl} - \overset{\overset{\displaystyle O}{\|}}{C} - C$$

(angle between the mean planes : 6°)

We report here the first results of the crystal structure determination of the β-cyd•benzil complex. Space group : $C222_1$; a = 19.58 Å, b = 24.00 Å, c = 32.84 Å, Z = 8. The unit cell parameters are simply connected with those a', b', of β-cyd•p-nitroacetanilide complex (7) by the approximate relation :

$$\vec{a} = \vec{b}' + \vec{a}' \qquad \vec{b} = \vec{b}' - \vec{a}'$$

These parameters fit a general pattern of similarities between

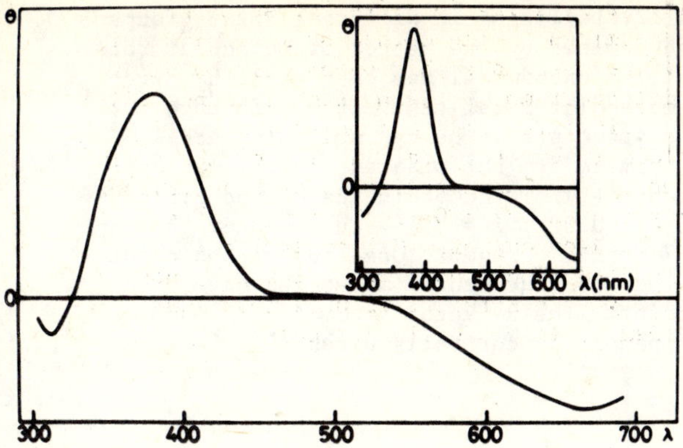

Fig. 3. Cd spectrum of bv•βcyd, pH = 10.2, at a molar ratio of βcyd to bv of 100 mol/mol, θ → 1cm ~ 5000 deg.cm^2dmole^{-1}.
Inset shows the cd spectrum of 1/1 complex bv-human serum albumin, pH = 7.4 (6)

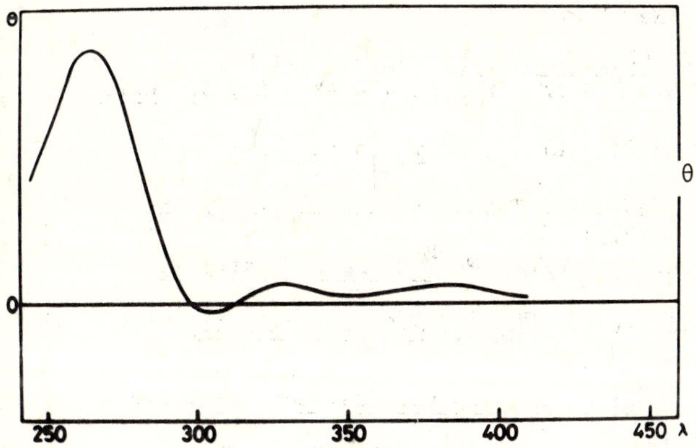

Fig. 4. Cd spectrum of benzil-βcyd, pH = 7, at a molar ratio of βcyd to benzil of 1 mol/mol, θ → 1cm ~ 600 deg.cm^2dmole^{-1}.

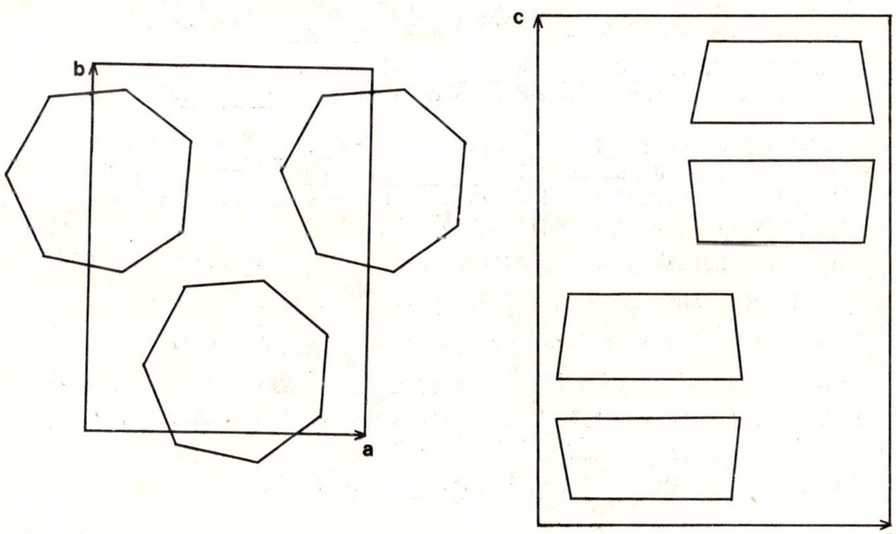

Fig. 5. c-axis projection of the structure (0 < z < 0.25) and a-axis projection of half the structure

structures of β-cyd complexes crystallizing in different space groups (8). The structure has been determined by the method of molecular rotation and translation function. The used model has been built by seven fold repetition of a glucose molecular model taken from reference (9). Fig. 5 shows an outline of the structure . The cyd molecules are associated in dimers, but the dimers are not stacked in columns as in the structure of cyd in reference (7). The coordinates of the cyclodextrin center are : x = 0.05 ; y = 0.32 ; z = 0.15. This center is defined as the mean plane of 7 bridge oxygens. One ethanol and about 10 water molecules have been located. The benzil molecules have not yet been identified with certainty. The R factor is 0.14 for 4700 non zero reflections. The refinement is currently pursued.

LITERATURE CITED

1. BENDER M.L. and KOMIYAMA M. (1978) Cyclodextrin chemistry, Springer Verlag, Berlin Heidelberg New-York

2. a. SENSSE K. und CRAMER F. Chem. Ber., 102, 509-521 (1969)
   b. THAKKAR A.L., KUEHN P.B., PERRIN J.H. and WILHAM W.L., J. Pharm. Sci., 61, 1841 (1972)
   c. IKEDA K., UEKAMA K., and OTAGIRI M., Chem. Pharm. Bull., 23, 201 (1975)
   d. UEKAMA K., OTAGIRI M., KANIE Y., TANAKA S. and IKEDA K., Chem. Pharm. Bull., 23, 1421 (1975)
   e. TAKENAKA S., MATSUURA N. and TOKURA N. Tetrahedron Lett. 26, 2325-2328 (1974)
   f. MATSUURA N., TAKENAKA S. and TOKURA N., J.C.S. Perkin II, 1419 (1977)

3. a. BLAUER G. and KING T.E., Biochem. Biophys. Res. Com., 31, 678 (1968)
   b. BLAUER G. and KING T.E., J. Biol. Chem., 245, 372 (1970)
   c. BLAUER G., HARMATZ D. and SNIR J., B.B.A., 278, 68 (1972)

4. a. SHELDRICK W.S., J.C.S. Perkin II , 1457 (1976)
   b. BONNETT R., DAVIES J.E. and HURSTHOUSE M.B., Nature, 262 , 326 (1976)
   c. MUGNOLI A., MANITTO P. and MONTI D., Nature, 273, 569 (1978)
   d. BECKER W. and SHELDRICK W.S. Acta Cryst., B34, 1298-1304(1978)
   e. LE BAS G., ALLEGRET A., MAUGUEN Y., DE RANGO C. and BAILLY M., Acta Cryst., B36, 3007-3011 (1980)

5. WOOLEY P.V., HUNTER M.J. and ARIAS I.M., B.B.A., 446, 115-123 (1976)

6. BLAUER G. and WAGNIERE G. J.A.C.S., 97, 1949 (1975)

7. HARDING M.M., MACLENNAN J.M., and PATON R.M., Nature, 274, 621 (1978)

8. HAMILTON J.A. and STEINRAUF L.K., Acta Cryst., B24, 1560-1562 (1968)

9. TOKUOKA R., ABE M., MATSUMOTO K., SHIRAKAWA K., FUJIWARA T. and TOMITA K., Acta Cryst., B37, 445-447 (1981)

10. BROWN C.J., SADANAGA R., Acta Cryst. 18, 158 (1965)

I.Int.Symp.on Cyclodextrins
Budapest, 1981 3.12.

INVESTIGATION OF CYCLODEXTRIN COMPLEXES BY X-RAY POWDER DIFFRACTION

K. SIMON,[1] Á. STADLER,[1] F. HANGE[2]

[1] CHINOIN Pharmaceutical-Chemical Works, Tó u. 1-5
H-1045 Budapest, Hungary
[2] Central Research Institute for Chemistry of the Hungarian Academy of Sciences, Pusztaszeri út 59-67
H-1025 Budapest, Hungary

SUMMARY

A great number of cyclodextrin complexes containing different type of pharmacons as guest molecule were prepared at the Chinoin Research Center. X--ray powder diffraction technique was used to prove complex formation in solid state. The dependence of the diffraction pattern on the solvent content was also tested. The patterns were indexed using structural information of complexes from the literature. An attempt was made to systematize the complexes using information of single crystal data and similarities in the powder diffraction patterns.

X-ray powder diffraction technique is a powerful tool for the justification of complex formation in solid state. In the last five years about 300 diffractograms of cyclodextrin complexes were recorded on a Philips powder diffractometer using Cu-K_α radiation. Mostly β-cyclodextrin complexes were studied, as guest molecules different pharmacons and aromas were applied. Some characteristic examples are represented in this paper.

There are several factors which has to be taken into consideration before reaching a decision on the complex formation, one is the water content of the sample. Powder diffraction patterns depend very strongly on the hydration rate as illustrated in Fig 1 (βCD water content 13 %) and in Fig 2 (βCD water content 6 %). Small guest molecules (i.e. HCl or HBr) form βCD complexes isomorphous with β-CD hydrate according to their similar unit cell data (1) and powder diffraction patterns (see βCD.HCl powder diagram in Fig 3).

The justification of complex formation is easier when the guest molecule is liquid. In this case a comparison of the diffraction diagram of the complex with that of the βCD is sufficient. We proved the complex formation for several aromas (i.e. active ingredient of garlic, onion, dill, cumin, camomile). The method is demonstrated with the powder diagram of βCD-anetol and βCD-benzaldehyde (Fig 4 and Fig 5).

When the guest compound is solid a comparison of the diffractograms of the assumed complex with a physical mixture of the guest compound and βCD is required. As the powder patterns of the βCD-Salicylic acid complex and βCD+ Salicylic acid physical mixture (containing analytically the same amount of salicylic acid and water as the complex) differ, the complex formation has taken place (Fig 6 and 7).

In contrary to that, the precipitate obtained after mingling up indomethacine dissolved in acetone with an aqueous solution of βCD turned out to be not a complex but a physical mixture, as the diffractogram (Fig 8) of this sample is basically the same as that of βCD except some peaks which correspond to the most intense reflections of indomethacin recrystallized from acetone (i.e. at $2\vartheta = 8.4^0$). A ternary complex sedimented, however, from a solution of βCD and indomethacine dissolved in dimethylformamide--water mixture (Fig 9).

2ϑ values and relative intensities of the powder patterns are summarized in Table I.

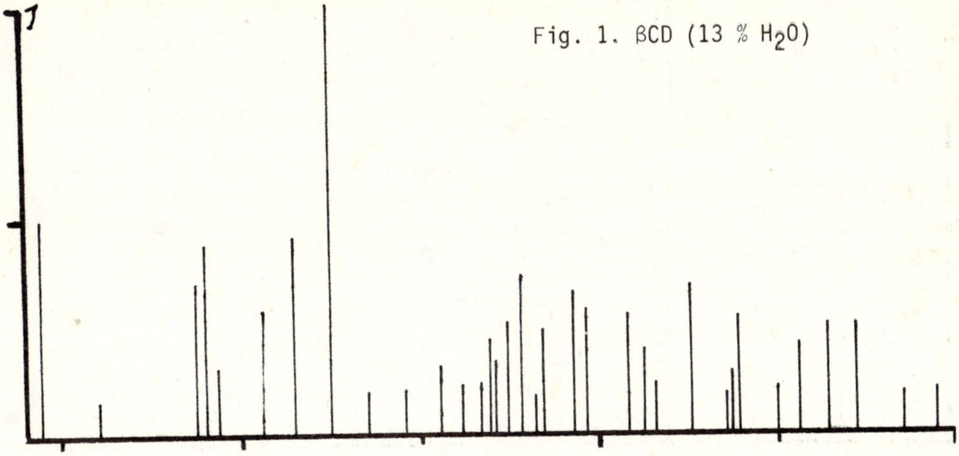

Fig. 1. βCD (13 % H_2O)

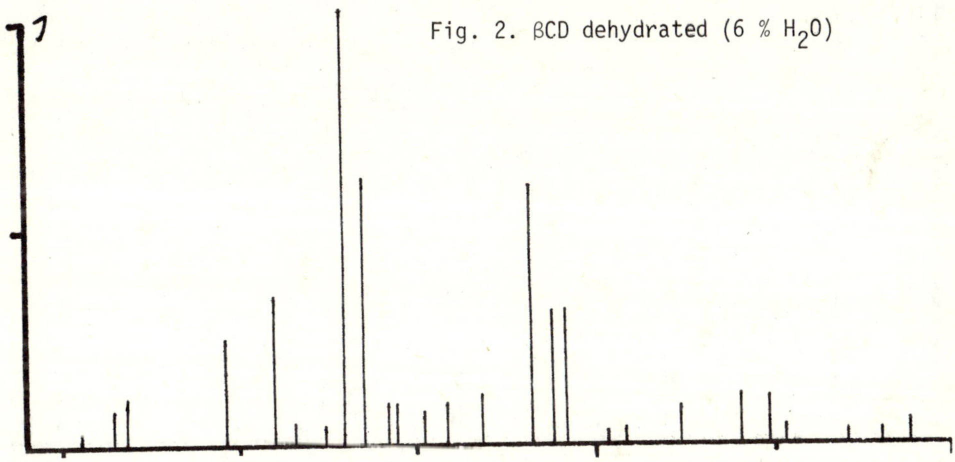

Fig. 2. βCD dehydrated (6 % H_2O)

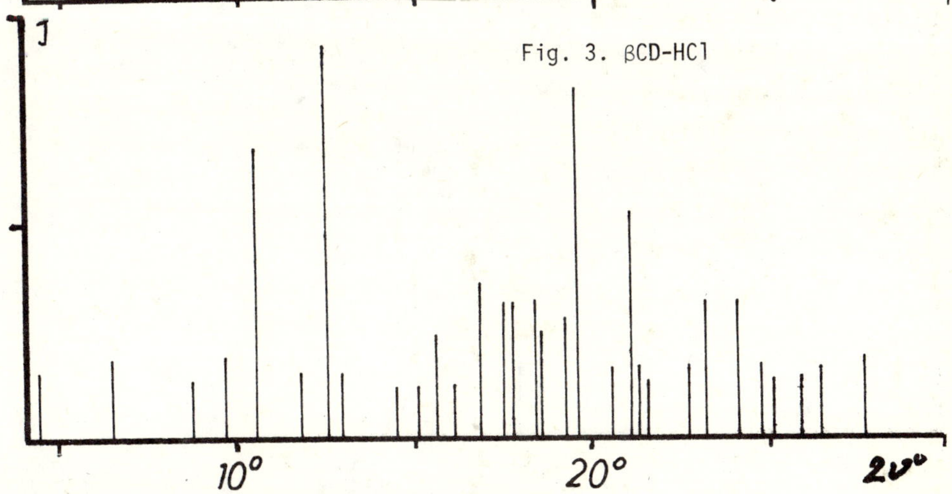

Fig. 3. βCD-HCl

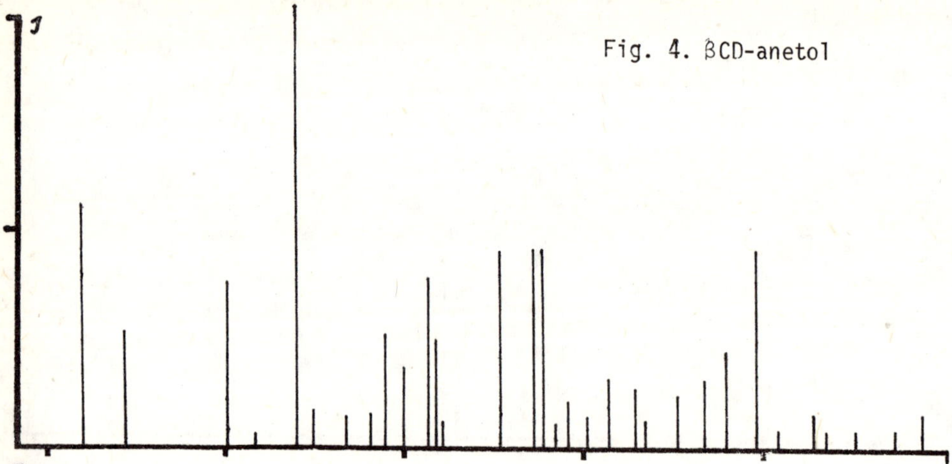

Fig. 4. βCD-anetol

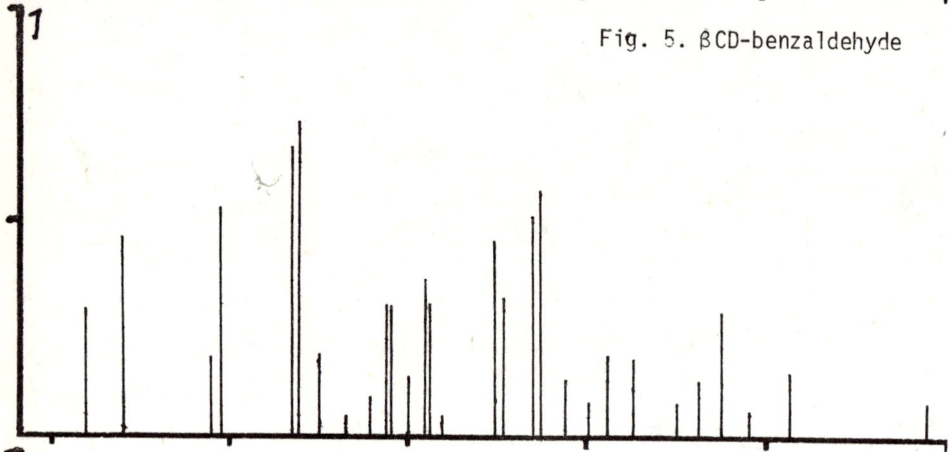

Fig. 5. βCD-benzaldehyde

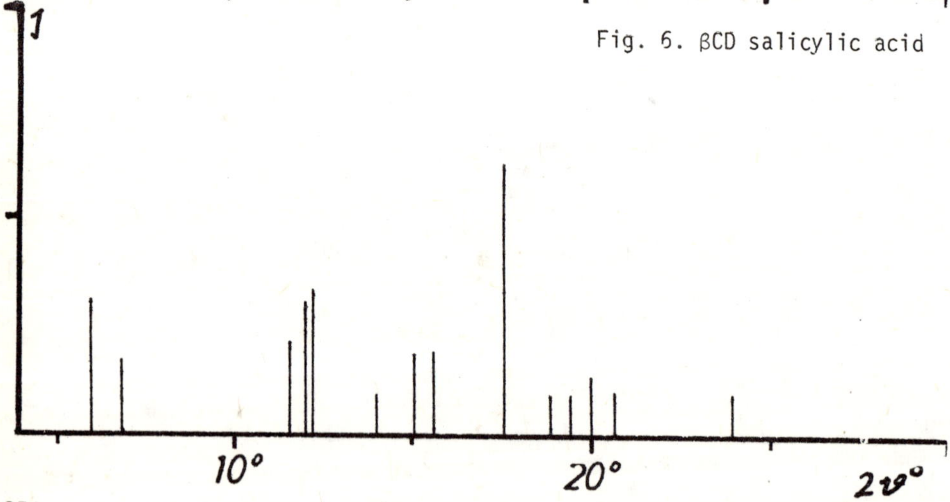

Fig. 6. βCD salicylic acid

On the basis of the powder diffraction data alone it is not possible to decide whether a given pattern represents a single phase or a mixture of several phases. Therefore, we indexed the powder diagrams of βCD, βCD.HCl and βCD.n-propanol using unit cell parameters from the literature (2, 3), the deviations of the observed and calculated ϑ values for the peaks between 2-15^{0} were less than 0.05^{0} in all cases. The powder diagrams of βCD.benzaldehyde and βCD.anetol are very similar and they are probably isomorphous with p-iodo-phenol.βCD, whose structure was reported by K.H. Jogun (3) as the powder diagrams (Fig 4 and 5) could be indexed using the cell parameter of p-iodo-phenol. Indexing was performed wiht the program POWDER (4).

While the positions of the peaks of the diffractograms are uniquely determined by the unit cell data, the intensities of the individual reflections depend on the relative position of the atoms in the unit cell. We calculated the powder diagram of βCD.n-propanol using atomic coordinates of the thermodinamically stable form of βCD.n-propanol complex (3), Fig 10. In order to reduce the number of parameters we determined the non-crystallographic symmetry which relates the two crystallographically independent βCD units in the cell. The relative coordinates of Mol 2 can be obtained with a good approximation from Mol 1 using the following equation:

$$\begin{bmatrix} x_2 \\ y_2 \\ z_2 \end{bmatrix} = \begin{bmatrix} -1.0 & 0.133 & -0.058 \\ 0.0 & 0.837 & -0.809 \\ 0.0 & -0.366 & -0.837 \end{bmatrix} \begin{bmatrix} x_1 \\ y_1 \\ z_1 \end{bmatrix} + \begin{bmatrix} 0.966 \\ 0.483 \\ 1.100 \end{bmatrix}$$

The deviations between the calculated and refined positions are less than 0.3 Å except the O6 and C6 atoms of the glucose units. Therefore we think that an averaged model of βCD which takes into account the sevenfold symmetry of the molecule can be used for generation of calculated βCD powder diagrams. Calculations were performed by the programs LAZY-PULVERX written by Yvon and his coworkers (5).

Acknowledgments

We are grateful to Dr. J. Stezowsky and Dr. K. H. Jogun for the coordinates of several unpublished βCD complexes. Many thanks are due to Dr. L. Párkányi for the adaptation of some programs used by us. We thank to Prof. Szejtli for his suggestions and encouragement.

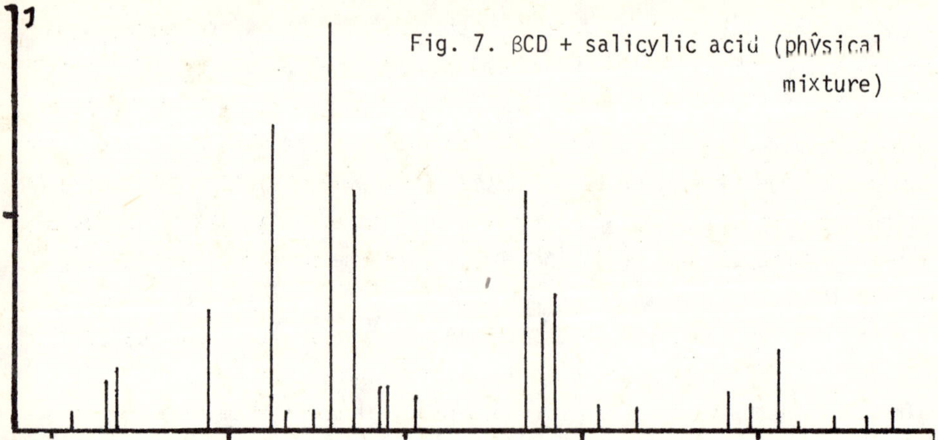

Fig. 7. βCD + salicylic acid (physical mixture)

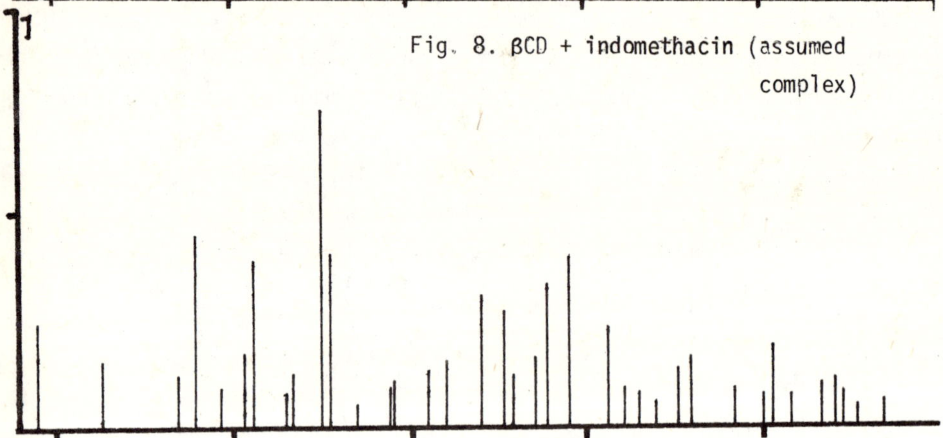

Fig. 8. βCD + indomethacin (assumed complex)

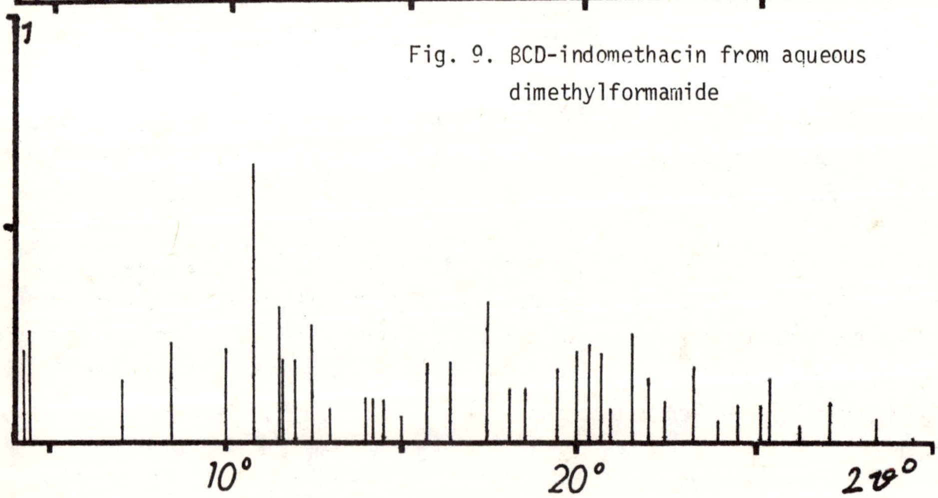

Fig. 9. βCD-indomethacin from aqueous dimethylformamide

Table I. 2θ and relative intensity (I) values of βCD (1), βCD dehydrated (2), βCD-HCl (3), βCD-anetol (4), βCD-benzaldehyde (5), βCD--salicylic acid complex (6), βCD + salicylic acid (physical mixture (7)), βCD + indomethacin (assumed complex) (8), βCD--indomethacin from aqueous dimethylformamide (9).

1		2		3		4		5		6		7		8		9	
2θ	I	2θ	I	2θ	I	2θ	I	2θ	I	2θ	I	2θ	I	2θ	I	2θ	I
4.6	46	3.4	7	4.6	34	6.0	37	6.0	45	6.0	47	5.6	26	4.5	43	4.4	40
6.3	10	5.5	10	6.4	40	7.2	22	7.1	62	6.6	32	6.5	32	6.2	25	7.0	25
8.8	35	6.5	15	8.7	34	10.0	25	9.6	37	11.5	42	6.7	35	8.4	22	8.5	36
9.1	45	6.7	18	9.7	40	10.9	8	9.9	63	11.8	45	8.7	23	8.9	48	10.0	31
9.7	17	7.8	6	10.6	84	12.0	99	11.9	75	12.1	51	9.0	22	9.7	20	10.8	74
10.7	30	8.7	11	11.7	34	13.3	9	12.0	80	13.8	28	9.6	43	10.3	27	11.6	38
11.8	42	9.0	11	12.5	99	14.5	18	12.5	24	15.1	36	11.1	90	10.6	51	12.0	32
12.6	99	9.6	32	12.9	36	15.0	16	13.3	16	17.6	78	11.5	21	11.4	21	12.5	40
13.6	15	11.0	38	14.4	30	15.5	22	14.5	39	18.9	22	12.4	24	11.6	25	13.0	20
14.7	15	11.5	13	15.1	28	15.7	17	15.4	47	19.3	23	13.0	99	12.5	83	14.0	26
15.4	20	12.4	12	15.5	52	16.0	9	15.6	41	20.0	30	13.5	69	12.7	50	14.3	24
16.1	15	12.9	99	16.0	32	17.6	25	15.9	18	24.1	26	14.2	26	13.5	14	14.6	25
16.7	15	13.5	99	16.8	60	18.5	25	17.4	57			14.5	26	14.7	22	15.0	20
17.1	25	14.2	17	17.4	56	18.8	25	18.5	61			15.3	21	15.4	25	15.8	30
17.3	17	14.6	17	17.9	56	19.2	9	18.7	65			15.7	23	16.1	28	16.5	30
17.7	27	15.2	12	18.4	52	19.7	12	19.6	25			16.9	30	17.1	37	17.4	36
18.1	37	15.7	15	18.6	44	20.1	9	20.1	20			17.3	60	17.7	31	18.1	25
18.5	13	16.2	10	19.2	56	20.8	13	20.7	31			18.3	67	17.9	22	18.5	25
18.9	30	16.9	20	19.6	99	21.4	9	21.3	30			18.8	37	18.5	27	19.5	32
19.5	35	17.4	10	20.5	32	21.8	8	22.6	20			19.1	40	18.8	45	20.3	35
19.7	28	18.3	99	21.0	70	22.7	11	23.2	23			21.4	15	19.6	48	20.7	28
20.9	27	18.7	33	21.3	40	23.3	11	23.8	40			24.1	22	20.7	32	21.7	37
21.1	23	19.1	33	21.6	32	23.9	16	24.7	17			24.7	20	22.1	16	22.1	25
21.6	18	20.3	11	22.3	24	24.8	9	25.8	25			25.3	30	22.7	30	22.8	20
22.1	7	20.7	11	22.6	40	25.4	7	26.2	16			27.2	14	23.0	34	23.5	32
22.7	38	21.5	8	23.0	50	26.0	10	29.3	18			28.1	15	24.3	21	24.7	20
22.8	36	22.4	13	23.6	22	26.9	7					28.7	18	25.0	20	25.4	28
23.6	13	23.1	11	24.0	50	28.8	7					29.8	14	25.3	34	26.4	15
23.9	18	24.1	16	24.8	34	29.4	9							25.7	16	27.2	22
24.2	28	24.7	15	25.8	30									26.7	21	28.5	15
25.1	17	25.3	11	26.4	36									27.0	22		
25.7	24	27.1	10	27.6	32									27.6	13		
26.7	30	28.0	10	28.1	20									28.5	15		
27.2	28	28.7	12	28.6	22												
28.5	14	29.5	10	29.0	26												
29.6	12																

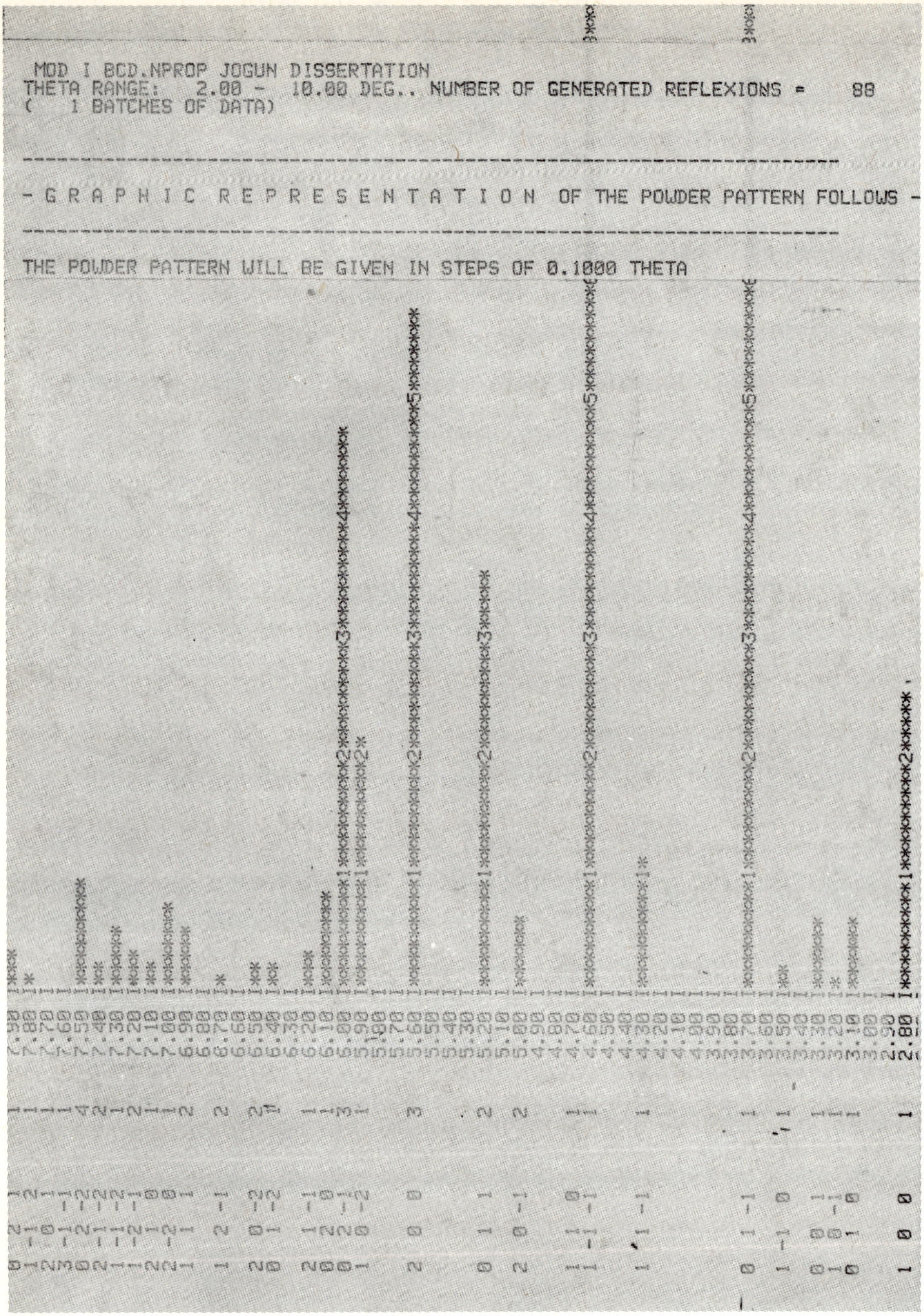

Fig. 10. Calculated powder diagram of βCD-n-propanol using coordinates published for the thermodinamically stable βCD-n-propanol complex (3).

Literature cited

(1) Szejtli, J., Budai, Zs.: Acta Chim. Acad. Sci. Hung. **94**, 383 (1977)

(2) Maclennan, J.M., Stezowski, J.: A.C.A. (Winter) 24, (1980) p. 123

(3) Jogun, K.H.: Dissertation 1979

(4) Evans, H.T., Aplleman, D.E., Handwerker, D.S.: POWDER program for cell refinement and indexing

(5) Yvon, K., Jeitschko, W., Parthe, E.,: LAZY-PULVERIX program written at Laboratoire de Cristallographie Aux Rayons-X, Univ. de Geneve, Switzerland

I.Int.Symp. on Cyclodextrins 3.13.
Budapest, 1981

TERNARY COMPLEXES OF β-CYCLODEXTRIN AND BENZOIC ACID

Á. BUVÁRI, L. BARCZA, J. SZEJTLI

Department of Inorganic and Analytical Chemistry, Eötvös Loránd University, Múzeum krt. 4/B
H-1088 Budapest, Hungary

SUMMARY

The solubility of benzoic acid in water is increased by the presence of β-cyclodextrin. The solubility vs. cyclodextrin concentration plots are practically linear but the slope is > 1, suggesting that not only simple binary but also ternary complexes are formed.

The solubility of benzoic acid was measured in solutions of β-CD and β-CD-benzoic acid complex, respectively, and reversely, the solubility of the 1:1 complex was measured in solutions of benzoic acid. All the three series could be interpreted in terms of complexes of 1:1 and 1:2 composition, with the following formation constants: for CDHB $\log \beta_1 = 2.9 \pm 0.15$ and for $CD/HB/_2$ $\log \beta_2 = 3.5 \pm 0.2$. Since two molecules of benzoic acid cannot penetrate into the β-CD cavity and the second molecule is bound rather weakly ($\log \beta_2 - \log \beta_1 = 0.6$), it must be attached to the cyclodextrin by hydrogen bonds.

The solubility increasing effect of the β-CD is decreased by SCN^- and ClO_4^--salts (because of competing reactions), it is increased by NO_3^- and SO_4^{2-}-salts (formation of mixed complexes), while NaCl is nearly indifferent.

I.Int.Symp. on Cyclodextrins
Budapest, 1981

POLAROGRAPHIC DETERMINATION OF GUEST-MOLECULES IN THE PRESENCE OF β-CYCLODEXTRIN

L. DARUHÁZI, L. BARCZA, J. SZEJTLI

Department of Inorganic and Analytical Chemistry, Eötvös Loránd University, Múzeum krt. 4/B
H-1088 Budapest, Hungary

SUMMARY

I. β-cyclodextrin adsorbs on the surface of the mercury drop. The adsorption is potential-dependent, and the adsorbed CD layer partially inhibits reductions. It causes a decrease in the polarographic wave-height. Three methods were tried to eliminate this effect:

i) The use of surface-active polymers such as Triton X-100 and gelatine, in order, that they would adsorb instead of the CD;

ii) Guest molecules, that are inactive in polarography were added in excess, in order to reduce the free cyclodextrin concentration;

iii) The CD concentration was raised to reach maximum surface coverage on the mercury drop.

The first method proved to be insufficient, because the polymers form complexes with the CD; the complexes adsorb also, forming a more compact adsorption layer. The second and the third were effective, but in the second case the "antiguest" molecule must be selected carefully, because a competition exists between the active guest- and the inactive "antiguest" molecules.

II. Step-wise reductions were also found in some cases. In the presence of β-CD Cu^{2+} ion first forms $Cu/I/CD^{+}$ complex, which is insoluble and adsorbs strongly, then $Cu/I/CD^{+}$ is reduced to yield $Cu/Hg/_{x}$. Similar phenomenon occurs in the reduction of nitroglycerin. The glycerin-mononitrate is formed in a 4 electron reaction, then it gives glycerin and the three equivalent nitrite ions at a more negative potential.

I.Int.Symp. on Cyclodextrins
Budapest, 1981 3.15.

HYDROLYSIS OF PHOSPHOROCHLORIDOTHIONATES IN THE PRESENCE OF β-CYCLODEXTRIN

M. MIKOŁAJCZYK, J. DRABOWICZ

Center of Molecular and Macromolecular Studies, Polish Academy of Sciences, Department of Organic Sulfur Compounds [1]
Boczna 5
90-362 Łodz, Poland

SUMMARY

Hydrolyses of organophosphorus esters and halides in the presence of cyclodextrins are proposed to involve the formation of the phosphorylated cyclodextrins. The aim of the present study was to check this possibility by means of the ^{31}P-NMR and stereochemical experiments using chiral O-ethyl ethanephosphonochloridothiomate (1) as model compound.

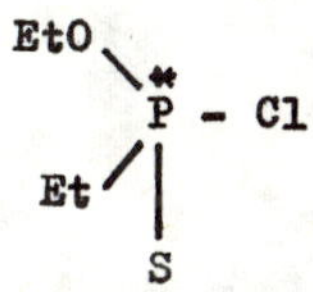

*- denotes optically active center

1

The results so far obtained concerning the hydrolysis of 1 in the presence of β-cyclodextrin do not support the formation of the corresponding O-phosphonyl β-cyclodextrin.

---

1 M.L. Bender, M.Komiyama, Cyclodextrin Chemistry, Springer Verlag, Berlin, Heidelberg, New York, 1978.

I.Int.Symp. on Cyclodextrins
Budapest, 1981

MOLECULAR DYNAMICS OF CD-COMPLEXES IN SOLUTION: A ^{13}C-N.M.R. STUDY

A. NESZMÉLYI

Central Research Institute for Chemistry of the Hungarian Academy of Sciences, Pusztaszeri ut 59-67
H-1025 Budapest, Hungary

SUMMARY

Molecular motions in various complexes formed by β-cyclodextrin have been studied by ^{13}C nuclear relaxation at 25.16 MHz. For medium sized molecules the results show that a/ the formation of the complex (the inclusion of the guest molecule) always increases the reorientation time and so it may be used for its detection and b/relaxation times (T_1-values) allow a detailed study of the dynamic rigidity of the system in solution. Usually the inclusion process does not hinder the free rotation of the guest around the symmetry axis of the macrocycle. If there are different subunits of the guest molecule then a gradient of T_1 values is observed which points towards the free end (uncomplexed site).

Using the 2,6-di-O-methyl derivative of β-cyclodextrin its high solubility (20-35% w/v) and the 1:1 host ratio ensures the possibility of precise T_1 determination in reasonable machine time.

Interfering signals of the CD molecule can be eliminated from the spectra taking advantage of the much shorter T_2 values. A Hahn type simple spin-echo has been used for that.

High resolution ^{13}C-n.m.r. is a very powerful method for the study of molecular motion is solutions. It allows the simultaneous observation of spectral lines belonging to many different residues of a complex molecule. The n.m.r. parameters of these lines all depend also on the reorientation time of that part of the molecule where the respective carbon is situated. Of particular interest is the longitudinal relaxation time (T_1). The set of T_1 values for all carbon atoms in a particular molecule, their distribution, offer detailed insight into molecular dynamics (1) provided their interpretation is unique.

For medium sized molecules the dipole-dipole relaxation is owerwhelming (2), i.e. for molecular weight below 2000. Then the measured T_1 may be interpreted as a local correlation time according to the usual formula

$$\tau \sim \frac{1}{T_1}$$

valid for isotropic motion. The overall distribution of T_1's rquires of course special consideration for a meaningful interpretation in case of macromolecules(1).

In case of inclusion complexes of cyclodextrin derivatives the overall tumbling motion is slowed down by the inclusion. The dynamic rigidity of the complex system shortens all T_1's of the substrate but does not change significantly the motion of the big macrocycle.

We understand the essentials by a simple example. The relaxation time of Benzyl β -D-glucopyranoside in D_2O at room temperature is around 1 sec. Looking at Fig. 1. we observe the relaxation time of a 1:1 complex of the same molecule. There is a factor of about 4 in the glucose moiety in the reduction of relaxation times as expected from the participation in the overall tumbling motion of the macrocycle. At same time, however, the presence of the methylene group mechanically decouples certain degrees of freedom, so nonbonded interactions between host and guest do not interfere (glucose vs. CD).

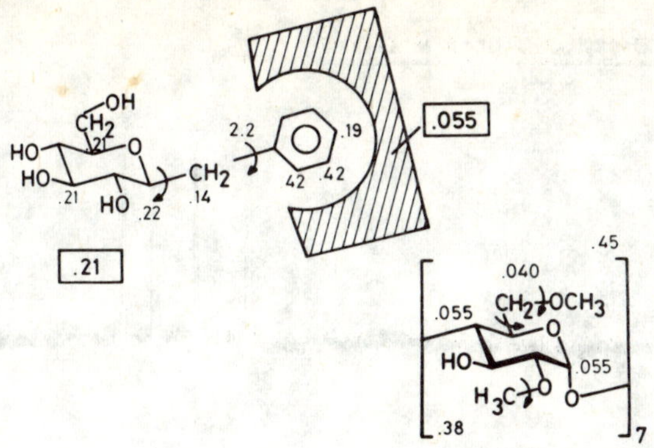

Fig.1. ^{13}C longitudinal relaxations times (in sec) for the 1:1 complex of 2,6,-di-O-methyl-β-cyclodextrin and Benzyl β-D-glucopyranoside (1.5 g in 4 ml of D_2O, room temperature) measured at 25.16 MHz.Mean T_1 values are enclosed in rectangles.

It is interesting to observe the anisotropy in the motion of the phenyl group: the ratio of the T_1 values belonging to meta or ortho carbons to the T_1 of the para carbon is as big as 2, clearly indicating the relativly free rotation of this group inside the cavity of the macrocycle.

The cyclodextrin itself shows varoius internal degrees of freedom too in its motion.E.g. the the O-methyl substituents have considerably higher T_1 values than the glucose core carbons.C-6 in CH_2 has a NT_1 of .08 sec,slightly higher than the average .055,showing rotation along this side-chain.

The conclusion has a high practical value:on the n.m.r. scale the inclusion complexes are stable and relativly flexible,however,the increase in reorientation time of the complex with respect of the guest alone in the solution,demonstrates the presence of the complex moiety.Of course a lot of other interesting details are available: which part of the guest is inside the cavity, which is the free end of a chain like molecule, and so on.

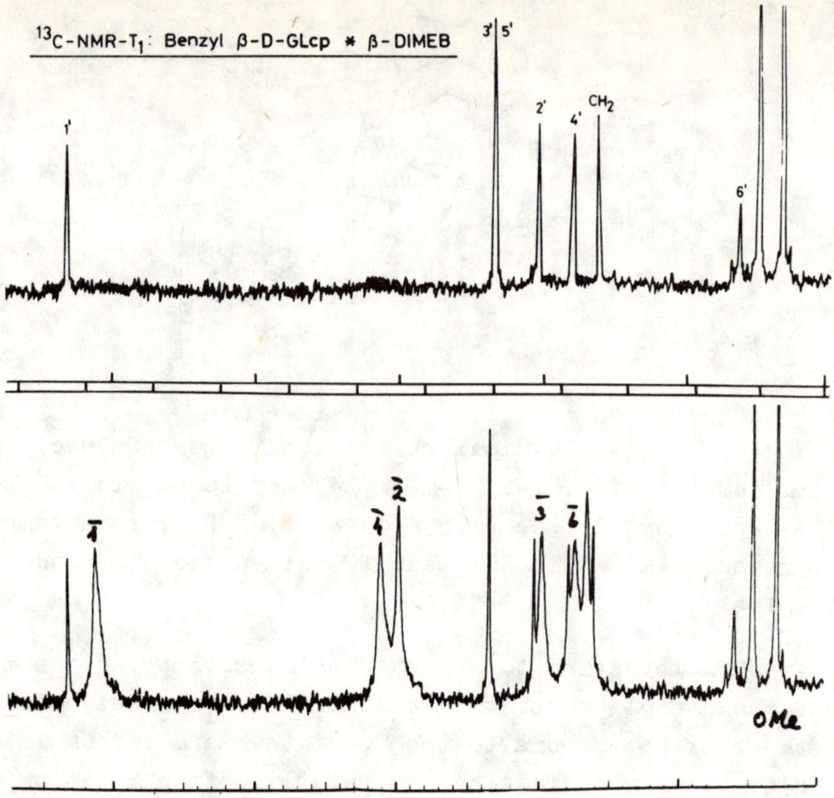

Fig.2. ^{13}C-n.m.r. spectra (aromatic lines are not shown here) for Benzyl β -D-glucopyranoside as a 1:1 complex with 2,6,di-O-methyl- β -cyclodextrin in D_2O at room temperature and 25.16MHz. Bottom trace: broadband decoupled spectrum. Upper trace: the same combined with a simple Hahne-echo (τ =56msec) to suppress the CD lines.

If overlapping of lines of the guest molecule and those of the CD itself happen, a good practical method is to exploit the differences in T_2 (transverse relaxation time). Because $T_2 \leqslant T_1$, a simple Hahn type spin-echo is enough to distinguish between the slow macrocyle's lines and those of the guest molecule. The pulse sequence, in our experience, may contain just two echoes to compensate pulse imperfections

$$\left[PD - \left(\frac{\pi}{2}\right)_x - \tau - (\pi)_x - 2\tau - (\pi)_{-x} - \tau - FID\right]_n$$

with τ equal to T_1 of CD. Fig. 2. shows the spectra with and without this "mass-filter" (3) in case of the same molecule of Fig. 1. Only the first 100 ppm is shown, so aromatic lines are missing. A nice cancellation of the CD lines can be realized this way.

Acknowledgement: Prof. J. Szejtli and dr. A. Lipták helped this work by discussions and submitting the samples.

Literature Cited:

(1) Ribeiro, A., Wade-Jardetzky, N.G., King, R., Jardetzky, O.: Applied Spectroscopy, 34, 299 (1980)

(2) Levy, G.C. ed.: Topics in Carbon-13 NMR Spectroscopy, Vol. 1., Wiley, 1974. See p. 74.

(3) Rabenstein, D.L.: J.Magn.Res. 34, 669 (1979).

CYCLODEXTRIN-DERIVATES

I.Int.Symp. on Cyclodextrins
Budapest, 1981

THE CHEMISTRY OF CYCLODEXTRIN DERIVATIVES

A. LIPTÁK,[1] P. FÜGEDI,[1] Z. SZURMAI,[1] J. IMRE,[1] P. NÁNÁSI,[1]
J. SZEJTLI[2]

[1] Institute of Biochemistry, Kossuth Lajos University
Egyetem tér 1
H-4010 Debrecen, Hungary
[2] CHINOIN Pharmaceutical-Chemical Works, Tó u. 1-5
H-1045 Budapest, Hungary

SUMMARY

The aim of the presentation is to give a comprehensive picture about the different methods used for the synthesis of chemically modified cyclodextrin derivatives with well defined and characterized structures. The efficiency of the different chromatographic and spectroscopic methods used for the structure elucidation of CD-derivatives are treated and estimated.

New and modified alkylation procedures for the preparation of fully and partially protected ß-cyclodextrin derivatives are treated and reviewed. The utilisation of the compounds are discussed in detail.

Reduction of 6-bromo-6-deoxy-ß-cyclodextrin derivatives, nucleophilic displacement of the bromo-compounds by sodium azide, reduction of the azides and also the introduction of new reagents in the field of CD-chemistry are demonstrated.

^{13}C-NMR spectra of the synthesized compounds demonstrate the structure postulated.

INTRODUCTION

The cyclodextrins are cyclic α-(1-4)-linked oligomers of D-glucose produced from starch by the enzyme of Bacillus macerans (1). Their ability to give inclusion complexes by insertion of various organic molecules into their hydrophobic cavity has received considerable interest. Two sets of their hydroxyl groups available for chemical modifications give the possibility to obtain derivatives having

a) increased complexation properties
b) higher solubility in different solvents or solvent systems
c) reactive functional groups.

RESULTS and DISCUSSION

The cyclodextrins have toroidal shape; the "top" side has six (α-), seven (ß-) or eight (γ-cyclodextrin) primary hydroxyl groups and the slightly wider "bottom" side is covered by the twelve (α-), fourteen (ß-) or sixteen (γ-cyclodextrin) secondary hydroxyl groups. Owing to the six-, seven- or eight-fold symmetry, the primary hydroxyl groups are equivalent from chemical and spectroscopic point of view. Similarly, the hydroxyl groups at positions 2 are equivalent but their reactivity considerably differ from that of the chemically equivalent hydroxyl groups situated at positions 3.

Theoretically, a single glucopyranosyl unit of the cyclodextrins can be considered as a 4-O-substituted α-D-glucopyranoside derivative, but in cyclodextrins this model is loaded by steric and statistic problems (vide infra). Due to these two effects, the 4-O-substituted α-D-glucopyranosides can not be considered as attractive models for selective substitutions.

Despite these difficulties, a number of modified cyclodextrins have been synthesized but the results of earlier studies have to be taken into account very critically (2). Now we can say that the synthetic and physico-chemical investigation methods have reached levels suitable for settling the configurational problems emerged during the selective modification of cyclodextrins.

Recently, a number of excellent books and articles (3-8) on cyclodextrins have become available including chapters on cyclodextrin chemistry. The present paper is aimed at the completion of the above literature and also to present the new results obtained in the author's laboratories. Here we wish to deal with the following subjects:

1) Alkyl derivatives of cyclodextrins,
2) acyl derivatives of cyclodextrins,
3) halogeno- and deoxy functions in cyclodextrin derivatives,
4) nitrogen containing analogues of cyclodextrins,
5) silylated ß-cyclodextrin derivatives.

The modified cyclodextrins treated in this paper are derivatives with retained or modified symmetry - as compared to the parent compound - and can be characterized by a x-fold symmetry axis.

Before detailing the above subjects we wish to give a theoretical review on the selective protection of cyclodextrins in the light of the excellent work of Lehn and his coworkers (2). Differentiation between the primary- and secondary hydroxyl groups can be done by two different ways:

1) direct approach: this method relies on the difference existing between the reactivity of primary and secondary hydroxyl groups. This approach involves two problems, statistical and steric ones. The decrease of the number of the sugar units in the macroring declines the selectivity due to the decrease of the steric crowding of the whole molecule. At the same time, however, the reactivity of the hydroxyl groups of the macroring also increasing and this effect causes the decrease of the selectivity, as well. This synthetic route is advantageous for ß-cyclodextrins but unfavourable for α-cyclodextrin. All earlier attemps used this approach for synthesizing partially substituted cyclodextrin derivatives.

2) indirect approach: this method is based on the nonselective reaction of all hydroxyl groups followed by selective deprotection of the primary hydroxyl functions. Being the "bottom" of a perbenzoylated α-cyclodextrin more crowded than the "top", the debenzoylation reaction selectively removes the primary benzoyl groups to give hexakis-(2,3-di-O-benzoyl)-α-CD. Generally, the indirect approach is beneficial for the preparation of α-cyclodextrin derivatives.

1. Alkyl derivatives of cyclodextrins

The first attempt for the preparation of methylated ß-CD was made by Irvine, Pringsheim and MacDonald (9). Correct physical data were reported by Freudenberg (10), using the Muskat method (11) for the methylation of α- and ß-cyclodextrins. This methylation was proceeded in liquid ammonia in the presence of metallic sodium and methyl iodide. Later Casu and his coworkers (12) applied the Kuhn's procedure (13-14) for the methylation of α- and ß-cyclodextrins.

We have found that the Brimacombe's method is a very useful procedure (15) for the methylation of α-, ß- and γ-cyclodextrins (16) and in each case the per-O-methylated derivatives were obtained in crystalline form. The ^{13}C-NMR spectra showed the six-, seven- and eight-fold symmetry and the presence of only six skeleton carbon signals in all three cases. It is necessary to underline that ^{13}C-NMR spectroscopy is an irreplaceable tool in the structure investigation of cyclodextrin derivatives. Similar results were published by Lehn et al. (2) for the methylation of α- and ß-cyclodextrins.

Casu et al. (12) and Staerk and Schlenk (17) reported for the first time, that the methylation of α- or ß-CD with Kuhn's procedure using BaO and $Ba(OH)_2 \cdot 8H_2O$ as base and dimethyl sulfate as the reagent resulted in hexakis-(2,6-di-O-methyl)-α- and heptakis-(2,6-di-O-methyl)-ß-CD.

We reinvestigated and optimalized the above method for the synthesis of heptakis-(2,6-di-O-methyl)-ß-CD (16). It has been found that the quantity of the base can be considerably diminished and by this way an easily treatable suspension was obtained. This is a very good and significant improvement, because the selective reaction of the fourteen 2,6-hydroxyl groups of ß-CD strongly dependent upon the temperature of the reaction mixture. With this method 2 kg of heptakis-(2,6-di-O-methyl)-ß-CD was prepared in our laboratory.

Another procedure was also worked out for the preparation of this latter dimethyl ether. Recently, Fügedi (18) has found that treatment of various carbohydrate derivatives in DMF or DMSO solution with alkylhaloids or with dialkyl sulfates in the presence of powdered KOH at room temperature resulted in complete alkylation within 20-30 min. Using this procedure for ß-CD, its 2,6-di-O-methyl derivative was prepared in 55-60% yield.

The selective 2,6-alkylation method has been extended for all three types of cyclodextrins and octakis-(2,6-di-O-methyl)-γ-CD has been synthesized most recently (19). The preparation of the following selectively alkylated cyclodextrin analogues is published: 2,6-di-O-ethyl-α- (19), 2,6-di-O-allyl-ß- (20), 2,6-di-O-propyl-ß- (21), 2,6-di-O-benzyl-ß-CD (21).

Heptakis-(2,6-di-O-allyl)-ß-CD gave possibility for the synthesis of heptakis-(3-O-methyl)-ß-CD (20), being the allyl groups, after isomerization (22-24), easily removable blocking groups. Similarly, the increased reactivity of OH-2 made possible the preparation of heptakis-(2-O-methyl)-ß-CD (25). For this synthesis heptakis-(6-bromo-6-deoxy)-ß-CD (26)(*vide infra*) served as starting compound. It was methylated with dimethyl sulfate in the presence of BaO/$Ba(OH)_2$. Nucleophylic displacement of the bromine and saponification of the resulting 6-benzoyl derivative gave heptakis-(2-O-methyl)-ß-CD.

Of the 6-O-methyl ethers only hexakis-(6-O-methyl)-α-CD was prepared by Lehn *et al.*(2). The starting compound was hexakis-(2,3-di-O-benzoyl)-α-CD which was methylated by diazomethane in the presence of BF_3-etherate. Removal of the benzoyl groups by saponification resulted in the target compound.

Here, we would like to present the synthesis of heptakis-

(2,3-di-O-methyl)-ß-CD. In this case the task is to protect all the primary hydroxyl groups. In the field of monosaccharide chemistry one of the most common procedure for this purpose is the tritylation of the primary hydroxyl function. Unfortunately substitution of all the primary hydroxyl groups of the cyclodextrin ring is impossible due to severe steric overlapping of the bulky substituents between the neighbouring glucose units. In the case of ß-CD it was found that maximally four trityl groups could be introduced (27). Recently, it was shown that α-CD gives a tri-O-tritylated α-CD derivative (28) which has three-fold symmetry indicating that every second primary hydroxyl group is tritylated.

These results clearly showed that the tritylation, the most frequently used procedure for the protection of the primary hydroxyl groups of monosaccharides (29), can not be used for the cyclodextrins, and the preparation of the 2,3-di-O-methyl ethers of ß-CD requires a new method, as follows.

The starting material for this synthesis was heptakis-(6-O-t-butyldimethylsilyl)-ß-CD (*vide infra*) which was methylated using the Brimacombe procedure to give heptakis-(6-O-t-butyldimethylsilyl-2,3-di-O-methyl)-ß-CD. Hydrolysis of this latter with dilute acid resulted in heptakis-(2,3-di-O-methyl)-ß-CD.

2. Acyl derivatives of cyclodextrins

Per-O-acyl derivatives of all the cyclodextrins were first synthesized and characterized by French (30) and their structure was proved by proton- (31) and carbon-13 NMR spectroscopy (32).

Perbenzoylated α- and ß-cyclodextrins were prepared by Cramer *et al*. (33). These compounds served as starting materials for the selective deprotection of C_6-OH in the work of Lehn and his coworkers (2) to prepare hexakis-(2,3-di-O-benzoyl)-α-CD in excellent yield. Application of the above method for heptakis-(2,3,6-tri-O-benzoyl)-ß-CD did not give satisfactory results.

Selective benzoylation of ß-CD in pyridine with benzoyl chloride at low temperature resulted in heptakis-(2,6-di-O-benzoyl)-ß-CD as major product and its structure was investigated by a degradative method (34) involving methylation with diazomethane. Ogawa and Matsui (35) used trialkylstannyl alkoxides to enhance regioselective nucleophylicity of the hydroxyl groups at positions 6 and 2 of α-CD. Using these derivatives for benzoylation hexakis-(2,6-di-O-benzoyl)-α-CD could be prepared with 95% yield. The structure of the benzoate was verified by ^{13}C-NMR spectroscopy.

Tosylation of the primary hydroxyl groups of hexopyranosides is a well documented procedure. Several groups (36-37) published the preparation of hexakis-(6-O-p-toluenesulphonyl)-α-CD and heptakis-(6-O-p-toluenesulphonyl)-ß-CD. In a substantial study Tsujihara *et al.* (38) showed that sulphonylation with p-toluenesulphonyl chloride did not proceed with sufficient selectivity and pure heptakis-(6-O-tosyl)-ß-CD was obtained only in 1.2% yield. Similar results were obtained also by Lehn *et al.* (2) with α-CD: they were unable to isolate the desired hexakis-(6-O-tosyl)-α-CD. In a patented procedure Kurita et al. (39) reported the selective sulphonylation of both α- and ß-CD with excellent yield using mesitylene chloride and 2,4,6-tri-isopropyl-benzenesulphonyl chloride, however, this reaction could not be repeated by Lehn (2). Lehn and his coworkers (2) reported the tosylation of hexakis-(2,3-di-O-benzoyl)-α-CD. These experiments indicate that the different sulphonyl chlorides are not suitable reagents for the selective protection of the primary hydroxyl groups of α- and ß-CD.

3. Halogeno- and deoxy-functions in cyclodextrin derivatives

The number of such derivatives is very limited. Takeo, Sumimoto and Kuge (26) reported the application of a Vilsmeier-type reagent for the preparation of hexakis-(6-bromo-6-deoxy)-α-CD, heptakis-(6-bromo-6-deoxy)-ß-CD and octakis-(6-bromo-6-deoxy)-γ-CD. The cyclodextrins were brominated with a mixture of methanesulphonyl bromide and N,N-dimethylformamide. In all three cases the respective 2,3-di-O-acetyl derivatives were also prepared and the structure of the synthesized compounds was verified by ^{1}H-NMR spectroscopy.

Heptakis-(6-bromo-6-deoxy)-ß-CD served as starting material for the synthesis of heptakis-(2-O-methyl)-ß-CD (25); the 6-bromo-derivative could be methylated without replacement of bromine atom.

For converting the 6-bromo-6-deoxy derivatives into the corresponding 6-deoxy analogues a number of methods and reagents ($LiAlH_4$, Raney nickel, Pd-on-charcoal and Adams catalyst) have been applied without success (26). However, the reduction of the acetylated halogeno derivatives with $NaBH_4$ in dipolar aprotic solvents was found to be useful for the preparation of deoxy analogues.

We repeated the bromination of ß-cyclodextrin using the above reaction and obtained similar results, but difficulties were observed with the reduction. Therefore we tried a reduction procedure known for very long time and worked out by E. Fischer (40) for the synthesis of methylose-type monosaccharides. Thus, the acetylated bromo-derivatives were treated with Zn-powder in acetic acid at 70° for 2 h, and the reduced compound could be readily isolated from the reaction mixture by extraction with dichloromethane.

These results indicate that the reduction of bromine satisfactorily occured by use of *in situ* generated hydrogen and this inexpensive and convenient method is quite suitable for the preparation of deoxy-cyclodextrin derivatives.

4. Nitrogen-containing derivatives of cyclodextrins

Of the representatives of this group the azido- and amino-derivatives are the most important, and generally, the azido compounds serve as starting material for the synthesis of the amino-cyclodextrins.

The first azido-cyclodextrin analogue was prepared by the

Umezawa group (<u>37</u>) using hexakis-(6-O-tosyl)-α-CD for the azidolysis. For similar purposes the mesitylene or tri-isopropylbenzenesulphonyl derivatives of cyclodextrins were applied by Kurita *et al*.(<u>39</u>).

A novel procedure, called "oxido-reductive substitution route" has been elaborated by the Lehn's group (<u>2</u>). Treatment of α-CD with equivalent quantities of triphenylphosphine and carbon tetrabromide and a large excess of lithium azide in DMF at room temperature led to mixture of azido-deoxy-α-CD derivatives. This substitution is very attractive since no formation of hindered hexa-O-phosphonium derivative is required and the reaction proceeds *via* sequential activation of the primary hydroxyl groups and *in situ* displacement resulting in the azido derivatives. The same compound was also prepared by the azidolysis of hexakis-(2,3-di-O-benzoyl-6-O-tosyl)-α-CD. Methylation of 6-azido-α-CD resulted in hexakis-(6-azido-6-deoxy-2,3-di-O-methyl)-α-CD in excellent yield. Similarly, the "oxido-reductive substitution route" proved to be very useful in the case of ß-CD and heptakis-(2,3-di-O-acetyl-6-azido-6-deoxy)-ß-CD was prepared in 57% yield.

We chose a new route for the preparation of azido-ß-CD. As mentioned replacement of the primary hydroxyl groups with bromine using methanesulphonyl bromide gave heptakis-(2,3-di-O-acetyl-6-bromo-6-deoxy)-ß-CD in very high yield, the reaction of which with NaN_3 in DMF resulted in heptakis-(2,3-di-O-acetyl-6-azido-6-deoxy)-ß-CD in high yield. Saponification of the acetate yielded crystalline 6-azido-ß-CD, which was methylated to give heptakis-(6-azido-6-deoxy-2,3-di-O-methyl)-ß-CD, a new representative in the ß-CD series.

The amino-cyclodextrins were prepared by the reduction of the azido compounds, and also, in same cases, by the nucleophylic displacement of sulphonyl esters with ammonia or primary or secondary amines.

The Umezawa group (<u>37</u>) and the authors of the cited Japanese patent (<u>39</u>) used platinum oxide for the reduction of azido cyclodextrin derivatives. For the same purpose the Lehn's group

(2)and others (28, 41, 42) applied the triphenylphosphine-ammonium hydroxide system, and the conversion was found to be quantitative.

For the reduction of azido compounds, primarily noncarbohydrate azides H_2S, $(NH_4)_2S$ and Na_2S have been successfully applied (43-48). Searching for such inexpensive reducing agent we have observed that the reduction of the hepta-(azido)-compound with Na_2S required a very short time and heptakis-(6-amino)-ß-CD was isolated as the hepta-hydrochloride in acceptable yield.

Several attempts have been reported for the direct nucleophylic displacement of the sulphonyloxy functions of cyclodextrins by ammonia, methylamine and dimethylamine. The synthesis of tris-(6-amino-2,3-di-O-methyl)-α-CD, having symmetrical structure, has been recently accomplished by using tris-(6-O-trityl)-α-CD as starting material.

5. Silylated ß-cyclodextrin derivatives

As shown in the previous chapters there are only a few protecting groups suitable for the selective substitution of cyclodextrin derivatives. Search for appropriate reagents and methods **is** now a progress of primary importance in the chemistry of cyclodextrins.

Tert.-butyl dimethylsilyl chloride (49) has been recently introduced in the field of carbohydrates (50-51). This relatively bulky reagent shows increased reactivity towards the primary hydroxyl groups. Using this reagent different solvent systems and catalysts were applied in order to find appropriate conditions for the selective protection of the primary hydroxyl groups of ß-cyclodextrin. Finally, pyridine was found to be the most useful solvent serving, at the same time, as proton acceptor. An amount of 1.1 mole/hexopyranosyl unit of the reagent was used and the reaction was monitored on t.l.c.

The product of the silylation reaction was isolated in form of the corresponding acetate obtained by acetylation. The

^{13}C-NMR spectrum of heptakis-(2,3-di-O-acetyl-6-O-tert.-butyl-dimethyl-silyl)-ß-CD clearly demonstrated the expected seven-fold symmetry of this derivative. Saponification of the di-O-acetate resulted in the crystalline hepta-O-silyl-derivative. Methylation of this latter gave the 2,3-di-O-methyl analogue of the silylated ß-CD, the ^{1}H-NMR spectrum of which (at 200 MHz) indicated the seven-fold symmetry of this compound.

We belive that it is not necessary to emphasize how important this silylated ß-CD is! This intermediate can be further derivatize and now such experiments are being in progress in our laboratories.

In summarization we have to underline that the selective polyfunctionalization of cyclodextrins is not a simple problem, but can be successfully accomplished by chosing the appropriate cyclodextrin molecules together with careful consideration of the steric and statistic factors which strongly influence the substitution reactions.

LITERATURE CITED

1. French, D.: Adv. Carbohydr. Chem., 12, 189 (1957)
2. Boger, J., Corcoran, R. J., Lehn, J. M.: Helv. Chim. Acta, 61, 2190 (1978)
3. Cramer, F.: Einschlussverbindungen, Springer Verlag, Berlin 1954
4. Cramer, F., Hettler, H.: Naturw., 54, 625 (1977)
5. Griffiths, D. F., Bender, M. L.: Adv. Catalysis, 23, 209 (1973)
6. Bender, M. L., Komiyama, M.: Cyclodextrin Chemistry, Springer Verlag, Berlin-Heidelberg-New York 1978
7. Szejtli, J.: Cyclodextrins and Their Complexes, Budapest **Akadémiai Kiadó, 1981.**
8. Saenger, W.: Angew. Chem., 92, 343 (1980)
9. Irvine, J. C., Pringsheim, H., MacDonald, J.: J. Chem. Soc., 125, 942 (1924)
10. Freudenberg, K., Meyer-Delius, M.: Ber., 71, 1596 (1938)

11. Muskat, I.: J. Am. Chem. Soc., 56, 693 and 2449 (1934)
12. Casu, B., Reggiani, M., Gallo, G. G., Vigevani, A.: Tetrahedron, 24, 803 (1968)
13. Kuhn, R., Trischmann, H., Löw, I.: Angew. Chem., 67, 32 (1955)
14. Kuhn, R., Baer, H. H., Seeliger, A.: Ann., 611, 236 (1958)
15. Brimacombe, J. S., Jones, B. D., Stacey, M., Willard, J. J.: Carbohydr. Res., 2, 167 (1966)
16. Szejtli, J., Lipták, A., Jodál, I., Fügedi, P., Nánási, P., Neszmélyi, A.: Stärke, 32, 165 (1980)
17. Staerk, J., Schlenk, A.: 149th A. C. S. Meeting, Detroit, Spring 1965, Abstract 11C, No. 22
18. Fügedi, P., Nánási, P.: Acta Chim. Acad. Sci. Hung., in press
19. Pitha, J.: Life Sci., 29, 307 (1981)
20. Bergeron, R. J., Meeley, M. P., Machida, Y.: Bioorg. Chem., 5, 121 (1976)
21. Bergeron, R. J., Machida, Y., Bloch, K.: J. Biol. Chem., 250, 1223 (1975)
22. Gigg, J., Gigg, R.: J. Chem. Soc., 82 (1966)
23. Boss, R., Scheffold, R.: Angew. Chem. Internat. Edn., 15, 558 (1976)
24. Ogawa, T., Matsui, M.: Carbohydr. Res., 62, C1 (1978)
25. Takeo, K., Kuge, T.: Stärke, 28, 226 (1976)
26. Takeo, K., Sumimoto, T., Kuge, T.: Stärke, 26, 111 (1974)
27. Cramer, F., Mackenson, G., Sensse, K.: Ber., 102, 494 (1969)
28. Boger, J., Brenner, D. G., Knowles, J. R.: J. Am. Chem. Soc., 101, 7630 (1979)
29. Helferich, B.: Adv. Carbohydr. Chem., 3, 79 (1948)
30. French, D., Levine, M. L., Pazur, J. H., Norberg, E.: J. Am. Chem. Soc., 71, 353 (1949)
31. Takeo, K., Kuge, T.: Agr. Biol. Chem., 34, 1416 (1970)
32. Takeo, K., Hirose, K., Kuge, T.: Chem. Lett., 1233 (1973)
34. Kondo, Y., Takeo, K.: Carbohydr. Res., 52, 232 (1976)
35. Ogawa, T., Matsui, M.: Carbohydr. Res., 56, C1 (1977)

36. Lautsch, W., Wiechert, R., Lehmann, H.: Kolloid-Z., 135, 134 (1954)
37. Umezawa, S., Tatsuta, K.: Bull. Chem. Soc. Jpn., 41, 464 (1968)
38. Tsujihara, K., Kurita, H., Kawazu, M.: Bull. Chem. Soc. Jpn., 50, 1567 (1977)
39. Kurita, H., Kawazu, M., Urawa, S., Takashima, K.: D. Pat., 2425663 (1975)
40. Fischer, E., Zach, K.: Ber., 45, 3761 (1912)
41. Mungall, W. C., Green, G. L., Heavner, G. A., Letsinger, R. L.: J. Org. Chem., 40, 1659 (1975)
42. Hata, T, Yamamoto, I., Sekine, M.: Chem. Lett., 977 (1975)
43. Adams, R., Blomstrom, D. C.: J. Am. Chem. Soc., 75, 4305 (1953)
44. Forster, M. O., Rav, K. A. N.: J. Chem. Soc., 1943 (1926)
45. Ugi, I., Perlinger, H., Behringer, L.: Ber., 91, 2330 (1958)
46. Rathsburg, H.: Ber., 54, 3183 (1921)
47. Barger, G., Ewins, A. J.: J. Chem. Soc., 97, 2253 (1910)
48. Curtis, T.: J. Prakt. Chem., 52, 210 (1895)
49. Corey, E. J., Venkateswarlu, L.: J. Am. Chem. Soc., 94, 6190 (1972)
50. Kraska, B., Klemer, A., Hagedorn, H.: Carbohydr. Res., 36, 398 (1974)
51. Franke, F., Guthrie, R. D.: Austral. J. Chem., 30, 639 (1977)

I.Int.Symp. on Cyclodextrins
Budapest, 1981 **4.2.**

CYCLODEXTRIN DERIVATISATION: DIRECTED REACTION OF SILYLATED INTERMEDIATES

R.L. WIFE,* D.E. REED,** H.C. VOLGER

Shell Biosciences Laboratory
Sittingbourne, England

SUMMARY

Methods for the complete or partial (selective) silylation of β-cyclodextrin are described that provide intermediates for further derivatisation. The procedure is useful in directing reaction to previously functionalised centres affording material that is soluble in organic solvents, facilitating chromatographic separations and making possible structural assignments for the product macromolecules by proton NMR spectroscopy.

The method is applied to the sequential attachment and attempted capping of β-cyclodextrin by a porphyrin template. The product conjugate was found to be hydrolytically unstable. Representative proton NMR spectra for a series of silyl derivatives and intermediates are analysed to demonstrate further the advantage of the method.

---

* Current address: Koninklijke/Shell-Laboratorium, Amsterdam (Shell Research B.V.) Badhuisweg, Amsterdam, The Netherlands.

** Current address: School of Molecular Sciences, University of Sussex, Brighton, Sussex, England.

INTRODUCTION

The host-guest capability of the cyclodextrins has resulted in their extensive employment as models for receptor binding and enzymatic catalysis, either as the parent cage molecule [1-3] or after derivatisation [4-7]. Their chemical modification for the incorporation of catalytic centres of other functionalities is complicated by the inherent physical and chemical properties of the oligosaccharide class. Thus they are hydrolytically unstable and are soluble only in relatively polar solvents [8].

Equally important is the probable lack of selectivity achievable by partial functionalisation of the cage molecule because of positional isomerism. Cage modification has to proceed at the 2,3- and 6-hydroxyl groups of the glucose molecules by derivatisation and/or displacement reactions, and in the absence of a directing influence a large (statistical) number of products can result. There have been many spectacular claims for selective modifications of the cyclodextrin molecule, and the more respectable and proven methods are summarised in Figure 1. The principal modifications to the cyclodextrin cage shown in Figure 1 can be classified as:

1. Alkylation (formation of O-C or O-Si derivatives)
2. Sulphonation or acylation (O-SO_2R or O-COR)
3. Substitution (by nucleophiles to give -NR_2, -SR, -X, -N_3 functionalities) of sulphonyl (e.g. tosylate) derivatives.

Practical problems encountered in these transformations can be summarised as being due to:

1. Use of aqueous or polar solvents (DMF, DMSO etc.) to solubilise parent or partially substituted cyclodextrins.
2. Restrictive choice in chemical functionalisation methods imposed by solvent, and cyclodextrin hydrolytic instability.
3. Comparatively few techniques to direct partial functionalisation.
4. Tedious and scale-limited chromatographic purifications on reaction product mixtures.
5. Assessing the degree of reaction and location of reacted centres on the cyclodextrin skeleton by nmr (^{1}H or ^{13}C or by mass spectrometric methods.
6. Escalating costs in sequential transformations.
7. Product instability [proximity of underivatised hydroxy groups accelerates hydrolysis of (e.g.) acyl derivatives].
8. Misleading claims in the literature of selective or efficient derivatisation.

FIGURE 1 : LISTING OF AUTHENTIC DERIVATIVES OF ALPHA AND BETA CYCLODEXTRINS

I

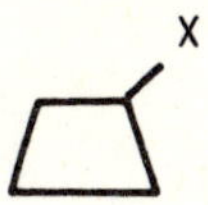

X = OTs (N_3, NH_2, Cl, Br, I, NR_2, $\overset{+}{N}R_3$, NH NH$_2$, NH NH NH etc..
SR, SFe_4S_4, OR)

= O.CO.R

II

X = OMs (N_3, NH_2) n=7
= OTs (NHR) n=7

III

X = OMe n=6,7
= $OSiHMe_2$ n=6
= OBz n=6
= OAc n=6,7,8

IV

X =	-	OMe	OMe	OTs	N_3	N_3	N_3	n=6
Y =	OBz	OBz	-	OBz	-	OMe	OAc	m=12

X = Tr (3Tr+3OMe) (3NH_2+3OMe)
Y = - 12OMe 12OMe

X = allyl n=7
Y = allyl m=7 OMe m=7

V

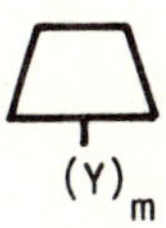

Y = OMe m=7

VI

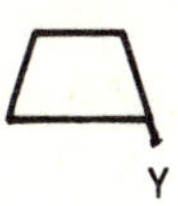

Y = OCOR, OCH_2COZ (Z = OR, NHR, N(OH)R)
= OTs (I, NH_2, NHR)

Fig. 1 (contnd)

VII

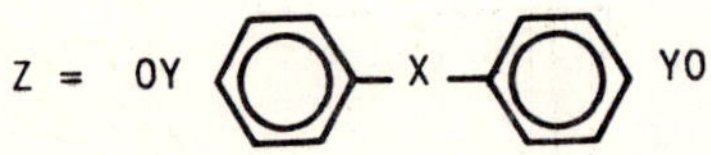

X = O, CH_2, Y = SO_2
= N=N = CO
= CO = CO
= CH=CH = SO_2

'capped cyclodextrin'

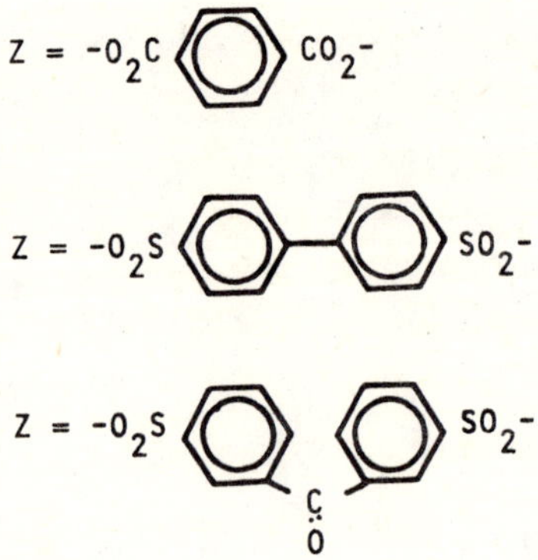

VIII

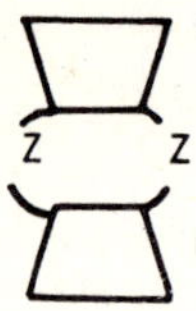

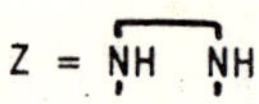

'duplex' cyclodextrin

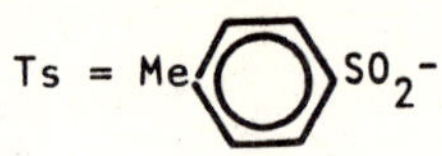

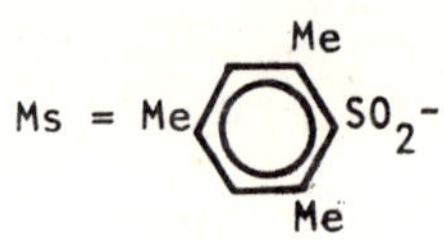

Bz = CO.Ph

Ac = $CO.CH_3$

Factors that have been shown to help accomplish selective derivatisation are as follows:

1. Inherent primary (6-O-hydroxyl) and secondary (2,3-O-hydroxyls) rate selectivities in reaction; steric and internal hydrogen-bonding effects further reduce the 3-hydroxyl derivatisation rate and the order 6-OH> 2-OH> 3-OH is established [9].
2. Blocking the primary face (e.g. with a trityl group) prior to derivatisation of the secondary face (Figure 2a) [10].
3. Preformation of an inclusion complex with the reagent, specific for secondary face derivatisation (Figure 2b) [11,12].
4. Transface capping by a spaced bifunctional reagent (Figure 2c) [13].

We sought to cap the cyclodextrin cage molecule, but with a rigid metal chelate and in this way to prepare a catalyst-cage configuration analogous to the active site of a metallo-enzyme. The selection of component materials and the attachment procedures were designed with an awareness of and an appreciation for the factors described above, in an attempt to make practical the handling, purification and identification of the macromolecules that resulted. Schematically the procedure is shown in Figure 3.

The 6-amino-6-deoxy-β-cyclodextrin (β-CD.NH_2) is readily obtainable and identifies one centre on the primary face for specific mono-attachment. The β-CD.NH_2 is made conveniently from the β-cyclodextrin monotosylate (β-CD.OTs) by azide displacement and then catalytic hydrogenation [14]. (The monotosylate prepared by controlled (selective) reaction of the primary hydroxyl with tosyl chloride as first described by Chao [15] and later by Matsui [16] is an extremely efficient procedure).

A tetraphenylporphyrin (TPP) with four carboxylic acid groups was chosen as the metal template with the eventual goal of achieving a cytochrome P-450 model enzyme configuration.

Without the step to protect the free hydroxyls with a blocking group, direct attachment of the TPP-tetraacid to β-CD.NH_2 (DCC coupling or through the tetraacid chloride) resulted in non-selective amide and ester bond formation. Complete silylation of the cage molecule was expected to confine reaction to the amino functionality exclusively (Step 1). Silylation would also produce macromolecules that could be handled in ordinary organic solvents, but serve as protecting groups removable by mild acid hydrolysis without rupture of the amide bond or the cage (Step 3).

FIGURE 2 : STRATEGIES FOR SELECTIVE DERIVATISATION OF CYCLODEXTRIN

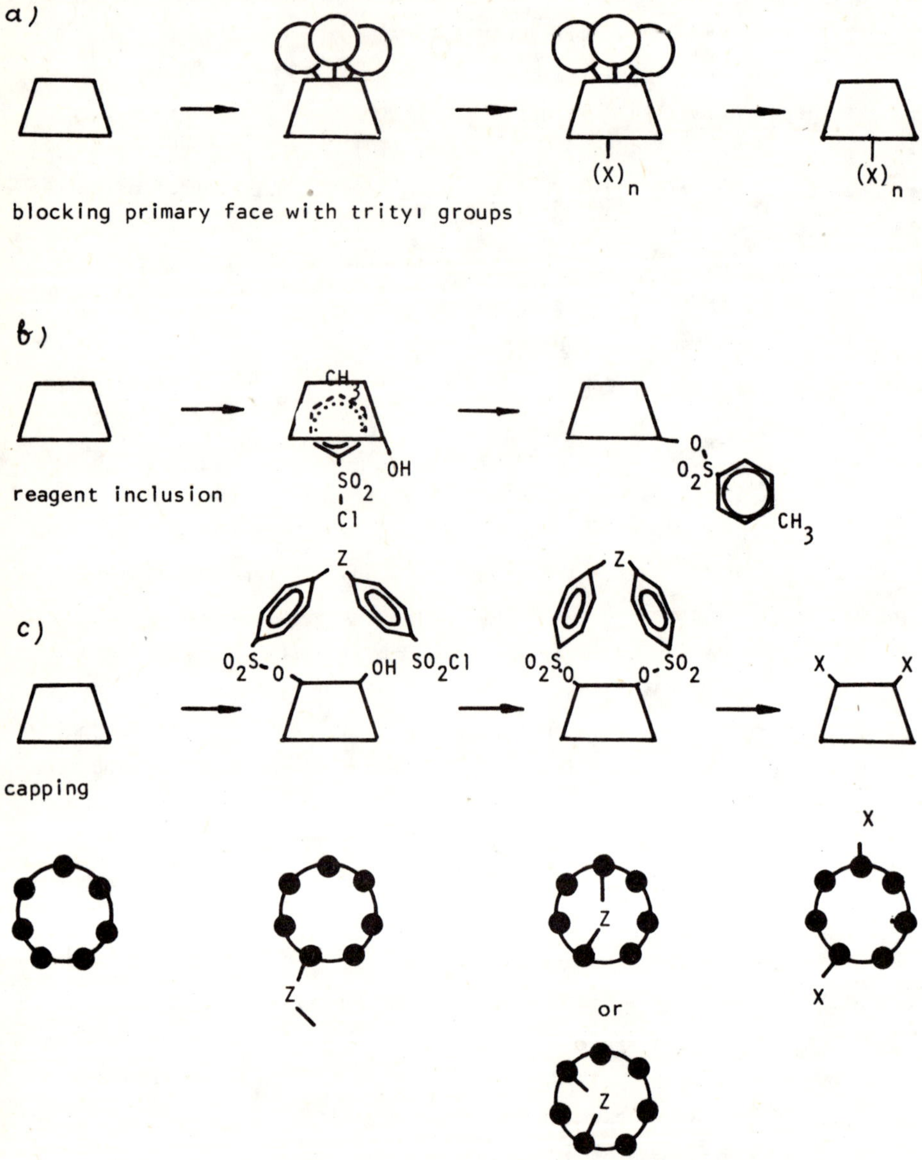

FIGURE 3 : SCHEMATIC CONJUGATION BETWEEN CHELATE (PORPHYRIN) AND CAGE (CYCLODEXTRIN)

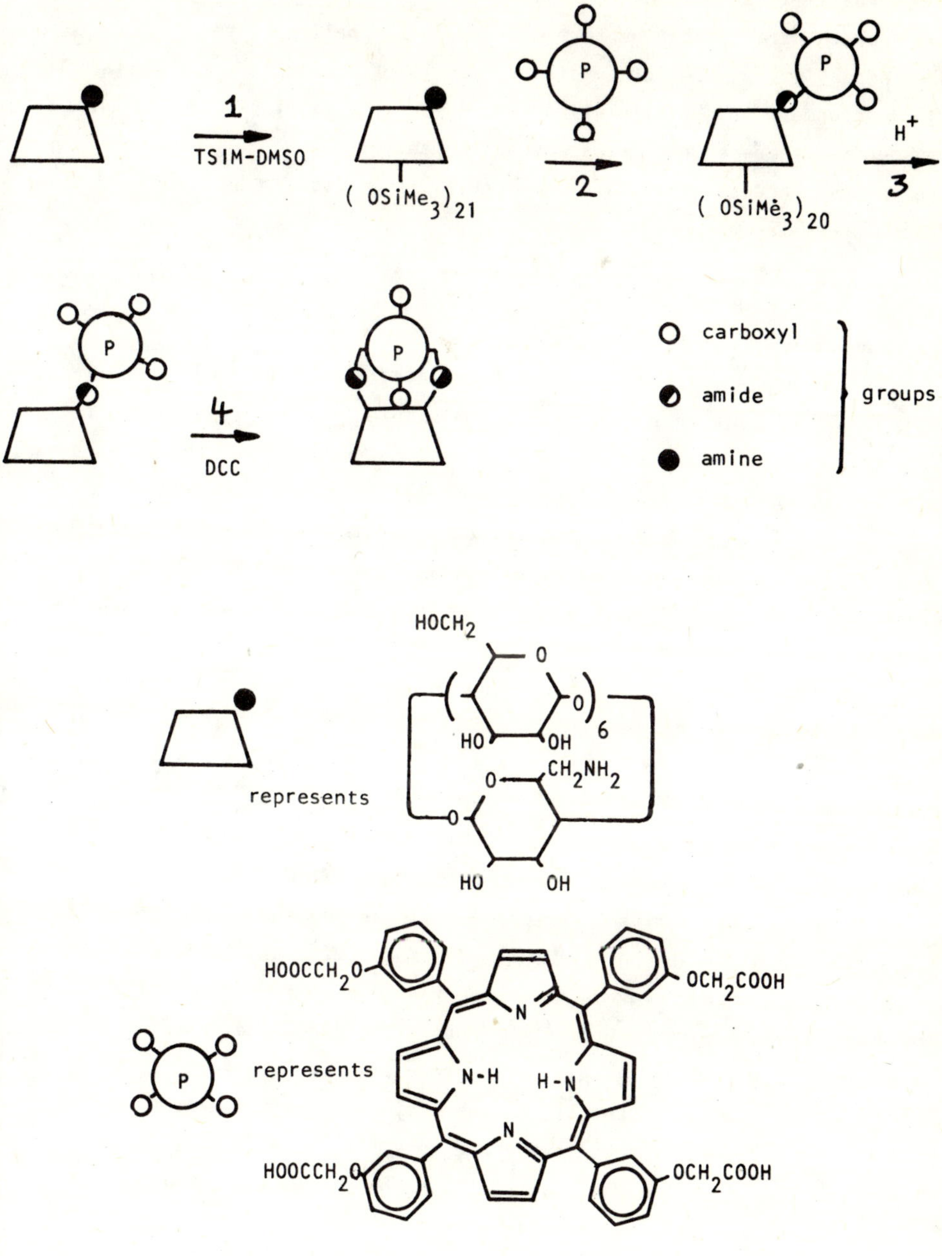

To achieve capping of the primary face, internal cyclisation (Step 4) constitutes the final step in the sequence.

Crucial in this sequence is the masking of the free hydroxyl groups in order to direct reaction to the amino functionality, (Step 2), and to obtain readily soluble intermediates by silylation. Model experiments in the development of the silylation method, in establishing unambiguous assignments for reaction centres and attachment procedures are described in the following section. [Full experimental methods are to be found in the Poster presentation].

RESULTS AND DISCUSSION

Trimethylsilylation: method and analysis

While complete dimethylsilylation of cyclodextrin has been described, pertrimethylsilylation is reportedly impossible [17]. As recommended to us [18] trimethylsilylimidazole (TSIM) in dimethyl sulphoxide (DMSO) was shown to give either complete (2,3,6-O-) or partial (2,6-O-) trimethylsilylation under extremely mild conditions. Product analysis by ^{1}H-nmr (a fuller discussion with sample spectra is given in poster form) is facilitated by the distinct resonances for each hydrogen on the saccharide ring, and clear signals for TMS groups attached at the 2- and 6-positions.

Unambiguous nmr assignments are essential for determining reaction selectivities. That, for example, monotosylation occurred at the 6-hydroxyl is entirely consistent with high-field proton spectra for β-CD.OTs and from the carbon spectrum for the derived β-CD.NH_2 material.

Monosubstitution at C-6 reduces the overall symmetry of the cyclodextrin and complete proton nmr assignments are not possible. Nevertheless, using signal shapes derived from model acylated and silylated derivatives the substitution sites can easily be verified. Silylation of β-CD.NH_2 (TSIM-DMSO) and product extraction into hexane gave the fully silylated component for mono-attachment to the porphyrin, and could be used without purification. The solubilisation of cyclodextrin and its derivatives in this way made it easy to confirm homogeneity by LH-Sephadex (LH-20 and LH-60) chromatography. (Preparative and analytical columns made up in chloroform and linked in a polarimeter-UV detector for eluent monitoring were routinely used for product analysis.)

Direct amination of β-CD.OTs by displacement reactions with primary alkylamines gave β-CD.NHR and β-CD.$NHCH_2CH_2NH_2$. (Conditions for these and similar amine displacements have now appeared in the literature [19].) Persilylation followed by N-acylation proved an intirely selective method for exclusive derivatisation at nitrogen. Dissolution in tetrahydrofuran (THF) and brief treatment with mineral acid (HCl) precipitated the 'free' cyclodextrin N-acyl derivative cleanly.

Preparation of the TPP-tetracarboxylic acid (TPP-$COOH_4$)

Rothemund condensation of pyrrole with m-formylphenoxyacetic acid (from m-hydroxybenzaldehyde and ethyl bromoacetate) in refluxing propionic acid gave, following chromatography, the TPP-tetraester in modest yield (10-12 %). (This porphyrin with meta-carboxylate groups on the phenyl substituents was judged from CPK models to give the best possible fit for attachment to the primary face of the cyclodextrin). The free acid functionality was readily obtained by base hydrolysis. (That this step is not scale-limited and the yield is comparable to simple Rothemund condensations makes this reaction acceptable as a one-step access to the desired TPP.)

TPP-CD attachment methods

In what was found to be the best procedure, TPP-$COOH_4$ was made into the tetraacid chloride (with oxalyl chloride) and admixture with the fully (TMS) silylated β-CD.NH_2 in dichloromethane over solid K_2CO_3 gave rapidly a crude conjugate mixture. This was purified by separation on LH-60 Sephadex. In this manner we separated and identified (following methylation of the free carboxylic acid groups with diazomethane) TPP-CD conjugates that showed two cage (CD) residues for each porphyrin template. The (unmethylated) 2:1 conjugates were expected to be capable of undergoing subsequent cyclisation that would afford a sandwich-like macromolecule. However, after deprotection with dilute mineral acid and treatment with dicyclohexylcarbodiimide (DCC) in DMSO, both conjugates gave as sole product what by nmr was shown to be a 4:1 cyclodextrin: porphyrin material. Co-elution of the β-CD.NH_2.TMS component with the mono-attached conjugates on LH-Sephadex and thus the presence of β-CD.NH_2 in the DCC-DMSO cyclisation reaction would account for the unexpected formation of this trivial product.

In an attempt to remove β-CD.NH_2 before this last step, the conjugate was deprotected and then rechromatographed on ion-exchange carboxymethyl cellulose (CM 52). The fast moving red conjugate band evidently decomposed rapidly during elution, and deposited free porphyrin tetra-acid. The apparent sensitivity of TPP-CD conjugates to hydrolysis in what was a slightly basic buffer (ammonium bicarbonate) was seen to increase on Sephadex or CM series columns. The proximal hydroxyl groups on the periphery of the cage are implicated in a transacylation reaction that facilitates hydrolysis of acyl linkages, and it is not unreasonable to suppose that supports that contain weakly mucleophilic carboxyl and hydroxyl groups will reinforce this reaction. It became impossible to free the conjugate satisfactorily from unreacted cyclodextrin amine. Thus, while it is evident that trimethylsilylation is effective in directing exclusive N-acylation, the product cage molecule is, after deprotection, susceptible to facile hydrolysis. It is further evident that N-derivatisation by a hydrolytically stable linkage could also be directed by prior silylation, to give a satisfactory (stable) conjugate.

Tertiary butyldimethylsilylation: method and analysis

The per-(2,6-O-trimethylsilyl)cyclodextrin derivatives hydrolyse readily and so cannot be chromatographed on silica. It is known that the t-butyldimethylsilyl (TBDM-Si) ethers are more stable as synthetic intermediates, withstanding quite severe reaction conditions [20], yet easily removable by potent nucleophiles (e.g. naked fluoride anion). The bulky nature of the silyl substituents in this reagent led us to believe that complete (cf TMS) silylation would be difficult, indeed that steric effects would direct partial silylation and leave free hydroxyls for the attachment of smaller functionalities.

Treatment of β-cyclodextrin with TBDM-SiCl (in DMF using equivalent amounts of imidazole as base) proceeded at 60° to give the per-(2,6-O-silyl)cyclodextrin but together with a material identified as retaining one free hydroxyl group at C-2. These two cage derivatives could be readily separated by short column chromatography [21] on silica without apparent hydrolysis, and they dissolved readily in common organic solvents. Partial TBDM-silylation afforded complex mixtures and no additional positional selectivity, while exhaustive treatment with the silylation reagents showed no derivatisation of the (inner) C-3 hydroxyl. The steric crowding imposed by silylation at 2- and 6-centres, as is apparent from CPK models, would make further attachment impossible. The material with one free hydroxyl was further derivatised by either acylation or tosylation. Thus it has proved possible to attach even moderately sized reagents to this free centre. The nmr spectra for these derivatives were entirely consistent with attachment of residues at C-2. (Poster presentation).

The mono-substituted primary face amines were also fully silylated by the same procedure, but could not be analysed satisfactorily by proton nmr, or further derivatised through attachments to the amino group.

CONCLUSIONS

From this brief evaluation of silyl derivatives as useful intermediates for cyclodextrin functionalisation the following is concluded:

1. TMS derivatives can be easily prepared and used to solubilise cyclodextrins in regular organic solvents for convenient handling, reaction and chromatographic (LH-Sephadex) separations.
2. Existing functionalities on the cage, such as amines, can be cleanly derivatised since the TMS groups effectively protect the majority of the hydroxyl sites from unwanted reaction.
3. Mild acid hydrolysis removes TMS groups and the free cyclodextrin is conveniently recovered by precipitation from the organic solution.
4. TBDM-silylation gives stable (towards hydrolysis) soluble derivatives that can be chromatographed on LH-Sephadex and silica selective secondary face derivatisation is possible, using the one C-2 hydroxyl remaining after this treatment.
5. The proton nmr spectra of the silyl derivatives are, in general, easier to interpret than those of the free cyclodextrin, and positional residue assignments can be made.

ACKNOWLEDGEMENT

We thank Professor C. Eaborn and Dr. D.R.M. Walton (University of Sussex) for invaluable advice and encouragement in the method of silicon chemistry.

LITERATURE CITED

Except where explicitly relevant, references to work contained in the two comprehensive review publications (Refs. [1] and [2] are NOT given.

1. Bender, M.L., Komiyama, M.: Cyclodextrin Chemistry, Springer-Verlag, Berlin-Heidelberg-New York 1978.

2. Saenger, W.: Angew. Chem. Int. Ed. Eng. 19 344, /1980/
3. Horikoshi, K.: Process Biochem. 26, /1979/
4. Breslow, R.: Isr. J. Chem. 18 187, /1979/
5. Boger, J., Brenner, D.G., Knowles, J.R.: J. Am. Chem. Soc. 101 7631, /1979/
6. Tabushi, I., Kuroda, Y., Mochizuki, A.: J. Am. Chem. Soc. 102 1152, /1980/
7. Breslow, R., Hammond, M., Lauer, M.: J. Am. Chem. Soc. 102 421, /1980/
8. Siegel, B., Breslow, R.: J. Am. Chem. Soc. 97 6869, /1975/
9. Bergeron, R.J., Meeley, M.P.: Bioorg. Chem. 5 121, /1976/
10. Boger, J., Corcoran, R.J., Lehn, J.M.: Helv. Chim. Acta. 61 2190, /1978/
11. Iwakura, Y., Uno, K., Toda, F., Onozuka, S., Hattori, K., Bender, M.L.: J. Am. Chem. Soc., 97 4432, /1975/
12. Onozuka, S., Kojima, M., Hattori, K., Toda, F.: Bull. Chem. Soc. Jpn. 53 3221, /1980/
13. Tabushi, I., Kuroda, Y., Yokota, K., Yuan, L.C.: J. Am. Chem. Soc. 103 711, /1981/ and references therein.
14. Melton, L.D., Slessor, K.N.: Carbohydr. Res. 18 29, /1971/
15. Chao, Y., PhD Thesis, Columbia University, New York 1973
16. Matsui, Y., Okimoto, A.: Bull. Chem. Soc. Jpn. 51 3030, /1978/
17. Beadle, J.B.: J. Chromatogr. 42 201, /1969/
18. Walton, D.R.M., Private communication.
19. Tabushi, I., Shimizu, N., Sugimoto, T., Shiozuka, M., Yamamura, K.: J. Am. Chem. Soc. 99 7100, /1977/; Matsui, T., Yokoi, T., Mochida, K.: Chem. Lett. 1037, /1976/
20. Corey, E.J., Venkateswarlu, A.: J. Am. Chem. Soc. 94 6190, /1972/
21. Hunt, B.J., Rigby, W.: Chem. & Ind. 1868, /1967/

I.Int.Symp. on Cyclodextrins
Budapest, 1981

4.3.

PREPARATIVE METHODS AND NMR ANALYSIS FOR SILYLATED DERIVATIVES OF CYCLODEXTRIN

R.L. WIFE,* D.E. REED,** D.P. LEWORTHY, D.M. BARNETT,
P.D. REGAN, H.C. VOLGER

Shell Biosciences Laboratory
Sittingbourne, England

SUMMARY

Methods for the complete or partial (selective) silylation of β-cyclodextrin are described that provide intermediates for further derivatisation. The method is applied to the sequential attachment and attempted capping of β-cyclodextrin by a porphyrin template. Representative proton NMR spectra for a series of silyl derivatives and intermediates are analysed to demonstrate further the advantage of the method.

----
* Current address: Koninklijke/Shell-Laboratorium, Amsterdam (Shell Research B.V.) Badhuisweg 3, Amsterdam, The Netherlands
** Current address: School of Molecular Sciences, University of Sussex, Brighton, Sussex, England.

INTRODUCTION

Our strategy for constructing a porphyrin-capped cyclodextrin has been described [1]. Crucial to this preparative method was a directed mono-attachment reaction that secured the component materials by an amide bond, with the provision for a subsequent cyclisation across the face of the cage molecule. The use of silylated intermediates, as protecting groups to direct reaction in the (previously functionalised) β-cyclodextrin amine, served also to solubilize the cage molecule in organic solvents and so facilitated ^{1}H-nmr analysis and product purification by chromatography.

Unambiguous nmr assignments are essential to determine reaction selectivities on the cyclodextrin ring, and are of further assistance in assessing compound homogeneity. We now describe proton nmr spectra recorded at 360 MHz on samples used in the preparation of the porphyrin-cyclodextrin conjugate, as well as on model compounds, together with experimental details for these procedures.

RESULTS AND DISCUSSION

Parent and fully derivatised cyclodextrins

The simplest and most easily assigned spectra are those of cyclodextrins where each derivatisable centre (C-2, C-3 and C-6) bears the same functionality (OH, $OCOCH_3$, $OSiMe_3$, OSi*t*-$BuMe_2$). The magnitude of discrete proton coupling constants is more or less independent of the functionality, and so each proton signal can be identified from a characteristic shape (Table I). Proton chemical shifts, however, are strongly influenced by the adjacent or even distant functionality.

The proton spectra for both α- and β-cyclodextrin have the anomeric hydrogen (H-1) at lowest field, with distinct high-field signals for H-2 and H-4 (the order is inverted for these two by switching from D_2O to d_6-DMSO as solvents) (Figure 1). Inbetween a four-proton envelope for H-3, H-5, H-6 and H-6' is located. Assignments made for β-cyclodextrin were further verified by decoupling, beginning with H-1, and further by sequential inclusion of d_6-benzene which resolved the spectrum partially, and decoupling of the modified spectrum, which identified H-5, H-6 and H-6'.

Mono-substitution imposes some asymmetry that complicates the spectra for these compounds. Complete assignments are no longer possible, but from the characteristic signal, key shifts can be identified that are consistent

Table I: Coupling constants (Hz) measured for α-, β-cyclodextrin and β-cyclodextrin peracetate*

α-cyclodextrin[a)]	H-1[c)] d (3.5)	H-2 dd (3.5, 10.5)	H-3 dd (10.5, 9.0)	H-4 dd (t) (9.0)
β-cyclodextrin[a)]	H-1 d (3.5)	H-2 dd (3.5, 9.5)	H-3 dd (t)(9.5)	H-4 dd (tbr) (9.5)
β-cyclodextrin[b)] peracetate	H-1 d (4.0)	H-2 dd (4.0, 9.5)	H-3 dd (9.5, 8.0)	H-4 dd (8.0, 9.3)
	H-5 dd (br) (9.3,4.5,1.5)	H-6 dd (4.5,12.5)	H-6'dd (1.5,12.5)	

a) Solution in D_2O

b) Solution in $CDCl_3$

c) Hydrogens are designated by the scheme shown:

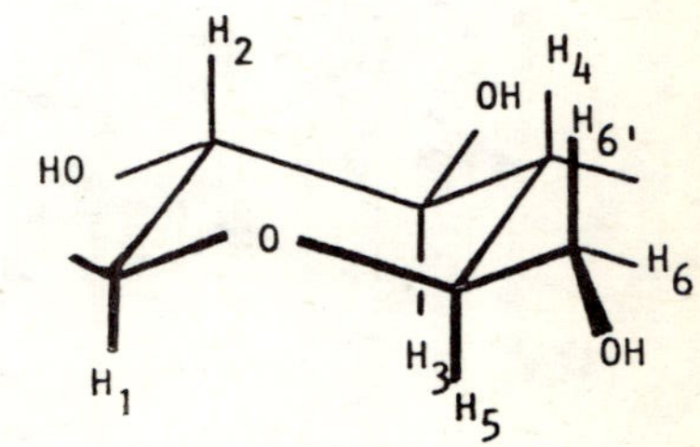

* d; doubled, dd; double doublet, t; triplet, br; broad

with what is anticipated from the nature of the substituent group and its position on the cyclodextrin frame.

In β-cyclodextrin monotosylate, the envelope of signals for H-2 to H-6 is clearly compressed (Figure 5). One anomeric hydrogen (H-1) is slightly shielded, but significantly, two signals for the methylene hydrogens that bear the tosyl group (H-6, H-6') are located downfield from the resonance position in the parent cyclodextrin. In confirmation, decoupling by irradiation at the position estimated for H-5 simplifies these two resonances. The downfield shift induced on H-6, H-6' is then about 0.8 ppm.

Peracetylation of the monotosylate helps resolve the multiproton envelope, and discrete proton groupings can be discerned (Figure 6). The cumulative deshielding by the acetyl groups, however, makes it no longer possible to locate the downfield-shifted methylene hydrogens in this spectrum.

With the bulky (TBDMS) groups on material that is almost completely derivatised, that the free hydroxyl group is located at C-2 is clear from the spectrum for the monoacetate, derived from the alcohol by acetylation (Figure 7). The H-2 proton is seen at low field, with a 1.25 ppm downfield shift from the other protons at the 2-position. The H-1 signal is also split, with one proton at lower field. This shift is larger for H-1 when a tosylate group is at this centre, but the signal corresponding to the H-2 cannot be located in this spectrum (Figure 8).

Cyclodextrin-Porphyrin conjugates

The spectra for these CD-P conjugates were as was expected for such macromolecules, significantly broadened, and no individual assignments are attempted. However, from the ratio of the porphyrin β-methine and amide signals to the envelopes for the cyclodextrin hydrogens the composition of the homogeneous (chromatographically) conjugates was established. The fully conjugated CD_4-P material had a comparatively more resolved spectrum, (Figure 10) but that for the CD_2-P material was extremely broad (Figure 9).

^{13}CARBON NMR SPECTRA

In order to establish that the monotosylate formed from β-cyclodextrin is 6-substituted, and so gives a 6-amino derivative either by direct substitution with a primary amine, or indirectly (azide reduction), the ^{13}C spectra for the amino-derivatives were examined.

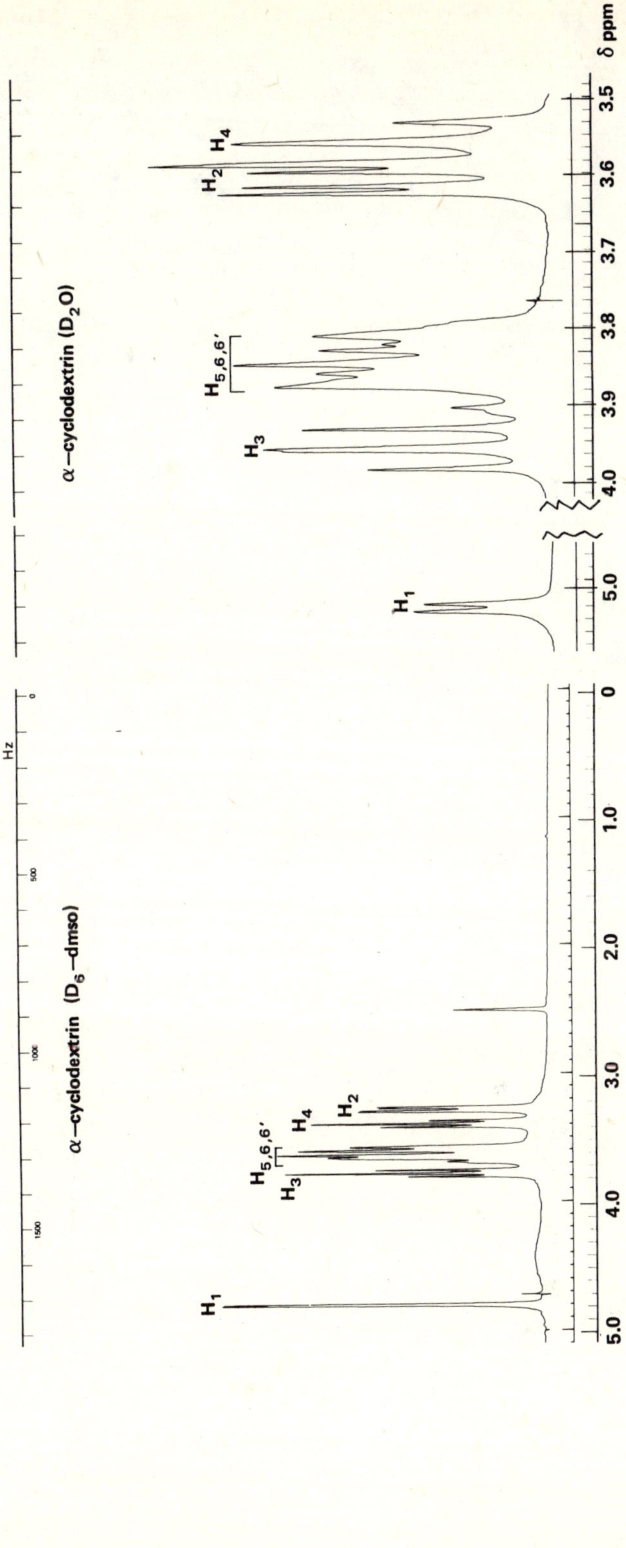

Figure 1a

E0658

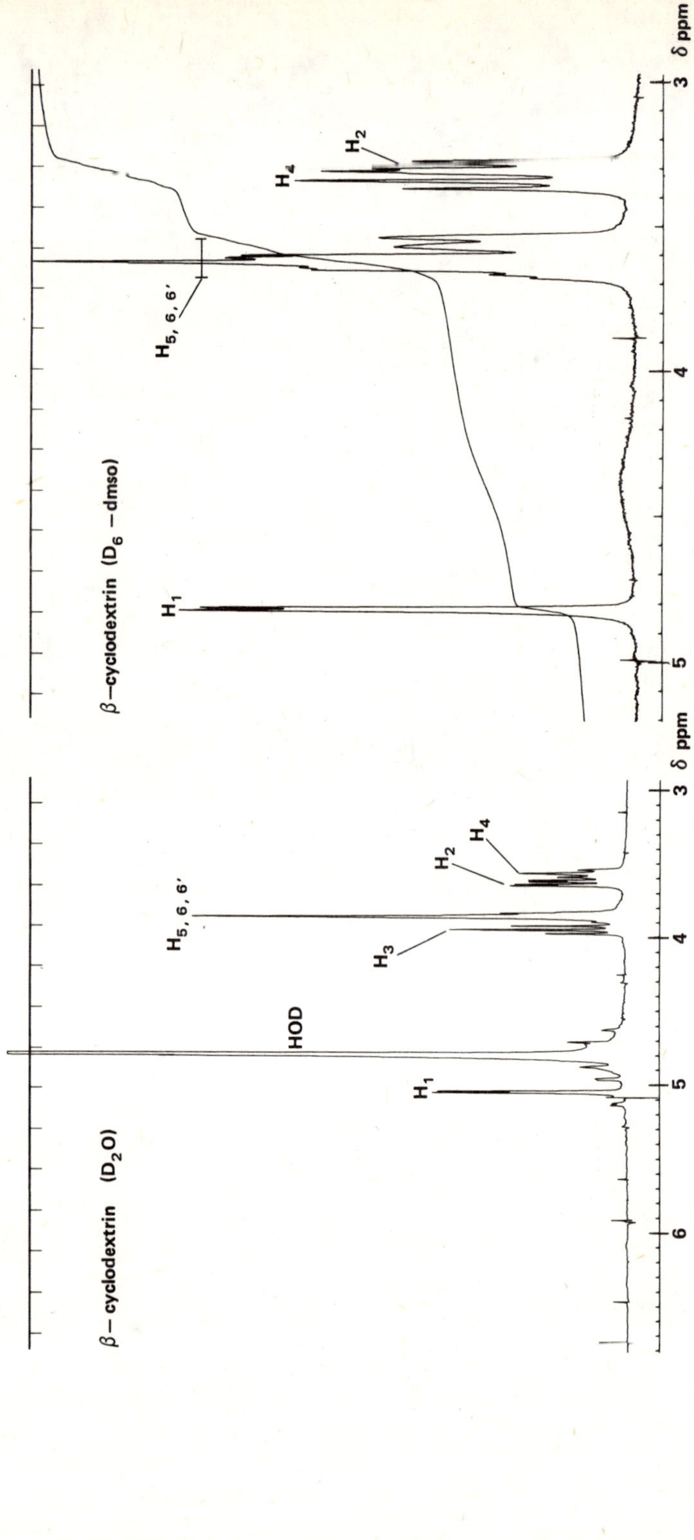

Figure 1b

E0658

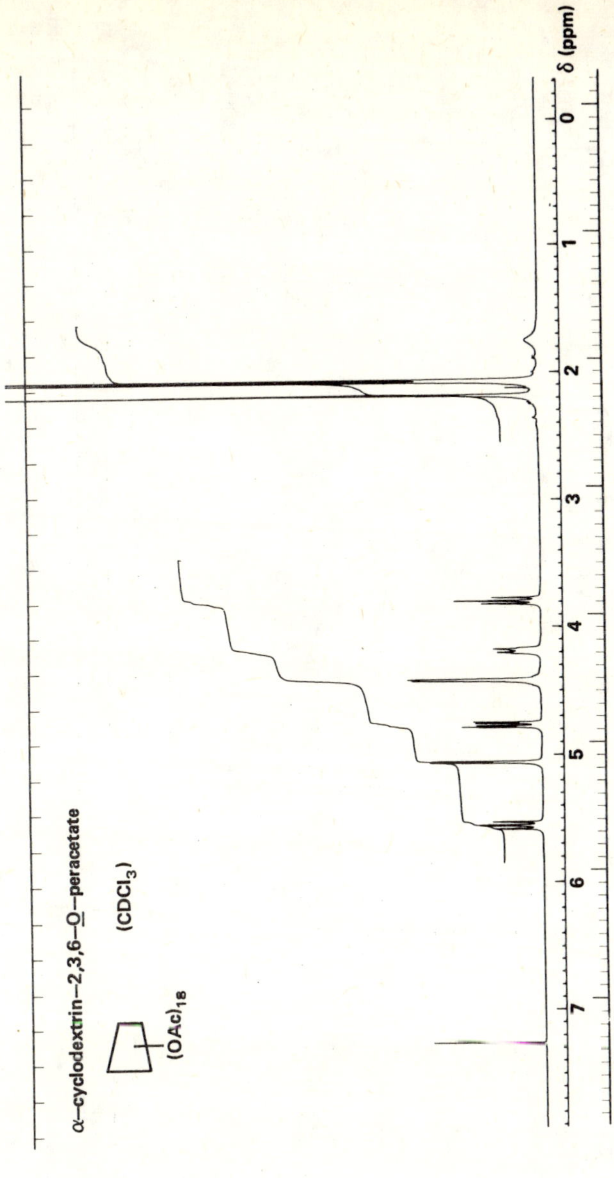

Figure 2a

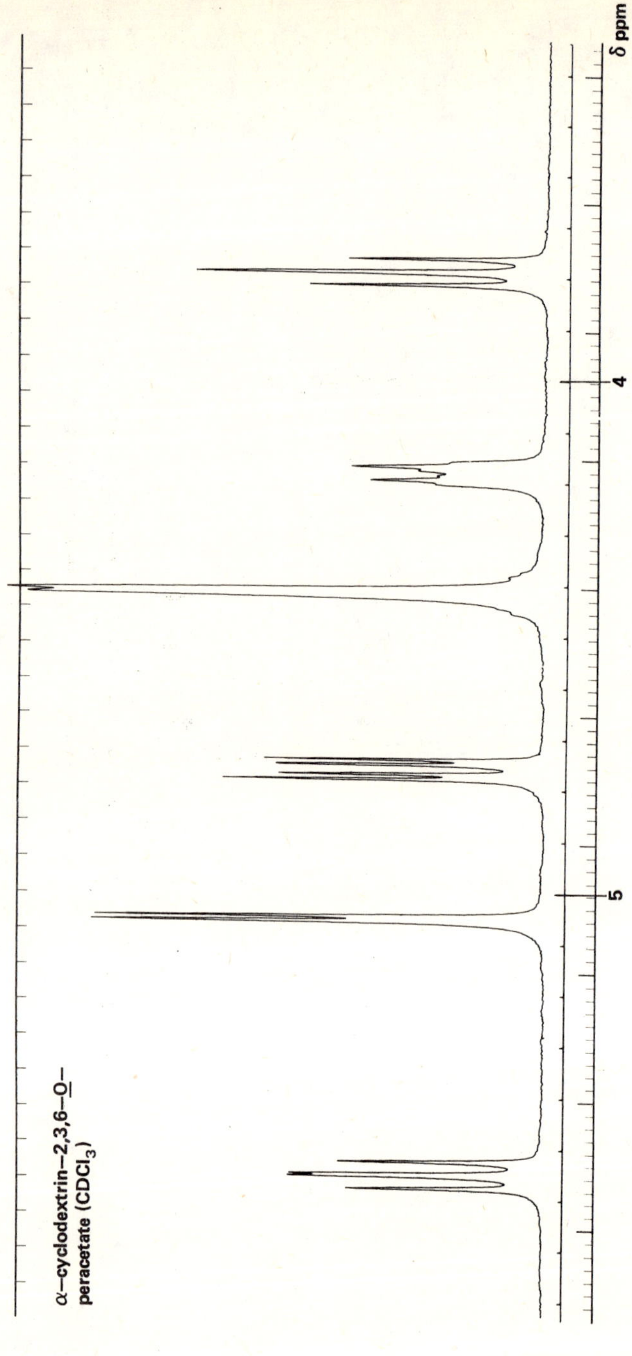

Figure 2b

E0658

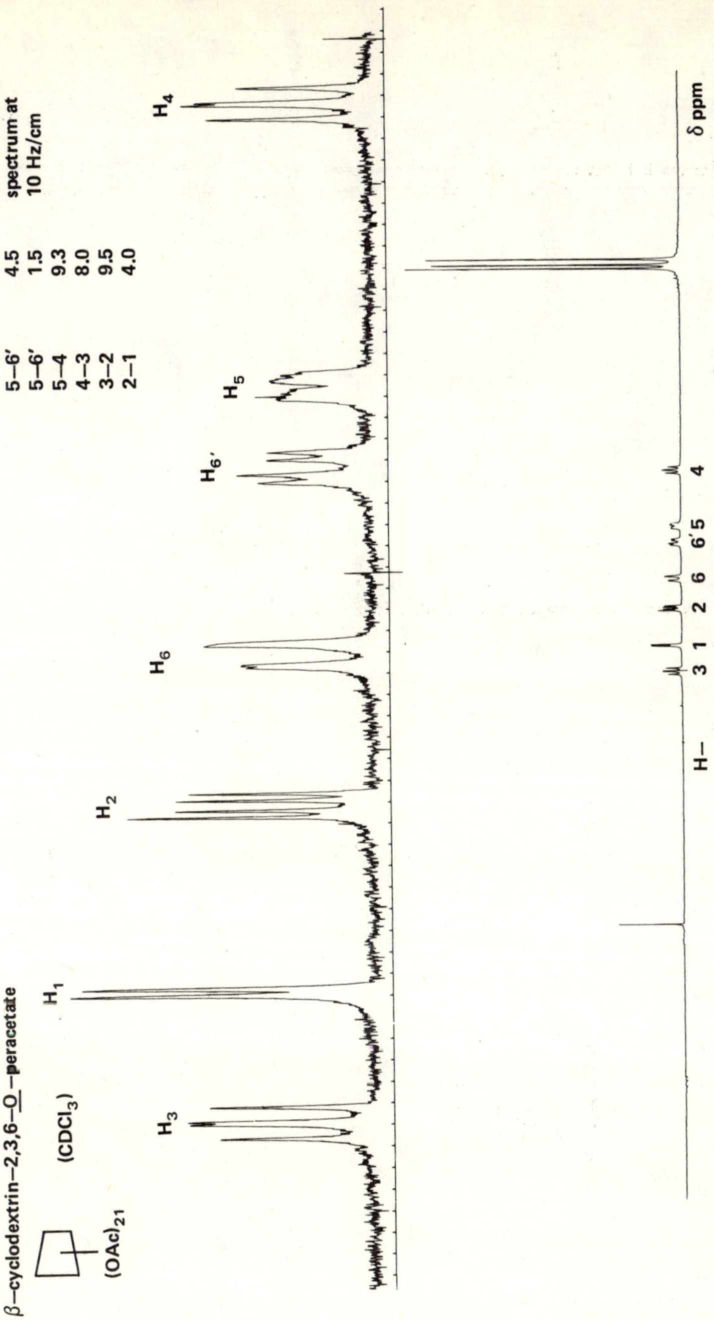

Figure 2c

E0658

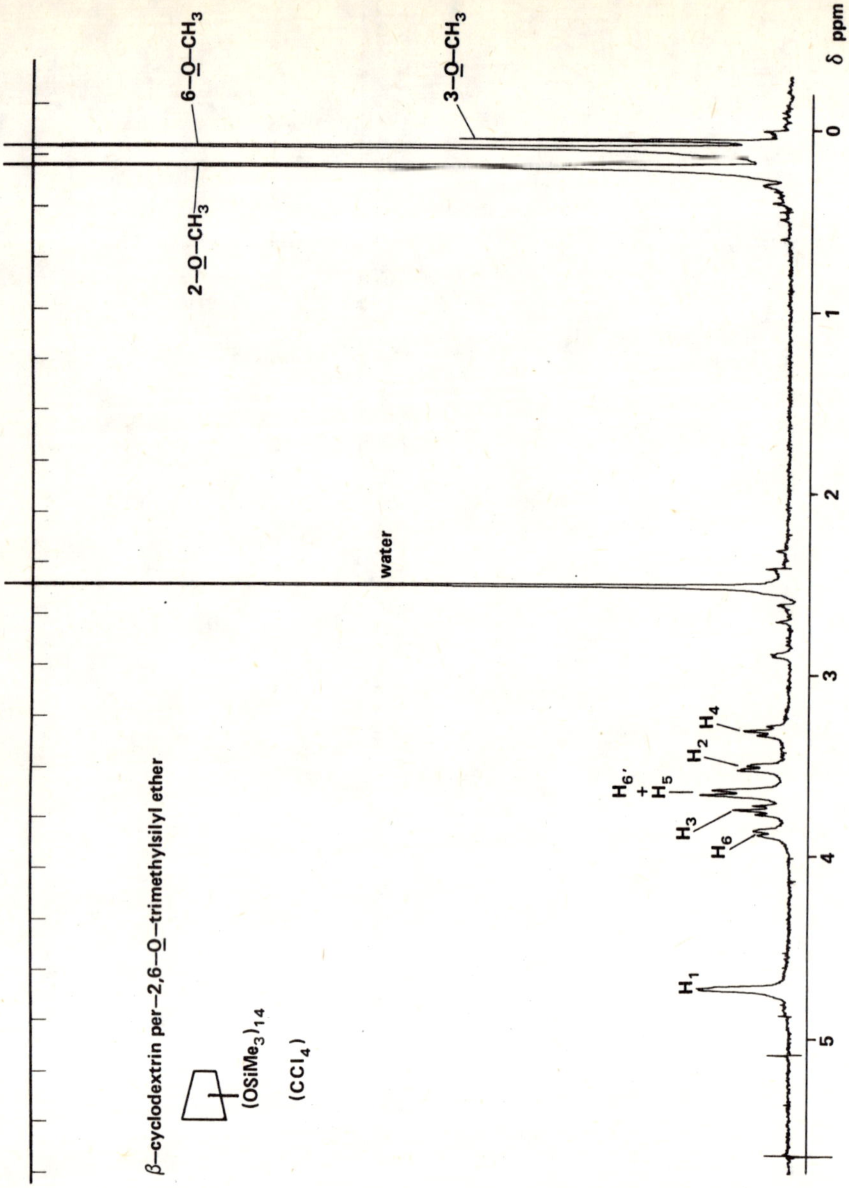

Figure 3

E0658

E0658

β–cyclodextrin per–2,6–O–t–butyldimethylsilyl ether

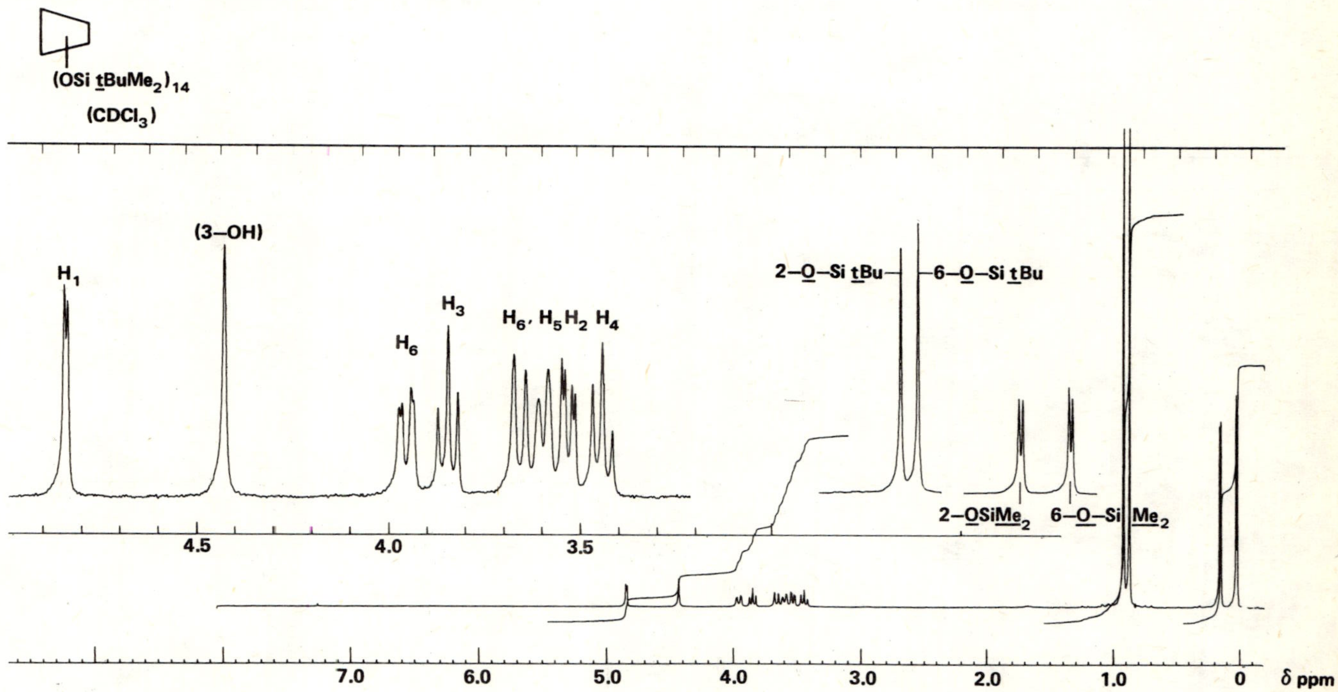

Figure 4a

E0658

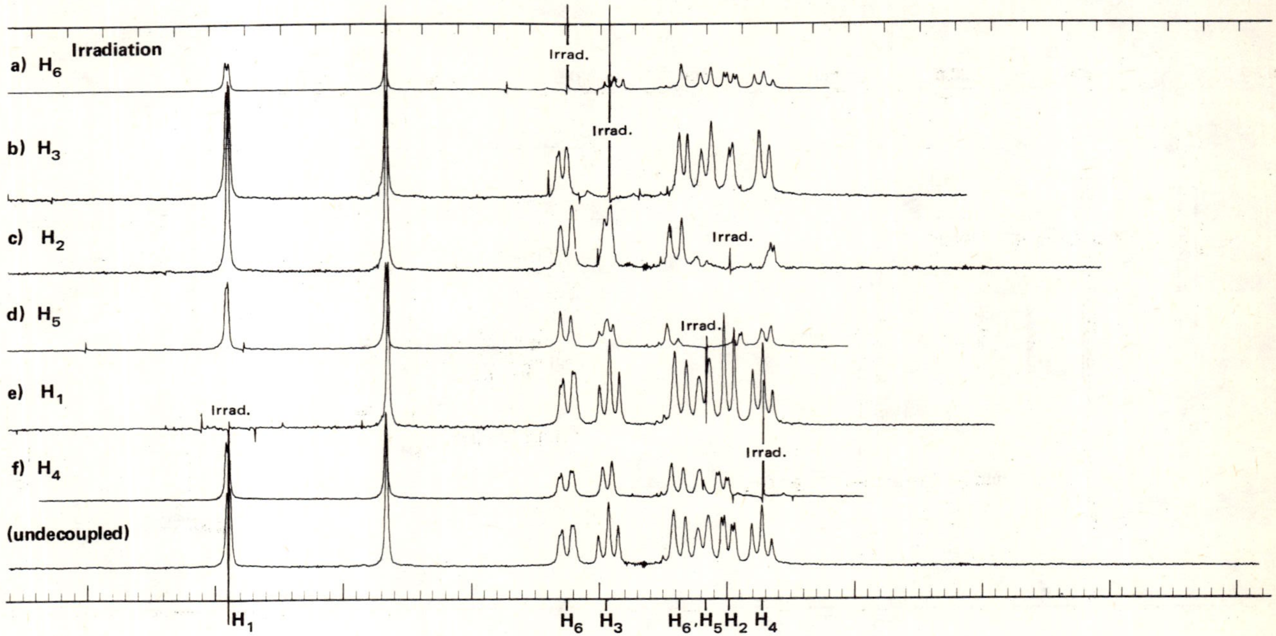

Decoupling experiment on β–cyclodextrin–2,6–O–per–t–butyldimethylsilyl ether

Figure 4b

E0658

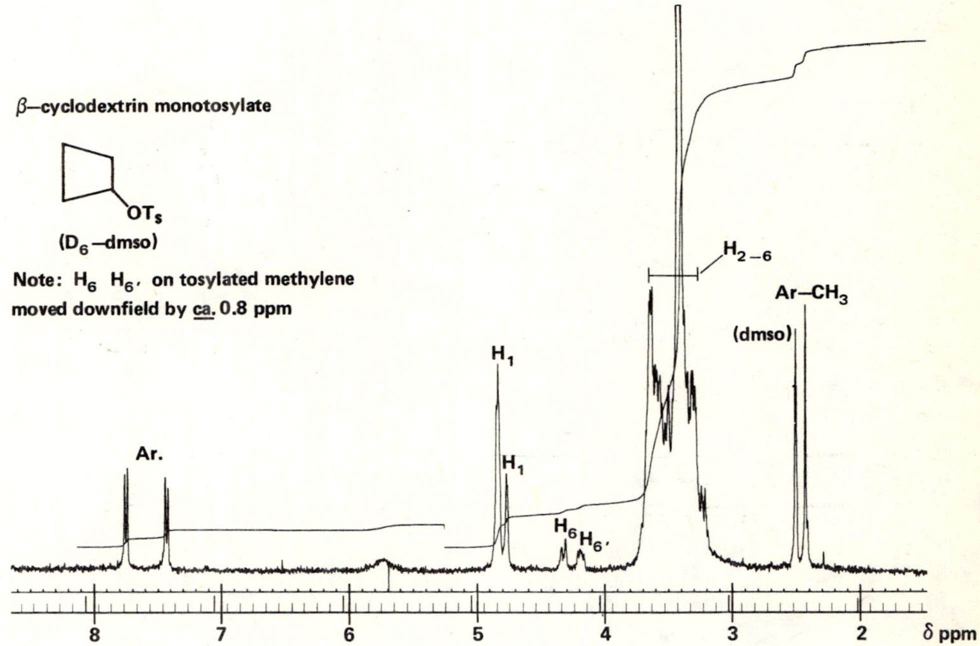

Figure 5a

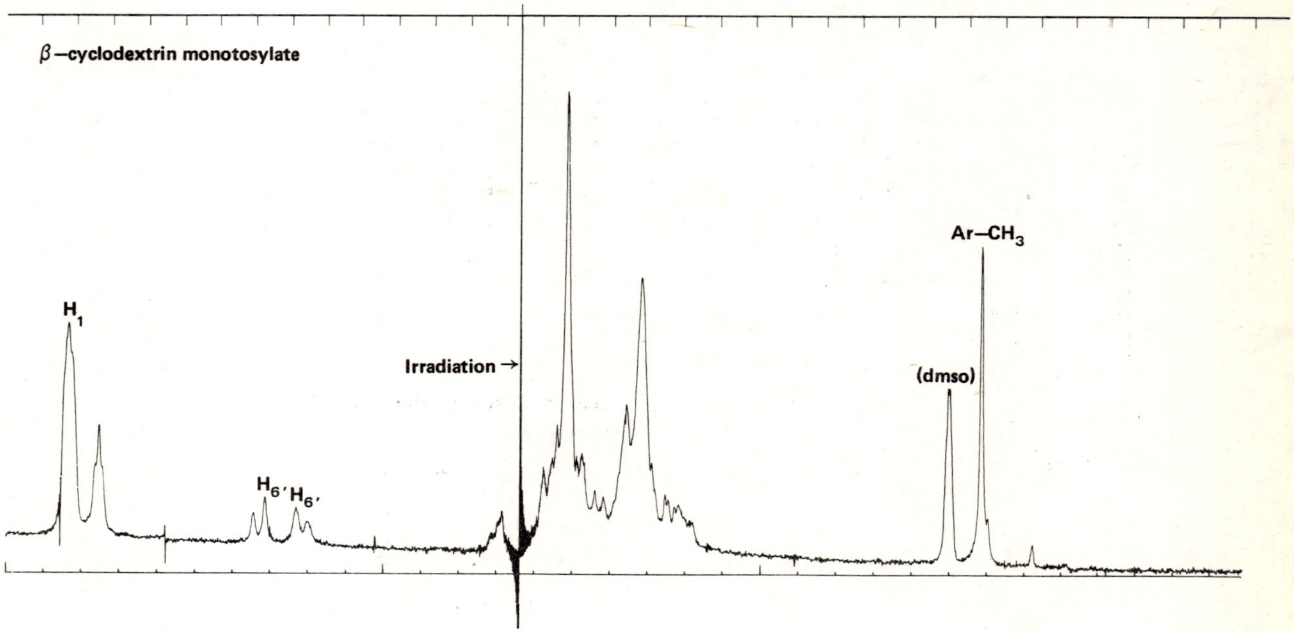

Decoupling experiment : Irradiation at H_5 , H_6 , and H_6 , sharpen to doublets confirming assignment

Figure 5b

E0658

E0658

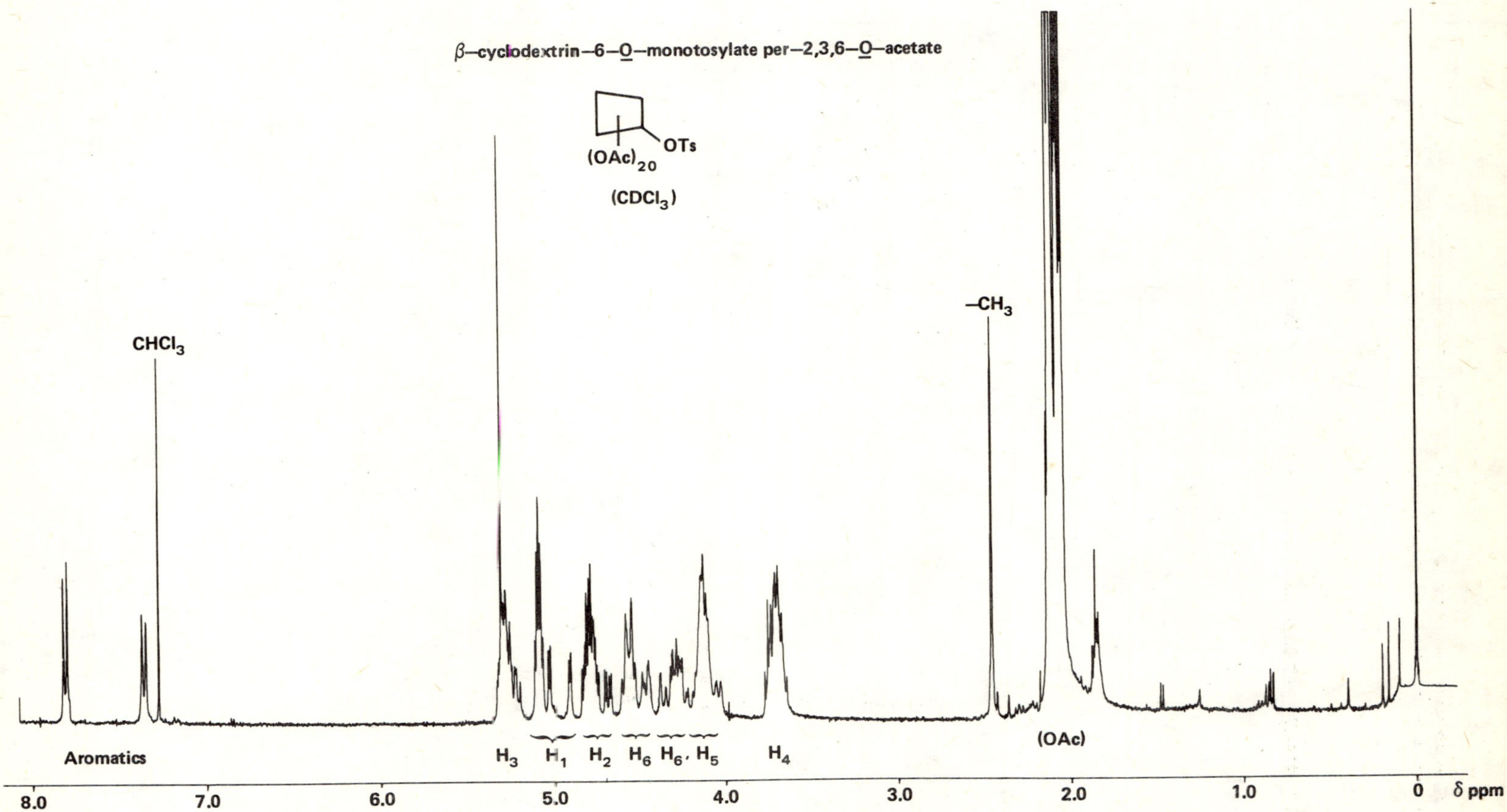

Figure 6

E0658

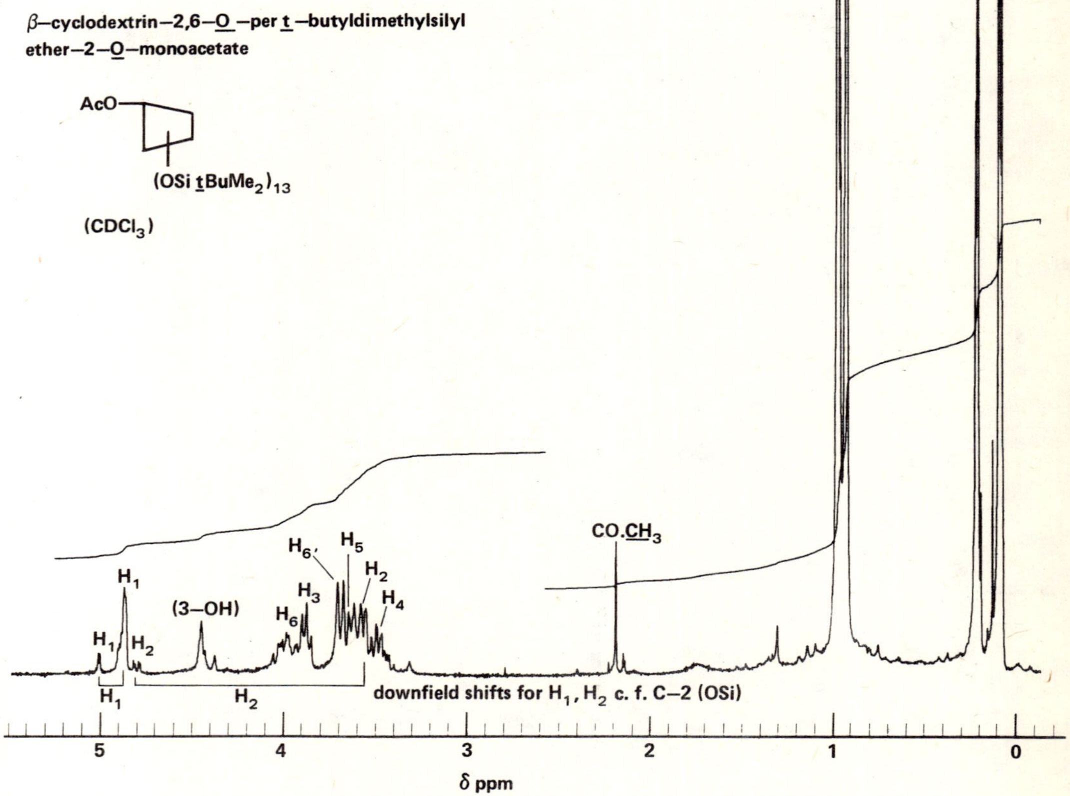

Figure 7

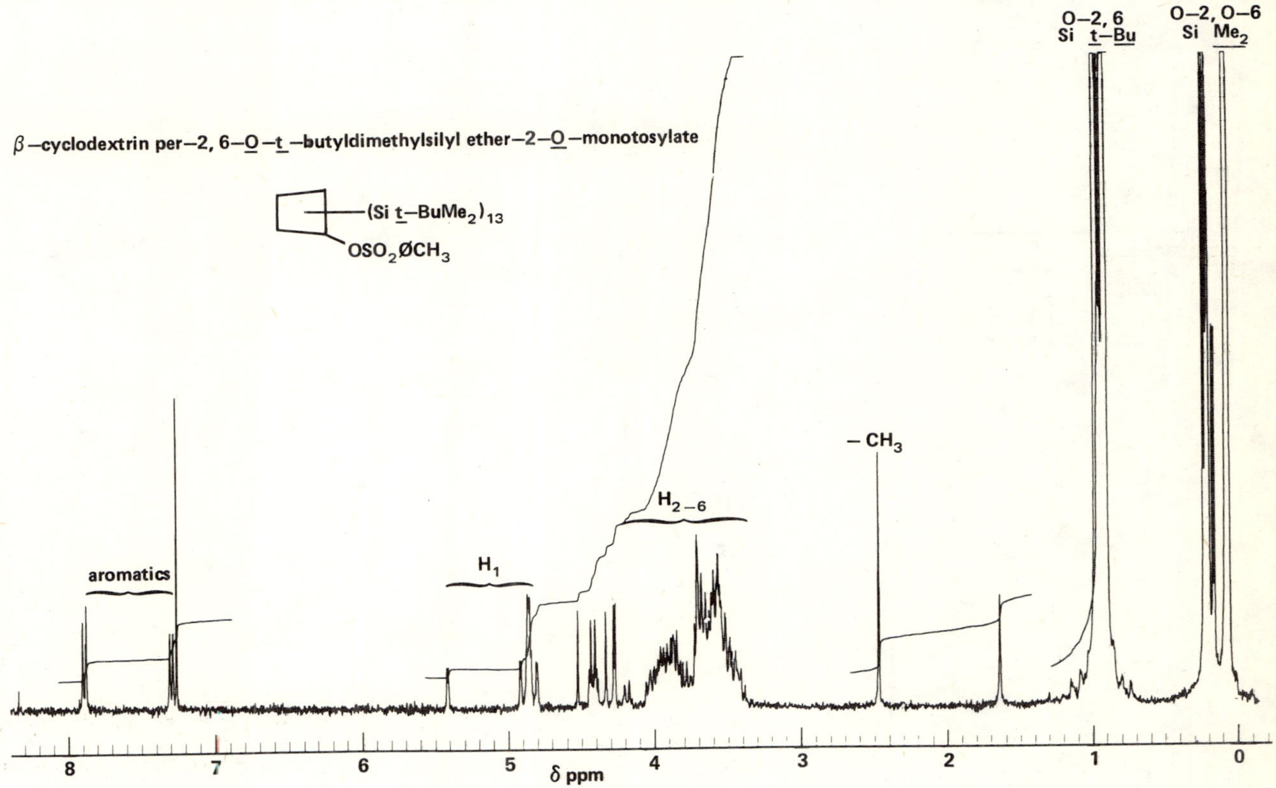
β-cyclodextrin per-2, 6-O-t-butyldimethylsilyl ether-2-O-monotosylate
(Si t-BuMe2)13
OSO2ØCH3
O-2, 6
Si t-Bu
O-2, O-6
Si Me2
- CH3
H2-6
H1
aromatics
8
7
6
5
4
3
2
1
0
δ ppm

Figure 8

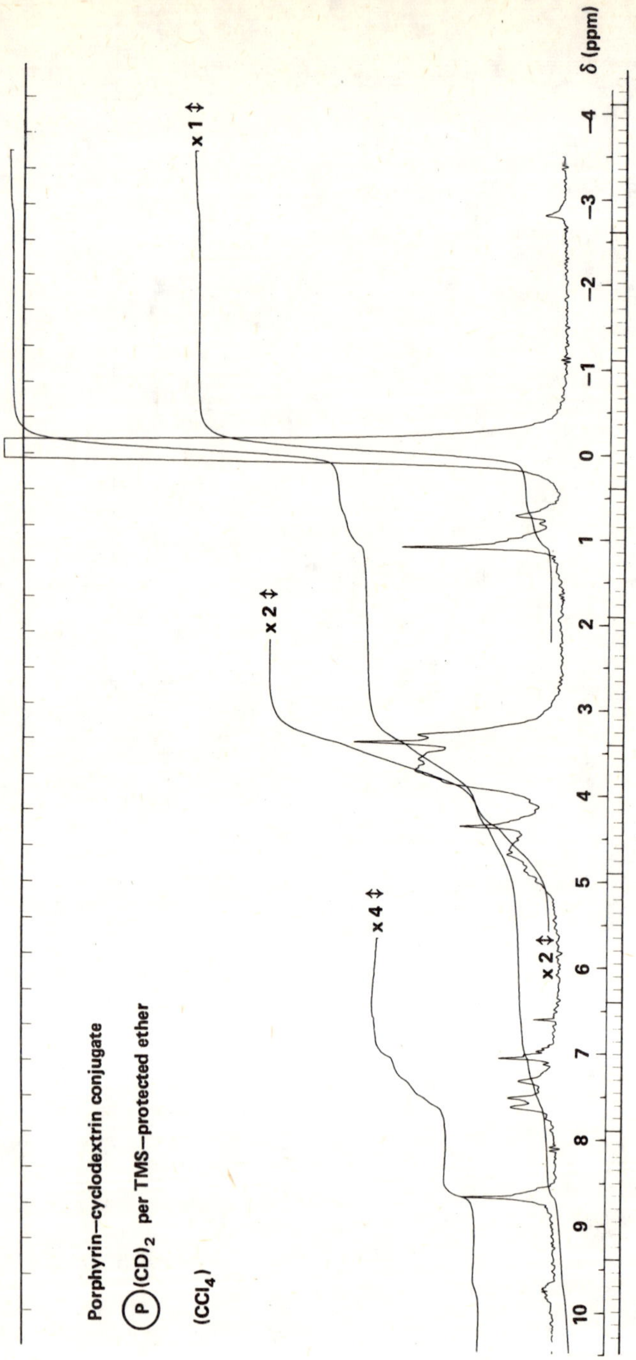

Figure 9

E0658

E0658

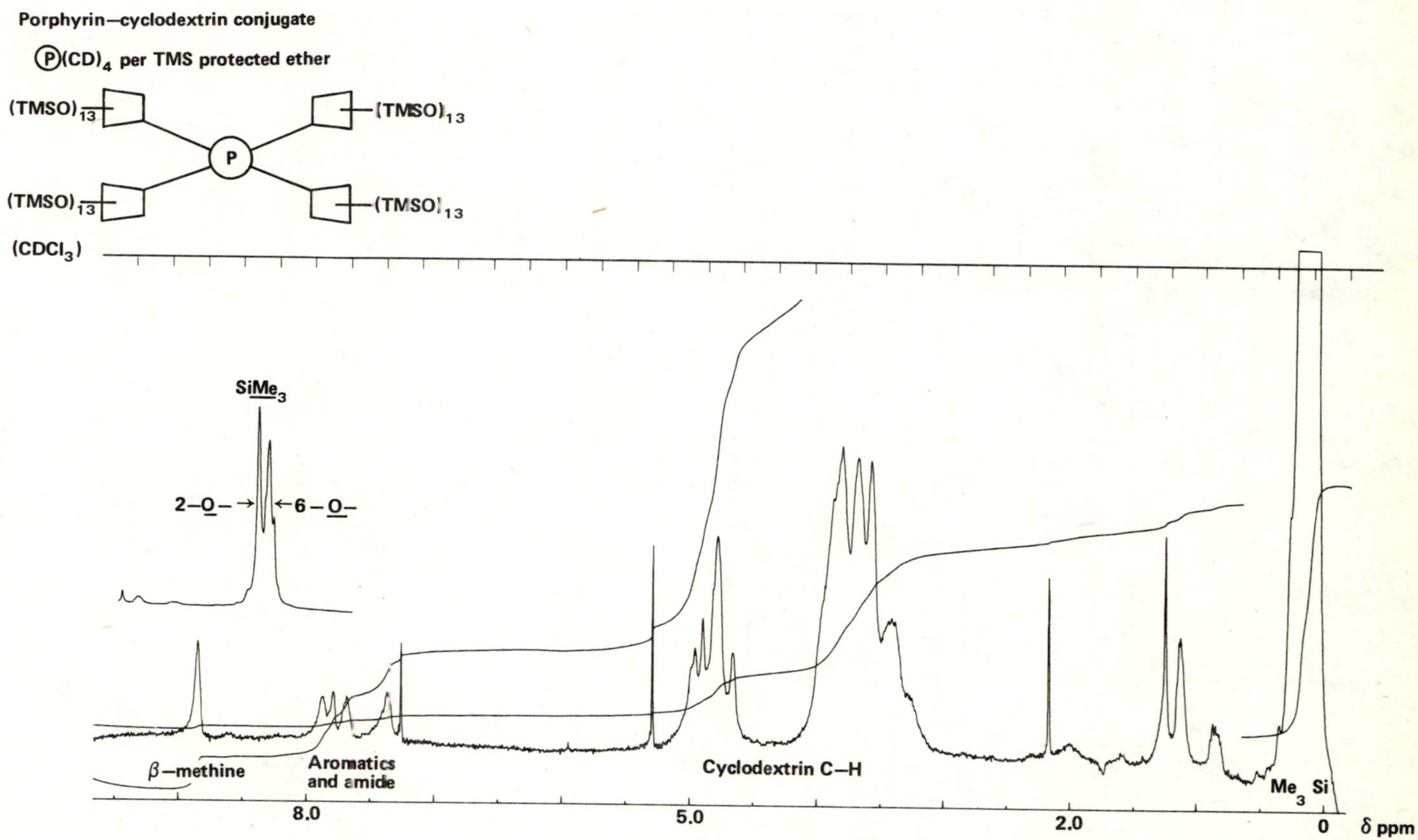

Figure 10

E0658

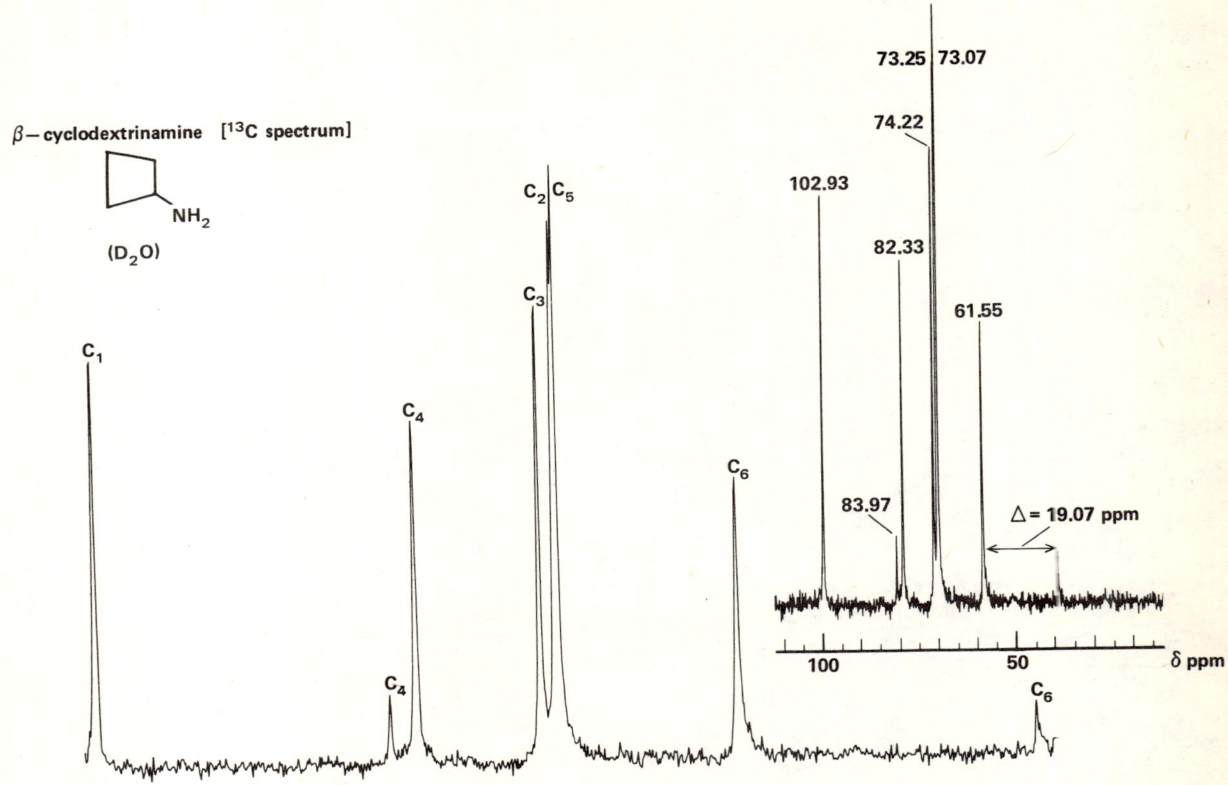

Figure 11

E0658

β-cyclodextrin methylamine [^{13}C spectrum]

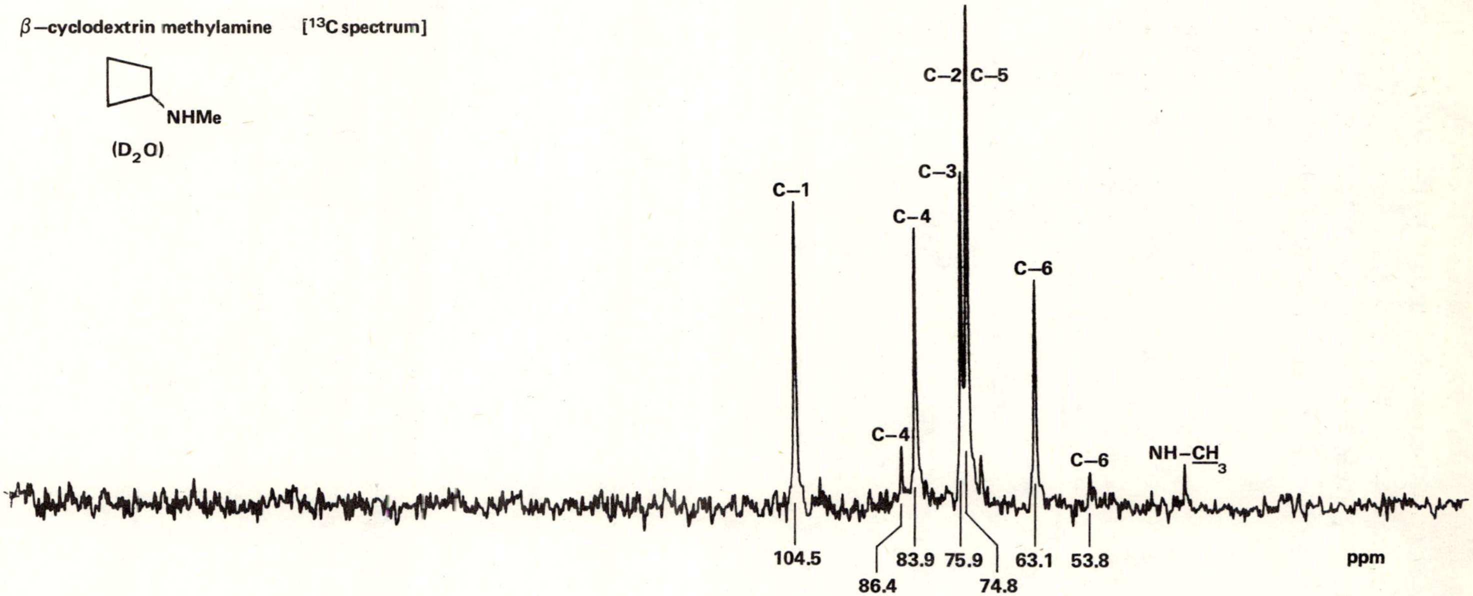

Figure 12

For both cyclodextrinamine and the N-methyl derivative, there is a clear signal upfield from the resonance position for the 6-carbon, assigned to the one 6-carbon that bears the less electronegative group (NH_2 or NHR cf O). Correspondingly, there is a small downfield shift for the H-4 signal on the glucose ring that contains the amino group which is precedented as a γ-effect.

EXPERIMENTAL

Solvents and sensitive reagents were purified by prescribed methods, and stored before use in an inert atmosphere. Other materials were used as supplied from commercial sources. Aldrich Chemical Co. supplied both α- and β-cyclodextrin which were dried by warming at 70 °C under high vacuum for at least 3 hours before use, and the silylating reagents N-(trimethylsilyl) imidazole and t-butyldimethylsilyl chloride.

Reactions were monitored by thin layer chromatography, on Merck Silica plates and visualised by acid-sprays (30 % sulphuric acid) and subsequent charring. Cyclodextrin mixtures were eluted using the solvent composition MEK: MeOH:H_2O (70:15:15) in thin layer analysis, porphyrin mixtures on silica (anal. and prep.) by chloroform or methanol/chloroform mixtures. Preparative separations on silica were as described for 'short column chromatography' [2], and on Sephadex G or Whatman cellulose CM 52 the commercial Pharmacia columns were used. Preparative separations on the Sephadex LH 60 or LH 20 materials were run in chloroform, using a custom-made glass/teflon column built to the Pharmacia design.

Eluent analysis from the columns described was achieved by monitoring through a Perkin Elmer 241 Polarimeter adapted for flow use, using a microcapillary flow cell as supplied, and through (in tandem) a UV detector as supplied from LKB both linked to a flat-bed recorder, and fractions indicated directly from the fraction collector (LKB). Connections between column and detectors were by teflon tubing, or nylon tubing for aqueous eluents. To avoid interference through the accumulation of air bubbles in the cells, all eluent solvents were rigorously degassed.

β-Cyclodextrin monotosylate and the monoamine were prepared as described elsewhere, [3-5] and without further modification.

Trimethylsilylation

i. β-Cyclodextrin per 2,6-O-silyl ether

Into a solution of β-cyclodextrin (0.287 g; 0.254 mmol) in dry DMSO (2 ml), N-(trimethylsilyl) imidazole (1 ml) was syringed. The reaction was stirred at 20 °C for 5 h in which time the homogeneity was lost and a thick cloudy suspension formed. The persilylated cyclodextrin product was extracted into dry carbon tetrachloride and the solution evaporated to give a white solid under high vacuum. This material was analysed directly by nmr after solution in either chloroform or carbon tetrachloride.

ii. β-Cyclodextrinamine per 2,6-O-silyl ether

In a similar manner, the monoamines of β-cyclodextrin were silylated in DMSO with TSIM and extracted into hexane, evaporated to give dry white powders and either used as such in conjugation experiments, or chromatographed on LH Sephadex columns.

Preparation of the tetraphenylporphyrin (TPP) tetracarboxylic acid

i. Ethyl m-formylphenoxyacetic ester

A solution of m-hydroxybenzaldehyde (25.0 g; 0.21 mol), ethyl bromoacetate (34.3 g; 0.20 mol) and sodium hydroxide (8.3 g; dissolved in a minimum amount of water) in ethanol (100 ml) was stirred at 0 °C overnight. Then o-mercaptobenzoic acid (6.5 g) in ethanol (100 ml) was added to remove unreacted bromoester and the reaction was left a further day. Following dilution with water (200 ml) the product was extracted into dichloromethane (200 ml) and the extracts washed with aqueous sodium carbonate (8 times 200 ml) and water (200 ml) before drying over sodium sulphate. In this way the desired ester (13.0 g) was obtained free from impurities.

ii. Rothemund condensation to give TPP-tetraester

A solution of the m-formylphenoxyacetic ester (7.0 g; 0.034 mol) in propionic acid (120 ml) was warmed to reflux temperature in an atmosphere of nitrogen, with the exclusion of light. To the solution was added pyrrole (2.25 g; 0.034 mol) by injection from a syringe, and the rapidly darkening reaction mixture was maintained at reflux temperature for a further half-hour. The mixture was cooled at ca 70 °C and chloroform (25 ml) was added. After further cooling to 0 °C the mixture was diluted with water (250 ml) and extracted with chloroform (250 ml). The extract was washed successively with sodium carbonate (4 times 250 ml) and water (250 ml), (Na_2SO_4) and evaporated to dryness. The crude product was chromatographed rapidly on Alumina (Brockmann G 111 neutral) and carefully on Silica (Woelm G 111; 120 g); elution with chloroform-diethyl ether (4:1) afforded the material that was finally purified by short column chromatography (ethyl acetate/toluene 1:19) to give the pure porphyrin tetraester (1.06 g; 12 % yield).

iii. Tetraphenylporphyrin tetracarboxylic acid

A solution of the tetraester (258 mg) in tetrahydrofuran (100 ml) was taken to reflux temperature under argon and treated with a solution of potassium hydroxide (0.78 g) in water (5 ml). After 4 h at this temperature, the mixture was cooled and left to stand. The organic layer was decanted and the aqueous layer diluted with water (50 ml) and methanol (50 ml), before acidification with hydrochloric acid (conc.) and extraction into chloroform. Methanol (ca 150 ml) was added to prevent precipitation of the acid, and the organic solution was then evaporated to dryness giving green crystals of the tetra-acid as the bis-hydrochloride salt with quantitative recovery.

TPP-CD Conjugation

The TPP-tetracarboxylic acid was converted to its acid chloride in methylene chloride by (excess) oxalyl chloride at 20 °C. The solvent and excess reagent were removed under high vacuum, and the residue redissolved in methylene chloride; this solution was then added to a solution of the persilylated cyclodextrinamine in methylene chloride over anhydrous potassium carbonate. Typically, in this manner the TPP-tetracarboxylic acid (1 equiv.), was reacted with the cyclodextrinamine (2 equiv.) over 1 h at 20 °C. After filtration of the red product solution through a glass sinter, the solution was concentrated. An aliquot was generally applied to a Sephadex (LH 60 in $CHCl_3$) column for analysis after treatment with diazomethane to derivatise the free remaining caboxylic acid functionalities. The major part of the reaction mixture was then hydrolysed by treatment of a solution in THF with hydrochloric acid (6 N; few drops) to obtain a green precipitate from the organic phase that was centrifuged to obtain a pellet after decanting of the supernatant (almost colourless). Application of this material to a Whatman CM 52 column in ammonium bicarbonate buffer and elution gave some resolution before decomposition of the conjugate by hydrolysis. In subsequent experiments, the removal of the silyl groups was preceded by chromatography on LH 60 and fraction isolation; following deprotection, the green pellet was dissolved in DMSO and treated with DCC (excess). The reaction was followed by taking aliquots, treating with TSIM-DMSO and monitoring them on an analytical LH 60 column. When complete after overnight reaction, the crude product was treated with diazomethane in chloroform and applied to a preparative LH 60 column, where the TPP-CD_4 conjugate was isolated.

t-Butyldimethylsilylation

i. β-Cyclodextrin persilylation

To a solution of β-cyclodextrin (0.43 g; 0.38 mmol) in DMF (1 ml) t-butyldimethylsilyl chloride (1.5 g; 9.9 mmol. 26 equiv.) was added and the mixture was taken to ca 90 °C under argon. At this point imidazole (1.35 g; 19.7 mmol. 52 equiv.) was added which caused the evolution of fumes that subsided as the reaction was cooled to 20 °C. The reaction was terminated after one day at room temperature by evaporation to dryness and brief elution through a short filter column on silica gel (Woelm G 111) to remove imidazole hydrochloride. The crude product was then separated by short column chromatography (in chloroform) to give two major product fractions. The least polar was the 2,6-O-persilylated derivative, (435 g; 42 %) followed by the material that had one silyl group less at C-2. (365 mg; 37 %); both were obtained as crystalline solids from $CHCl_3$/MeOH.

Derivatisation of the material with a hydroxyl group at C-2 was achieved by reaction with either tosyl chloride in chloroform-pyridine (1:0.5 v/v) or acetic anhydride in pyridine, and product isolation from a LH 60 column. This material could be selectively obtained from β-cyclodextrin by reaction with reduced amounts of the silylating reagent. Thus when reacted with only 14 equivalents of t-butyldimethylsilyl chloride the reaction gave in 38 % yield the cyclodextrin with 13 silyl residues, and only trace amounts of the fully silylated product.

LITERATURE CITED

1. Wife, R.L., Reed, D.E., Volger, H.C., Symposium on Cyclodextrins, Budapest, 1981. p.
2. Hunt, B.J., Rigby, W.: Chem. Ind. 1868, /1967/
3. Melton, L.D., Slessor, K.N.: Carbohydr. Res., 18 29, /1971/
4. Chao, Y., PhD Thesis, Columbia University, New York, 1973.
5. Matsui, Y., Okimoto, A., Bull. Chem. Soc. Jpn. 51 3030, 1978/

I.Int.Symp. on Cyclodextrins
Budapest, 1981 4.4.

CYCLODEXTRIN POLYMERS: TYPES AND SPECIFIC PROPERTIES

B. ZSADON,[1] É. FENYVESI[2]

[1] Department of Chemical Technology, Eötvös Loránd University, Múzeum krt. 6-8
H-1088 Budapest, Hungary
[2] CHINOIN Pharmaceutical-Chemical Works, Tó u. 1-5
H-1045 Budapest, Hungary

SUMMARY

By incorporating cyclodextrins (CDs) into polymer structures, the ability of CDs to form inclusion complexes can be usefully combined with specific polymer characteristics. There are many types of these so-called cyclodextrin polymers (CDPs). The most popular ones are produced by connecting CD molecules with suitable multifunctional agents e.g. epichlorhydrin or diepoxides. CDs have been polymerized also in the presence of poly (vinyl alcohol) using ethylene glycol bis (epoxypropyl) ether as the cross linking agent to give specific α-CD, β-CD and γ-CD copolymers in the form of differently swelling beads. These proved to be utilizable for separation purposes e.g. such as the stationary phases in chromatographic separations. Thus, a good separation of aromatic amino acids from one another and from the aliphatic amino acids could be achieved by inclusion chromatography on a column packed with β-CD-PVA copolymer. An other example is the inclusion chromatography of some indole alkaloids. Also the partial resolution of mixtures of optical antipode alkaloids has been observed.

INTRODUCTION

Up to the present, a large number of cyclodextrin derivatives - among them also polymer products - have been synthetized. In these polymers the inclusion complex forming ability of cyclodextrins is combined with specific properties derived from the polymer structure and this enables their practical uses first of all in different fields of separation technics. Probably the most important field is the so-called inclusion chromatography where these polymers can be used as stationary phases. Also in this case, the separation is mainly but not exclusively based on different stability of inclusion complexes reversibly formed with various compounds.

Several papers and patents concerning the preparation, properties and practical uses of these polymers have been published. These works have been in details reviewed recently by Saenger [1] and by us [2].

TYPES

Four types of polymer products containing cyclodextrin (CD) have been synthetized:

1. Most popular ones are the so-called cyclodextrin polymers which are produced by linking (or cross-linking) particular CDs, mixtures of CDs or mixtures of CDs and other carbohydrates with proper linking agents. The primary and secondary hydroxy groups on carbon 6, 2 and 3 are available as starting points of structural modifications without eliminating the central cavity of CD molecules. Bi-or polyfunctional compounds (for example epichlorhydrin, diizocyanates, diepoxides) can be used as linking agents.

By linking two or a few CD molecules, watersoluble derivatives can be produced, which form very good soluble and stable complexes.

On the other hand, by cross-linking CD molecules insoluble polymers can be prepared in form of resin, film, fibre or beads. These can be good swelled by water and form gel structures, which are permeable for aqueous solutions and make possible the complex formation between solutes and the polymer phase. As a consequence, these CD polymers (mainly in bead form) can be favourably used in separation technics.

2. Cyclodextrin-containing polymers of other type have been synthetized by polymerizing vinyl derivatives containing CDs (for example acrylic esters).

3. Cyclodextrins have been immobilized by binding those with covalent bonds on polymer supports, for example on

Sepharose, Bio Gel or cellulose. There are known examples for their use in the chromatography.

4. Polymer products containing CDs have been prepared also by incorporating them into synthetic polymer matrices by physical methods.

DISCUSSION OF EXPERIMENTAL RESULTS

For this once, I should like to deal with those of investigations going on in our laboratory, the goal of which was the preparation of bead-formed cyclodextrin polymers cross-linked by diepoxides and studying the possibilities of controlling their composition and characteristics.

We reacted cyclodextrins with ethyleneglycol-bis (epoxy-propyl) ether in the presence of polyvinylalcohol, in alcaline aqueous solution which was dispersed by continouos intensive stirring in water-immiscible indifferent solvent (for example in toluene) in order to get bead-formed product [3].

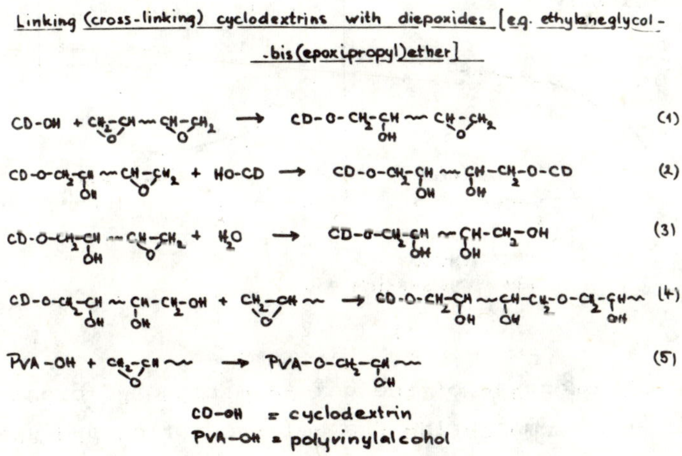

Figure 1.

On the Fig. 1. some basic reactions are demonstrated. One of the two epoxy groups of the linking agent can react, for example, with a hydroxy group of CD and forms a side chain through an ether bond, which holds an active epoxy group on its end. The latter can react with an other CD molecule and so two CD molecules will be bound through an epoxide bridge unit. Otherwise, the epoxy group can react with water, too, and this reaction yields diol groups on the end of the side-chain which either don't react further or react with a new epoxy group yielding longer binding unit or side chain. Epoxy groups can react also with hydroxy groups of polyvinylalcohol present and in this way polyvinylalcohol can be built in the cross-linked polymer structure.

Statistically informing data about the composition and structure of the products could be obtained by means of several chemical analyses and of measuring swelling characteristics [4].

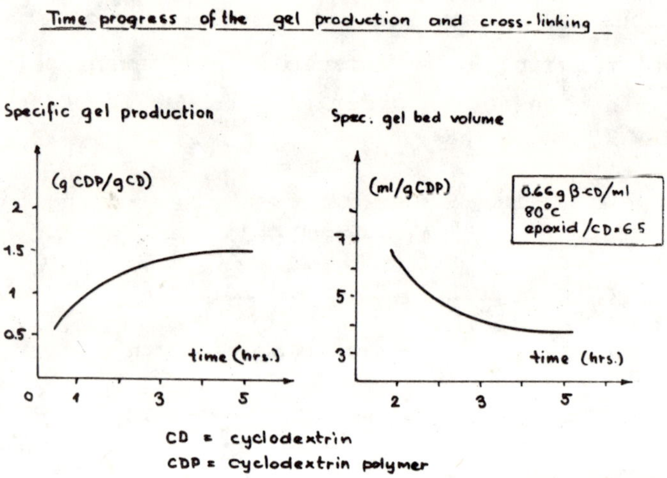

Figure 2.

The time-progress of the gel formation and cross-linking can be followed by measuring the gel production and swelling characteristics from time to time. As can be seen on the Fig. 2. the gel production could be detected only about half an hour after starting and the cross-linking arrived its maximum in three to four hours.

The swelling capacity of CD polymers, which has an

important role in practical uses, can be controlled by some reaction parameters peculiarly by the cyclodextrin concentration in the aqueous phase of reaction medium. The Fig. 3. shows the dependence of specific swelling characteristics, namely the gel bed volume and waterregain, within a series of our experiments.

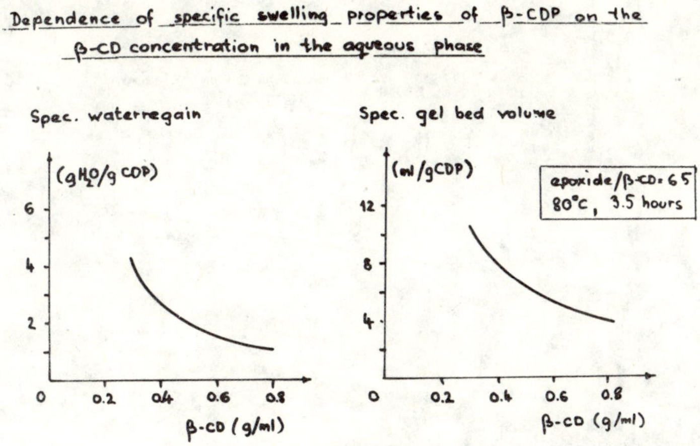

Figure 3.

The increase of swelling capacity was limited, namely no cross-linked polymer could be prepared in diluted solutions, e.g. in these experiments by CD-concentrations less than 0.3 g/ml.

Cross-linked polymers of much greater swelling capacity can be produced in diluted solutions in the presence of suitable polymer additives such as polyvinylalcohol. The Fig. 4. shows that swelling capacity of cyclodextrin-polyvinylalcohol copolymers somewhat decrease by increasing the polyvinylalcohol concentration. On the other hand, in the presence of this additive the CD-concentration can be considerably reduced and, as a consequence, cross-linked polymers can be produced even by a CD-concentration 0.14 g/ml, the specific gel bed volume of which can achieve 10-20 ml/g, too. In addition, also the elasticity and mechanical stability can be favourably controlled by building polyvinylalcohol into the polymer beads.

Dependence of the specific gel bed volume of bead-formed β-CDP on the PVA concentration at different β-CD concentrations (both in the aqueous phase of the reaction mixture)

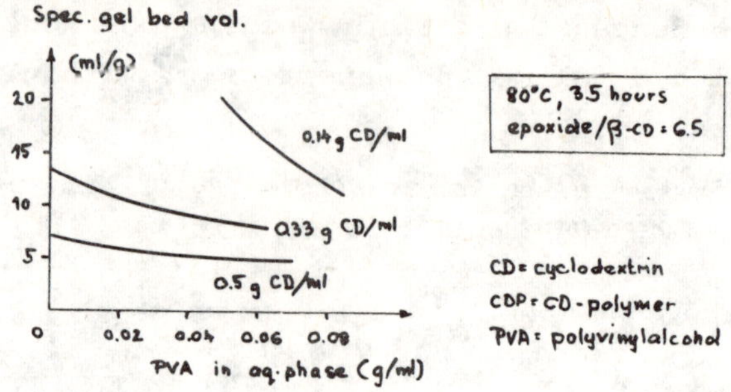

Figure 4.

Specific gel bed volume and cyclodextrin content of β-CDP plotted against the molar ratio of diepoxide and β-cyclodextrin

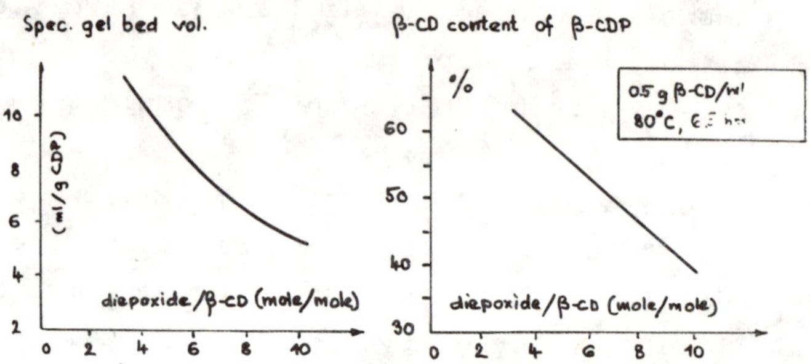

Figure 5.

The swelling capacity and, of course, the CD-content of the products can be controlled by the molar ratio of the cross-linking agent to CD in the reaction medium. As shown on the Fig. 5., in this way the CD-content could be altered within wide limits (40-70 %), while the swelling capacity altered a little, too. However, the reduction of the molar ratio and herewith the increase of the CD-content are limited: minimum three moles diepoxyde per one mole CD are necessary to get cross-linked polymer of proper mechanical stability.

For practical uses, the particle size and distribution are also important. They can be controlled, for example, by the size of the reaction vessel and the stirrer, by the speed of stirring, by the volume of the reaction mixture, by the ratio and viscosity of both phases and by the use and kind of any detergent [3].

On the basis of experiences outlined above, we developed a process for preparation of cross-linked CD-polymers in bead form [5]. By this process, polymers containing α-, β- or γ-CD can be reproducibly prepared. These polymers quickly swell in water, keep their form and elasticity in swollen state, are stable against heating and microorganisms and their main characteristics don't alter even after several drying and swelling procedures. They proved to be well utilizable for separation purposes especially as stationary phases in the chromatography. It is to be noted that over complex formation also secondary effects, for example gelpermeation and weak adsorption can appear here. In fortunate cases, these effects jointly increase the chromatographic resolution.

CHROMATOGRAPHIC BEHAVIOUR OF CD-POLYMERS

To demonstrate the chromatographic behaviour of our polymers, let me present the separation of natural amino acids. It is to noted that α-, β- and γ-CD-polymers were tried and the β-CD-polymer proved to be the most suitable for this purpose [4, 6].

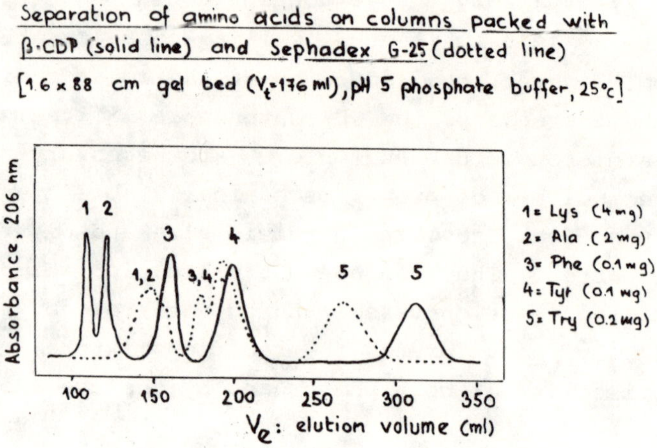

Figure 6.

The Fig. 6. shows the complete separation of five amino acids(those were lysine, alanine, phenylalanine, tyrosine and tryptophane) on a column packed with β-CD-polymer (the solid line). In addition, further fifteen non-aromatic natural amino acids were chromatographed, the peaks of which all appeared between or together with the peaks of alanine and lysine, therefore these couldn't be well separated from each other.

In comparison, the former five amino acids were chromatographed under identical conditions also on an other carbohydrate-type stationary phase; that is Sephadex G-25 (the dotted line). As can be seen, the sequence of elution was the same on both columns, but the resolution was significantly greater on β-CD-polymer than on Sephadex, which latter binds these amino acids only by weak adsorptive strengths.

Probably, for the most perspective field of practical uses of CD-polymers can be considered their use in the separation of a wide variety of structural, positional and optical isomeric compounds.

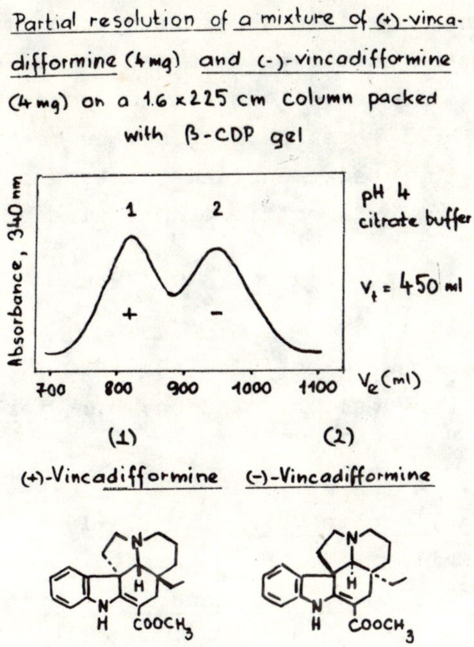

Figure 7.

For example, several enantiomers were observed to have different retentions on stationary phases containing optically active cyclodextrin and this provided the possibility of separating optical antipodes by chromatography. The Fig. 7. shows an example: the partial resolution of indole alkaloid antipodes (+) and (-) vincadifformine on bead-formed β-CD-polymer gel bed [8].

PERCPECTIVE

Scientific publications and also our experimental results allow to forecast the perspective of CD-polymers mainly in solving specific separation problems. Research works in the future should be directed toward profoundly studying the basic interactions between potential „guest" compounds and CD-polymers, developing the methods of preparation, enlarging the assortment and the use not only for analytical but also for specific preparative purposes.

REFERENCES

1. Saenger, W.: Angew. Chem. 92, 343 (1980).
2. Szejtli, J., Fenyvesi, É., Zsadon, B.: Stärke, 30, 127 (1978).
3. Fenyvesi, É., Zsadon, B., Szejtli, J., Tüdős, F.: Ann. Univ. Budapest, Sectio Chim. 15, 13 (1979).
4. Zsadon, B., Szilasi, M., H. Otta, K., Tüdős, F., Fenyvesi, É., Szejtli, J.: Acta Chim. Acad. Sci. Hung. 100, 265 (1979).
5. Szejtli, J., Fenyvesi, É., Zoltán, S., Zsadon, B., Tüdős, F.: Hung. Appl. CI-1845 (1978); Hung. Patent 177,419 (1981); Belg. Patent 877,653 (1980); US Patent 4,274,985 (1981); Ger. Offen 2,927,733 (1980).
6. Zsadon, B., Szilasi, M., Tüdős, F., Fenyvesi, É., Szejtli, J.: Stärke 31, 11 (1979).
7. Zsadon, B., Szilasi, M., Tüdős, F., Szejtli, J.: J. Chromatogr. 208, 109 (1981).

I.Int.Symp. on Cyclodextrins 4.5.
Budapest, 1981

PRELIMINARY STUDIES ON TABLET DISINTEGRATION BY SORPTION CALORIMETRY

E. KŐRÖS,[1] I. DÉKÁNY,[2] É. FENYVESI[3]

[1] Institute of Inorganic and Analytical Chemistry,
Eötvös Loránd University, Múzeum krt. 4/B
H-1088 Budapest, Hungary
[2] Institute for Colloid Chemistry, József Attila University
H-6700 Szeged, Hungary
[3] CHINOIN Pharmaceutical-Chemical Works, Tó u. 1-5
H-1045 Budapest, Hungary

SUMMARY

The heat of sorption of some materials - including cyclodextrin derivatives - used also as tablet disintegrators and of artificial tablets was measured in an LKB sorption microcalorimeter. A correlation between the disintegration rate and the heat of sorption is sought.

INTRODUCTION

Some cyclodextrin derivatives seem to have rather advantageous tablet disintegrating properties /1/ and it is likely that in the not-too-distant future they will be mixed in tablets to enhance the rate of disintegration. In order to qualify these cyclodextrin derivatives and other materials widely used at present as disintegrators, we have decided to reveal whether there exists any correlation between the disintegration rate of the tablets and their heat of sorption. Here we report on our first results.

EXPERIMENTAL

Measurement of the heat of sorption

The heat of sorption was measured in an LKB sorption microcalorimeter. The calorimeter is provided with a two-part reaction vessel where the chemical reaction takes place. Each part of the reaction vessel is in thermal contact with a thermoelement couple and with a heat-absorbing unit. Whenever, due to some kind of chemical interaction heat is evolved or absorbed, the heat flow induces a temperature difference on the thermocolumns, which are located between the reaction vessel and the heat absorber, and an electromotive force proportional to the temperature difference is produced. The output voltage of the thermocolumn is amplified and fed e.g. into a recorder. The calibration of the microcalorimeter /see below/ can be performed with the aid of a built-in heater.

Our liquid /water/-solid sorption experiments were performed as follows. About 100 mg accurately weighed sample was placed into the reaction vessel and kept at 90 ^{o}C for 24 hours in vacuum /at 1.3 pascal $\approx 10^{-2}$ torr/. After cooling, the wetting liquid /0.5 cm^3 water/ was pipetted into the upper part of the reaction vessel. Then approximately 0.7 cm^3 silicone oil was introduced into the vessel-holding block and the sorption vessel was placed into the calorimeter. Time /2-3 hours/ was allowed the system to reach thermal equilibrium, the set-in of which was indicated by the stability of the base line. Then the wetting liquid was allowed to contact the sample and the heat change accompanying the sorption processes was recorded. After the trace line returned to the base line electric calibration was performed. In the knowledge of the weight of the sample using the calibration value the heat of sorption referred to 1 g sample was calculated.

In the case of the sorption experiments of tabletted samples first tablets of the composition 94.0% AVICEL pH 102, 1.5% talcum, 2.5% magnesium stearate and 2.5% disintegrator were prepared by thorough mixing and applying a pressure of 9.8×10^5 kPa /10^4 kp/cm^2/. The tablets were of 75-79 mg in weight. Otherwise the procedure was the same as described above. The time of heat effect was 8-10 minutes with all samples.

A typical curve is shown on Fig. 1.

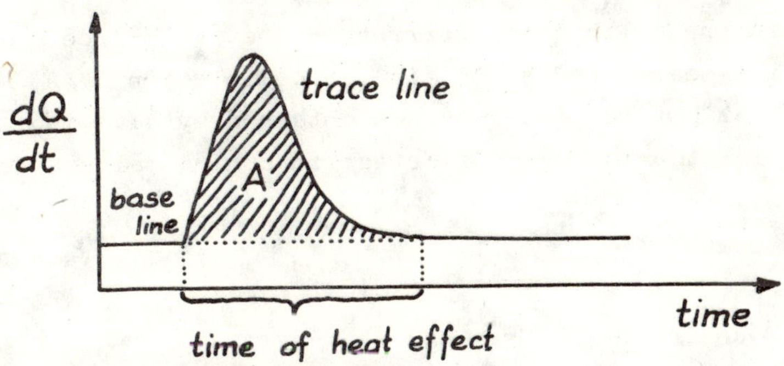

Fig. 1.

Calibration

The total heat evolved, Q, during the chemical interaction is directly proportional to the area under the trace line:

$$Q = \alpha A$$

where α is the calibration constant.

The heat change per unit time /dQ/dt/ is proportional to the deflection of the recorder, D:

$$\frac{dD}{dt} = \alpha D$$

Integrating the equation between t_1 /the time of initiation of the reaction/ and t_2 /the termination of heat evolution/ we obtain

$$Q = \alpha \int_{t_1}^{t_2} D dt = \alpha A$$

where A is the area under the trace line.

The calibration was performed electrically. The calibration heat, Q_{cal}, was calculated from the following equation:

$$Q_{cal} = I^2 Rt \text{ joule}$$

where I is the calibration current, R the resistance of the calibrating heater and t the time of calibration.

The calibration constant was obtained by dividing Q_{cal} with the integrated recorder defection during calibration, A_{cal}:

$$\alpha = \frac{Q_{cal}}{A_{cal}} \text{ joule/unit area}$$

If high precision is required, and this is the requirement in our investigations, the value of the calibration heat should be as close as possible to that of the heat evolved in the experiment. Thus it is advisable to make the calibration after performing the experiments. During calibration a known electric current was allowed to pass through the system for 50 seconds.

RESULTS AND DISCUSSION

The heat of sorption in water of different materials used either as ingredients /I/ or as disintegrators /D/ is given in Table 1.

Table 1.

	Sample	Heat of sorption joule/g
I	Microcrystalline cellulose /AVICEL pH 101/	50.6
I	Microcrystalline cellulose /AVICEL pH 102/	53.4
D	Cross-linked carboxymethyl cellulose /Ac-di-sol/	110.8
D	Formaldehyde casein /Esma Spreng/	27.6
D	Corn-starch /Amylum majdis/	118.5
D	Polyvinylpyrrolidone /Polyplasdone XL/	160.5
D	Cyclodextrin derivative A	70.1
D	Cyclodextrin derivative B	72.3
D	Cyclodextrin derivative C	68.2

The heat of sorption in water of tablets containing different disintegrators and the disintegration time are given in Table 2.

Table 2.

Disintegrator in the tablet	Heat of sorption joule/g	Δ joule/g	Disintegration time, min.
Corn-starch /Amylum majdis/	53.6	3.4	15
Formaldehyde casein /Esma Spreng/	54.5	4.3	10
Polyvinylpyrrolidone /Polyplasdone XL/	56.3	6.1	5.5
Cross-linked carboxymethyl cellulose /Ac-di-sol/	56.3	6.1	3.0

Table 2. /contnd./

Disintegrator in the tablet	Heat of sorption joule/g	Δ joule/g	Disintegration time, min.
Cyclodextrin derivative A	55.6	5.4	5.5
Cyclodextrin derivative B	55.0	4.8	5.5
Cyclodextrin derivative C	54.7	4.5	6.5

Δ = /heat of sorption of the tablet/ -53.4x0.94

The heat of sorption of talcum and magnesium stearate, respectively, in water is negligible.

Disintegration time was measured by the USP-XIX standard method.

The heat of sorption in the 50-100 joule/g range can be measured with an accuracy of ±0.1 joule.

Our results, compiled in Tables 1 and 2, indicate that in the system studied so far there seems to be a correlation between the heat of sorption and the disintegration time. The lower is the Δ value /which is the heat of sorption of the disintegrator in 1 g tablet/ the slower is the disintegration of the tablet. However, in order to establish whether such correlation - between kinetic and thermodynamic data - exists at all large number of experiments should be performed and many factors so far disregarded /e.g. the influence of pressure/ should be considered and looked at.

It is also worth mentioning that the heat of sorption of a disintegrator does not run parallel with the Δ value. Compare e.g. the values for corn-starch and formaldehyde casein in Tables 1 and 2. The heat of sorption of a disintegrator measured alone and in unpressed state does not provide us with any reliable information regarding its disintegrating efficiency. The high pressure applied at the tablet preparation may considerably alter the sorption properties.

Further experiments are in progress and will be reported in due course.

ACKNOWLEDGEMENT

The authors are indebted to Prof. J. Szejtli /Chinoin Chemical and Pharmaceutical Works, Biochemical Research Laboratory/ for many valuable discussions.

REFERENCE

1. Szejtli, J., Fenyvesi, É., Dósa, É., Antal B., Wágner, P., Kállói, K.: Hung. Patent application 2586/81.

I.Int.Symp. on Cyclodextrins
Budapest, 1981 4.6.

WATER-SOLUBLE CYCLODEXTRIN POLYMERS AND THEIR COMPLEXING PROPERTIES

É. FENYVESI,[1] M. SZILASI,[2] B. ZSADON,[2] J. SZEJTLI,[1] F. TÜDŐS[2]

[1] CHINOIN Pharmaceutical-Chemical Works, Tó u. 1-5
H-1045 Budapest, Hungary

[2] Department of Chemical Technology, Eötvös Loránd University
Múzeum krt. 4/B
H-1088 Budapest, Hungary

SUMMARY

Cyclodextrin polymers extremely soluble in water were prepared. Influence of the reaction parameters on the overage molecular weight was investigated. Inclusion complexes of these type of cyclodextrin polymers with appropriate guest molecules which are also well soluble in water were studied and their stability constants were calculated. The increase in solubility of poorly soluble substances in aqueous solutions of cyclodextrin polymers seems to be important from practical point of view /enhancement of reaction rate, selectivity, separations, etc./.

Introduction

The most important property of cyclodextrins is their ability to form inclusion complexes with suitable "guest" molecules (1). In many cases, the practical utilization of this effect is unfavourably limited by the low solubility of cyclodextrins, especially of β-cyclodextrin. By appropriate substituting or crosslinking agents the poorly soluble β-cyclodextrin can be transformed into cyclodextrin derivatives or moderately high molecular weight cyclodextrin polymers which are characterized with high solubility in water also in presence of complex forming agents. An advisable way for preparing this sort of polymers is the crosslinking of cyclodextrin molecules with epichlorohydrin under proper conditions.

Though epichlorohydrin is not a diepoxide it reacts as if it were. It is capable to form ethers reacting with one of the alcoholic hydroxyl groups of cyclodextrin in alkaline aqueous solution. Through the epoxy group obtained in consequence of the cleavage of hydrochloric acid the deriving side chain can react further, coupling cyclodextrin rings together. Moreover, also side reactions can occur of which the most important is the hydrolysis of epoxy groups. With the hydroxyl groups formed another epichlorohydrin /or an epoxy group of a side chain coupled to a cyclodextrin/ can react. Consequently, the side-chains and the glyceryl bridges connecting the rings can be of different length(2). Thus, the cyclodextrin molecules are linked according to the following reaction scheme:

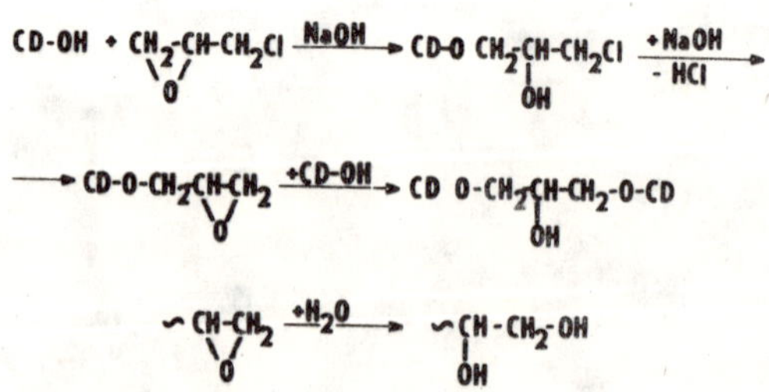

where CD-OH represents the cyclodextrin molecule.

Low degree polymerization results in water-soluble polymers (4) while a more extensive polymerization leads to a structure of three-dimensional network (3,4,5) i.e. to a water-insoluble polymer.

In the present paper our results are summarized in preparation of water-soluble β-and γ-cyclodextrin polymers and their complex forming ability is illustrated by some examples.

MATERIALS and METHODS

The chromatographically pure β- and γ-cyclodextrin /used for the preparation of polymers/ and Indomethacin /1-p-chlorobenzoyl-2-methyl-5-methoxy-indole-3-yl-acetic acid/ were the products of Chinoin Pharm.-Chem. Works, Budapest. Cholic acid was a product of Koch-Light Laboratories. The other guest substances are marketed by Reanal Fine Chemical Works, Budapest.

Preparation of water-soluble cyclodextrin polymers:

NaOH solution was weighted into a three-necked flask provided with a reflux condenser and a stirrer, and cyclodextrin was dissolved in it. Epichlorohydrin was added slowly dropwise at 40-80 °C under stirring, then the mixture was cooled, neutralized and desalted by dialysis or ion-exchange. The product was obtained by freeze drying as white powder.

Chromatographic analysis:

Molecular weight distribution of our products was characterized by gel chromatography of Ultrogel ACA 54 and 34 on polyacrylamide gels. The chromatographic equipment consisted of a K 16/100 Pharmacia column, a peristaltic pump, a Thorn NPL Automatic Digital Polarimeter and an LKB Flat Bed Recorder.

Narrow fractions of the polymer were separated, and their molecular weights were determined by HP 302 Vapour Pressure Osmometer. The chromatograms were then calibrated on the basis of the relationship between the relative elution volumes (V_e/V_t) and molecular weights /Fig.1./.

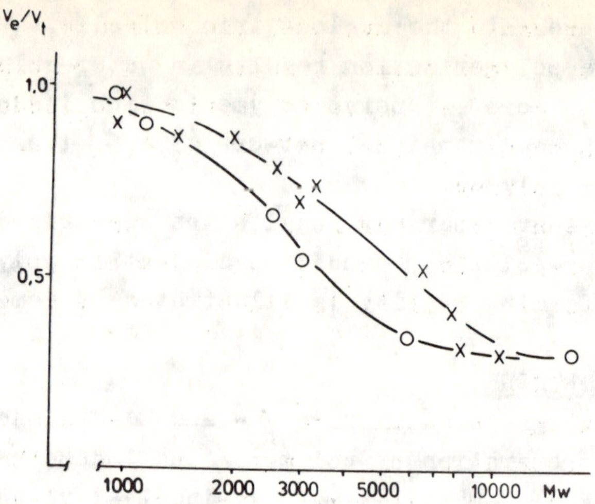

Fig. 1. Dependence of the relative elution volume (V_e/V_t) on the molecular weight. Ultrogel ACA 34(x), ACA 54(o).

The weight-average molecular weights of different cyclodextrin polymers could be better calculated from the chromatograms obtained on ACA 34 gel because this type of gel gives good resolution in wider range of molecular weight. Otherwise the chromatograms obtained on ACA 54 gels are more illustrative because one peak represents the derivatives consisting of only one substituted cyclodextrin ring and the other peak belongs to the polymers obtained by connenting two or more cyclodextrin rings /Fig. 2./.

Determination of increase in solubility:

Solid or liquid guest molecules, poorly soluble in water were added in excess to the aqueous solution of the cyclodextrin polymer, and after standing and occasional shaking for an hour the aqueous phase was separated from the undissolved rest. The concentration of the dissolved substance was determined by UV photometry /no substantial shiff of UV maxima were observed/ or in the case of cholic acid the amount of the undissolved material was weighted /since it has no usable UV absorption band/.

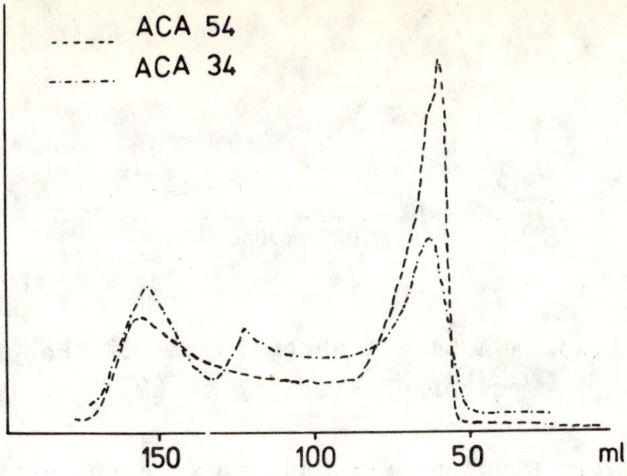

Fig. 2. ACA 54 and ACA 34 chromatogram of the same cyclodextrin polymer ($\bar{M}_w$ = 4600)

Determination of the stability constants of complexes:

Different amounts of cyclodextrin or polymer was added to 2×10^{-5} M/l solution of congo red /diphenyl-bis-azo-naphtionic acid/ in pH = 7,5 tris buffer, and the absorbance of the solution was measured at 520 nm. The stability constants were calculated by the Benesi-Hildebrand equation (6) assuming the formation of 1:1 complex.

RESULTS

Preparation of cyclodextrin polymers of different molecular weight:

Reproducing the method known from literature (4) for preparation of water-soluble cyclodextrin polymer and characterizing the obtained product by gel-chromatography on ACA 54 polyacrylamide gel /Fig. 3./ it proved to be a heterogeneously substituted cyclodextrin. The average molecular weight calculated on the basis of elution diagram is not more than $\bar{M}_w$ = 1800. Molecular weight of cyclodextrin is 1134. This fact suggests that mainly different kinds of cyclodextrin-glyceryl-ethers could be obtained this way.

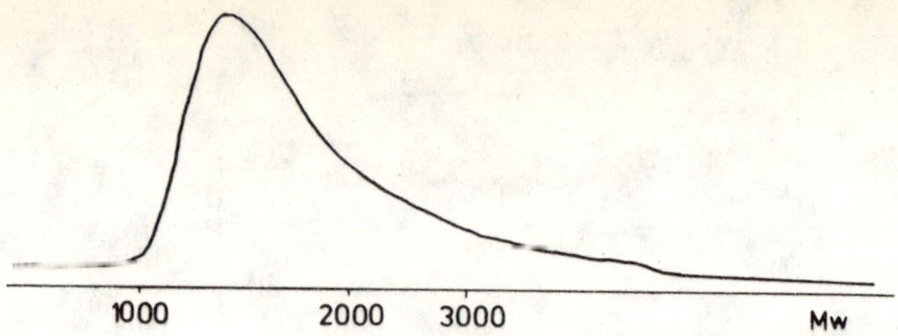

Fig.3. ACA 54 gel chromatogram of the product prepared according to (6) $\bar{M}_w$ = 1800.

Systematic studies were made to find the appropriate conditions for the coupling of cyclodextrin molecules, for the preparation of cyclodextrin polymers built up from 2 or more cyclodextrin rings per molecule.

By varying the reaction parameters it was observed that a product of higher molecular weight can be obtained when the undesired hydrolysis of epichlorohydrin is repressed. For this purpose a relatively high cyclodextrin concentration and a great excess of the crosslinking agent is needed, moreover this latter has to be added slowly, dropwise to the reaction mixture.

Figure 4. shows the chromatograms on ACA 54 gel of three products prepared from solutions of different cyclodextrin concentration with the same molar ratio of epychlorohydrin to cyclodextrin /10:1/. In diluted solutions /20 g/100 ml cyclodextrin/ mainly simple derivatives of average molecular weight $\bar{M}_w$ = 1500 were formed /Fig.4/a./. More concentrated solutions increase the ratio of those products which are excluded from the gel matrix /eluted near the void volume V_o/ because of their high molecular weight. The products obtained in the solutions containing initially 40 g/100 ml and 42 g/100 ml cyclodextrin have the average molecular weight

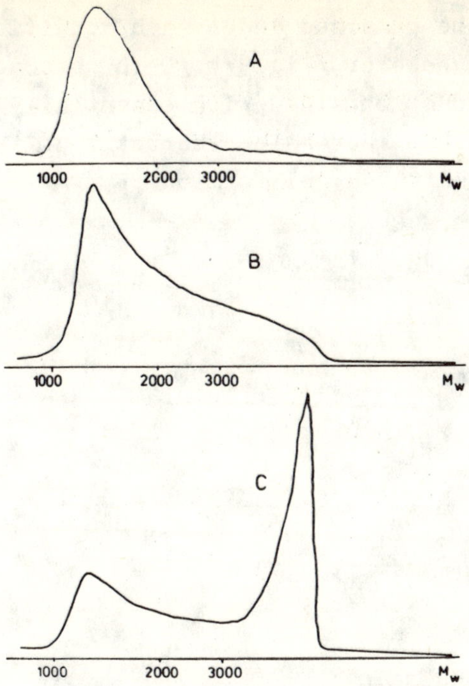

Fig.4. ACA 54 gel chromatograms of the products prepared with the initial β-cyclodextrin concentration of 20 g/100 ml (A), 40 g/100 ml (B) and 42 g/100 ml (C). Average molecular weights are 1500, 2500, 5600 respectively.

of $\bar{M}_w$ = 2500 /Fig.4/b./ and $\bar{M}_w$ = 5600 /Fig.4/c./ respectively. Increasing further the initial cyclodextrin concentration the product obtained cannot be dissloved in water, it swells only.

Similar relationship was found with varying the ratio of epichlorohydrin to cyclodextrin at constant cyclodextrin concentration /40 g/100 ml/.

According to the chromatogram /Fig.5/a./ of the product prepared with 2:1 molar ratio of epichlorohydrin to cyclodextrin, predominantly derivatives of low molecular weight

$/\bar{M}_w = 1400/$ could be obtained under such conditions. Applying 20:1 molar ratio, however /Fig. 5/b./, the average molecular weight of the product obtained with essentially higher yield was $\bar{M}_w = 5300$. Further increasing the ratio of the cross-linking agent resulted in gel forming.

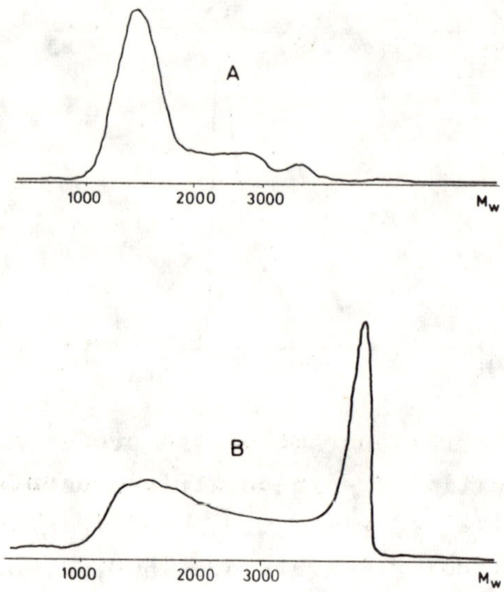

Fig. 5. ACA 54 gel chromatograms of the products prepared with 2:1 (A) and 20:1 (B) molar ratio of epichlorohydrin to β-cyclodextrin. Average molecular weights are 1400 and 5300 respectively.

Water-soluble polymers prepared by the above reaction and used to further investigations were white, powderlike materials. They had wide molecular weight distribution /from 1200 up to about 10000/, their average molecular weight ranged between 4-6000. The cyclodextrin content calculated from reducing power following acid hydrolysis was about 50 %. They were very well soluble in water already at room temperature and even 40-60 g/100 g solutions are of not high viscosity. In such solutions the concentration of bound cyclodextrin is about 160-360 g/l. It should be noticed that the solubility of β-cyclodextrin is not more than 18 g/l in water at room temperature (7).

The solubility increasing effect

The solubility of cyclodextrin complexes are usually around of 0,1 g/100 ml, therefore they crystallize readily from aqueous solutions. The complexes of water-soluble cyclodextrin polymers, however, are well soluble in water. This makes difficult to obtain them in solid state, but this is an important advantage when poorly soluble molecules has to be dissolved.

Some solubility data determined in water and in 10 w % aqueous solution of β-cyclodextrin polymer are illustrated in the following table.

	Solubility /g/1000 ml/		
Substance	in distilled water S_w	in 10 w % β-cyclodextrin polymer solution S_{CDP}	$\frac{S_{CDP}}{S_w}$
Benzene	0,7	4	5,7
Benzoic acid	4,0	13	3,2
Cholic acid	0,3	16	53,3
Indomethacin^{x}	0,9	3	3,3
Anethole^{xx}	0,7	2	2,8

^{x}determined in pH = 8,5 tris buffer
^{xx}determined in pH = 7,6 phosphate buffer

It is to be mentioned that even higher solubility can be achieved in more concentrated solutions. For instance, in 1000 ml of 40 w % solution of β-cyclodextrin polymer 14 g benzene /$S_w/S_{CDP} = 20$/ or 50 g benzoic acid /$S_w/S_{CDP} = 12,5$/ can be dissolved.

Stability of complexes

The high increase in solubility observed in many cases is probably due to complex formation of the cyclodextrin polymer and the guest molecule. To investigate the stability of complexes congo red was used as a model. The spectrum of congo red changes in consequence of complex formation: an increase in absorbance can be observed (8).

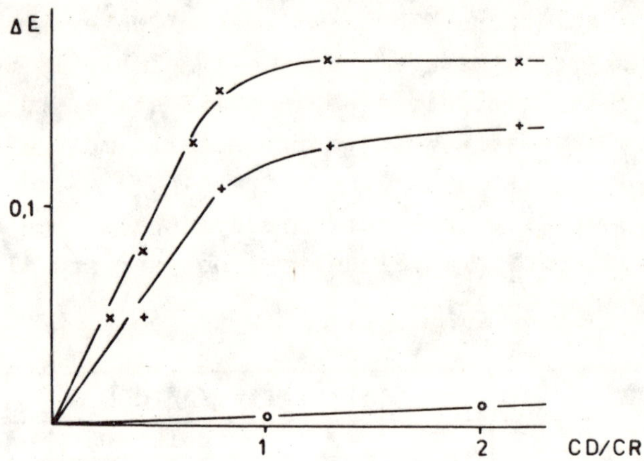

Fig. 6. Increase in absorption of congo red /CR/ solution at 520 nm in the presence of β-cyclodextrin /o/ and the fractions of β-cyclodextrin polymer having low /+/ and high /x/ molecular weight /$\bar{M}_W$ = 1500 and 8300 respectively./

Figure 6. shows the change in extinction of congo red solution as a function of molar ratio of cyclodextrin to congo red. The increase in absorbance was measured in presence of β-cyclodextrin and two types of its derivatives prepared in crosslinking reaction with epichlorohydrin. These derivatives were obtained by fractionating the product characterized by Fig. 4/c.

The curves, referring to the two fractions /Fig.6./ /i.e. to the cyclodextrin-glycerylethers of $\overline{M}_w$ = 1500 average molecular weight and to the compounds built up from 4-5 rings per molecule of $\overline{M}_w$ = 8300 average molecular weight/ were calculated on the basis of their cyclodextrin content.

The complex stability constant for β-cyclodextrin was $1{,}8 \times 10^3$ l/M, for the low molecular fraction $1{,}2 \times 10^5$ l/M, and even higher for the high molecular fraction: $3{,}8 \times 10^5$ l/M.

There seems to be a correlation between the enhancement of solubility and the complex stability. This is born out also by the fact that the well soluble γ-cyclodextrin /232 g/l; (8) / forms a complex with congo red /stability constant is $6{,}7 \times 10^4$ l/M/ which is more than forty times more stable than a similar complex with β-cyclodextrin; and it is only doubled /to $1{,}3 \times 10^5$ l/M/ when γ-cyclodextrin polymer /$\overline{M}_w$ = 5200, Fig.7./ is studied.

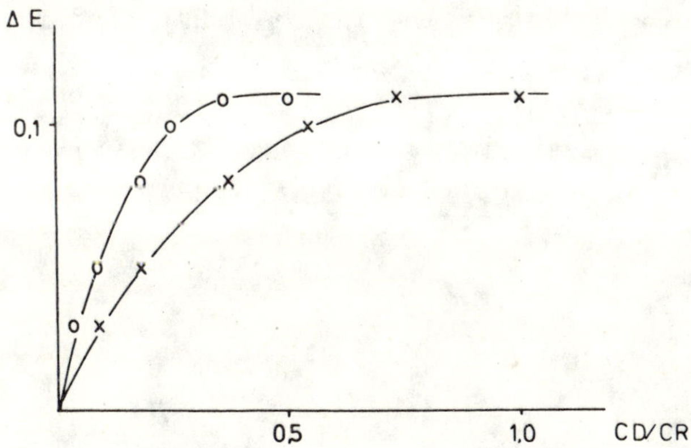

Fig. 7. Increase in absorption of congo red solution at 520 nm in the presence of γ-cyclodextrin /x/ and γ-cyclodextrin polymer /o/ /$\overline{M}_w$ = 5200/

In crosslinking reaction with epichlorohydrin cyclodextrin polymers were prepared which are extremely soluble in water, are characterized with well reproducible molecular weight distribution and with the ability to form water-soluble inclusion complexes with a number of appropriate guest molecules (9). With respect to practical use the increase in solubility achieved by complex formation seems to be important.

Literature Cited

1. Saenger, W.: Angew.Chem. 92 343 (1980)
2. Wiedenhof, N.: Stärke 21 163 (1969)
3. Nestlé Societé des Produits Neth Appl. 6, 505,361 (1965)
4. Wiedenhof, N., Van Panthaleon Van Eck, C.L., Lammers, J.: N.J.J. Brit 1,244,990 (1971)
5. Szejtli, J., Fenyvesi, É., Zoltán, S., Zsadon, B., Tüdős, F.: Hung.Pat. 177-419
6. Benesi, H.A., Hildebrand, J.H.: J.Am.Chem.Soc. 71, 2703 (1949)
7. French, D., Levine, M.L., Pazur, J.H., Norberg, E.: J.Am. Chem.Soc. 71,353 (1949)
8. Sensse, K., Cramer, F.: Chem. Ber. 102,509 (1969)
9. Szejtli, J., Zsadon, B., Fenyvesi, É., Szilasi, M., Tüdős, F.: Hung.Pat. CI-1960 (1980)

I.Int.Symp. on Cyclodextrins
Budapest, 1981

CYCLODEXTRIN POLYMERS AS SPECIFIC SORBENTS

K. OTTA,[1] É. FENYVESI,[2] B. ZSADON,[1] J. SZEJTLI,[2] F. TÜDŐS[1]

[1] Department of Chemical Technology, Eötvös Loránd University
Múzeum krt. 4/B
H-1088 Budapest, Hungary
[2] CHINOIN Pharmaceutical-Chemical Works, Tó u. 1-5
H-1045 Budapest, Hungary

SUMMARY

Cyclodextrin polymers which are insoluble in water or other solvents but have good swelling capacity can be prepared, for instance, by linking together the cyclodextrin rings with epichlorohydrin or diepoxy compounds in alkaline aqueous medium, or coupling the cyclodextrin to some kind of polymer support, for instance, to cellulose.

Cyclodextrin polymers bind compounds from aqueous solutions or vapor phase via inclusion complex formation. This selective sorption property makes possible to use these types of polymers e.g. for cleaning industrial waste waters.

INTRODUCTION

Cyclodextrins (CD) bind compounds from solutions, gas or vapor phase via inclusion complex formation. In the case of cyclodextrin polymers this ability of cyclodextrins is combined with the characteristic properties of the polymeric structure. Several patents have been published in connection with the production of cyclodextrin polymers [1., 2., 3., 4.]. For some purposes, for instance, for chromatography cyclodextrin polymers in bead form can be used advantageously which are insoluble in water or other solvents, but have good swelling capacity /3-20 ml/g in water). The characterization and chromatographic behaviour of these have been reported by Zsadon et. al. [5].

A method was developed also for immobilization of cyclodextrins on cellulose [6]. Because of their higher specific surface area deriving from the fibrous structure of the cyclodextrin-cellulose (CD-cell) are sorbents of higher capacity than the cyclodextrin polymers prepared in crosslinking reaction. In this case the cyclodextrin molecules are linked to the surface of the preparate and this way they are in easy to reach for the guest molecules. In this way the complex forming ability of the cyclodextrin are not modified. Therefore, the cyclodextrin-cellulose could be advantageously used for the solving of special purification, isolation problems.

MATERIALS AND METHODS

Pure α-CD, β-CD, γ-CD were immobilized on cellulose, regenerated cellulose or a fibrous material with cellulose content. Cyclodextrins were prepared by the CHINOIN (Hungary)

Substances used as linking agent: epichlorohydrin (REANAL), 1,2-bis-(2,3-epoxi-propoxi)-ethan (MERCK).

The linking reaction was performed in two steps , at 60°C. The products were washed with water to neutral pH and dried at 105°C. The cyclodextrin content of our white, fibrous products were measured after acidic hydrolysis with the determination of the liberated reducing sugars [5].

RESULTS

The cyclodextrins were linked to the cellulose by epichlorohydrin or diepoxy compounds in two steps. First the epichlorohydrin was linked to cellulose in weak alkaline solution (e.g. 0,2-3% w/v NaOH):

$$\text{Cell-OH} + \underbrace{CH_2\text{-}CH}_{O}\text{-}CH_2\text{-Cl} \rightarrow \text{Cell-O-}CH_2\text{-}\underset{\displaystyle OH}{CH}\text{-}CH_2\text{-Cl}$$

Usable cellulose-glucoseunit/ epichlorohydrin molar ratio was 1:1o-1:55. In the second step epoxy-groups were formed on the sidechains in more concentrated alkaline solution (e.g. 40 % w/v NaOH) and linked to the cyclodextrin molecules according to the following equations:

$$\text{Cell-O-CH}_2\text{-}\underset{\text{OH}}{\underset{|}{\text{CH}}}\text{-CH}_2\text{-Cl} \xrightarrow[\text{- HCl}]{\text{(Na OH)}} \text{Cell-O-CH}_2\text{-}\underset{\text{O}}{\underset{\backslash\ /}{\text{CH-CH}_2}}$$

$$\text{Cell-O-CH}_2\text{-}\underset{\text{O}}{\underset{\backslash\ /}{\text{CH-CH}_2}} + \text{HO-CD} \longrightarrow \text{Cell-O-CH}_2\text{-}\underset{\text{OH}}{\underset{|}{\text{CH}}}\text{-CH}_2\text{-O-CD}$$

Under these circumstances the epoxyde group reacts with water too resulting vicinal diol endgroups which also can react with epichlorohydrin and this way the cellulose and cyclodextrins can be linked by longer chains.

The complex formation ability of our CD-cell products was also investigated using benzene, phenol and p-cresol as model guest compounds. The quantities of substances retained from diluted aqueous solutions on columns packed with immobilized cyclodextrin on cellulose (mol/mol CD) are summarized in Table I.

Table I.

Substance conc.of solution	DDVP* (o,o5%)	Benzene (o,o5%)	Phenol (o,1%)	p-cresol (o,1 %)
α-CD-cell (CD cont:4o %	o,o5	o,15	o,21	o,16
β-CD-cell (CD cont:21 %)	o,o8	o,15	o,25	o,26
γ-CD-cell (CD cont:26 %)	o,o8	o,19	o,18	o,4o

*DDVP = dichlorovinyl-dimethyl-phosphate

These data show that the sorption is much more greater in the case of aromatics of good complex forming ability. In these experiment the capacity of our sorbents does not reach the maximum value because the quantity of the retained compound depends on the concentration of its aqueous solution. Fig. 1. shows the dependence of the retained phenol /β-CD molar ratio on the concentration of the aqueous phenol solution using immobilized β-cyclodextrin on cellulose as sorbent.

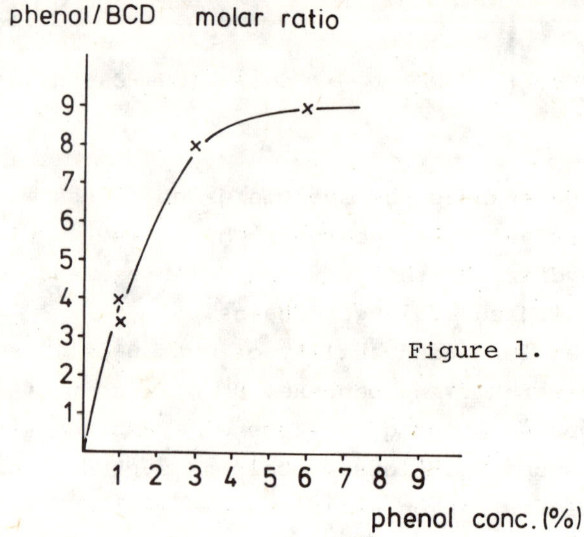

Figure 1.

This relationship was investigated for aqueous p-cresol solutions too. (Fig. 2.)

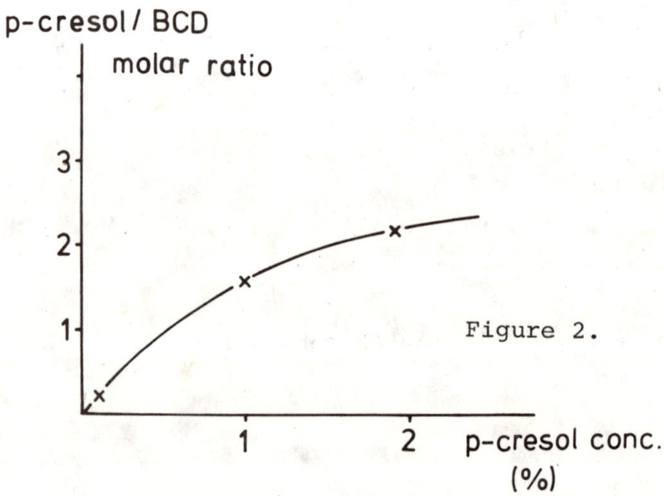

Figure 2.

The higher molar ratio values at more concentrated solutions suggest that inclusion complex formation and adsorptive effects are combined in the binding mechanism.

In the last Fig. (Fig. 3.) the cleaning of an industrial waste water (phenol content 0,12 %, measured by Koppeschaar) is shown. For column packing 5 g β-CD-cellulose (CD content 32 %) was used. The retained phenol/β-CD molar ratio was 0,43 mol/1 mol β-CD.

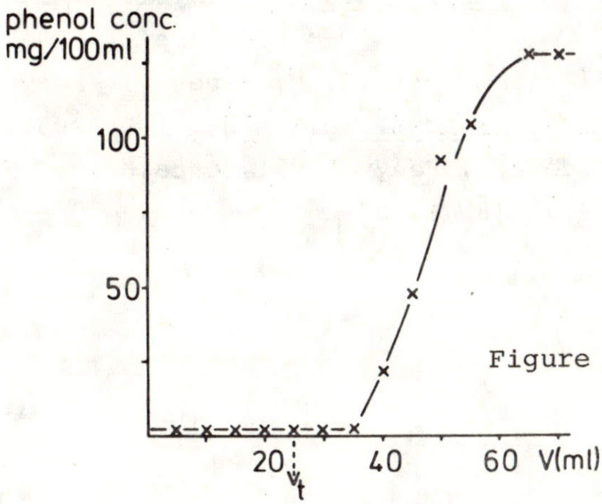

Figure 3.

During the investigation of cyclodextrins immobilized on cellulose we found that in these products the favourable properties of cellulose (adsorption, wettability, good swelling capacity) can be combined with the complex forming ability of cyclodextrins. These products are usable for the isolation and concentration of several compounds in separation and purification processes. This can be important especially from the view point of environmental protection. Further investigations are going on in our laboratory for refining the preparation these products and using them for solving different separation problems.

REFERENCES:

1. Wiedenhof, N., van Panthaleon, C., L., van Eck, B. and Lammers J., N., J., J. US Patent 1244990

2. Buckler, Sh., A., Martel, R., F., Moshy, R., J. US Patent 3,472,835

3. Société de Produits Nestlé S.A. Swiss Patent No 65o5361

4. Szejtli, J., Fenyvesi, É., Zoltán, S., Zsadon, B., Tüdős, F. Hung. Patent 177419(1981) Hung. Appl. (1978)

5. Zsadon, B., Szilasi, M., Otta, K., H, Tüdős, F., Fenyvesi, É., Szejtli, J. Acta Chim. Acad. Sci. Hung. 100. 265 (1979)

6. Szejtli, J., Zsadon, B., Fenyvesi, É., Otta, K., H, Tüdős, F. Hung. Appl. CI-1961/80 (1980)

I.Int.Symp. on Cyclodextrins
Budapest, 1981

CYCLODEXTRIN-CROWN ETHER COMBINATION

L. TŐKE,[1] B. ÁGAI,[1] I. BITTER,[1] G. SZABÓ,[1] M. KAJTÁR[2]
J. SZEJTLI[3]

[1] Institute for Organic Chemical Technology, Technical University, Gellért tér 4
H-1111 Budapest, Hungary
[2] Institute for Organic Chemistry, Eötvös Loránd University
Muzeum krt. 4/B
H-1088 Budapest, Hungary
[3] CHINOIN Pharmaceutical-Chemical Works, Tó u. 1-5
H-1045 Budapest, Hungary

SUMMARY

β-cyclodextrin derivatives containing crown ether units /I/ have been prepared by the reaction of diamino-β-cyclodextrin with isocyanato-benzo-15-crown-5.

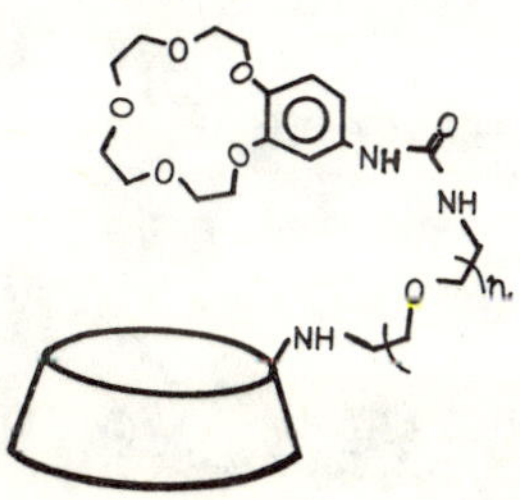

I

Compounds /I/ have much greater effect on the transport of salt-like pharmaceuticals through lipoid membran than β-cyclodextrin or crown ether separately. The measurements were run in Sartorius instrument. Constants of complex stability in water for Na^+ cation and p-nitro-phenolate anion were determined by electrochemical and CD methods, respectively. These values are in good agreement with our expectation on the possibility of "perfect molecular packing" of salts.

CYCLODEXTRINS IN PHARMACEUTICALS

CYCLODEXTRIN IN PHARMACEUTICAL INDUSTRY

KARL-HEINZ FRÖMMING

Institut für Pharmazie der Freien Universität Berlin
Königin-Luise-Str. 2-4
1000 Berlin 33, FRG

SUMMARY

The main reason for the great pharmaceutical interest in cyclodextrin (CD)-inclusion compounds compared to other inclusion compounds is due to the size of channel diameter, a good stability, and absence of toxicological side effects.

Solid CD-inclusion compounds can be used to enhance the chemical and physical stability of the guest compounds, to reduce or cover unpleasant smell and taste of drugs, to formulate liquid drugs, to avoid undesirable incompatibilities, or to reduce the volatility of drugs.

Especially interesting is the influence on the biopharmaceutical behaviour of the included drug. The release of heavily soluble drugs can be increased after formulation to a CD-inclusion compound, which can result in higher blood levels.

The main future pharmaceutical use of CD inclusion compounds is seen in the possibility to improve the biopharmaceutical properties and the stability of heavily soluble, low dosed drugs.

I want to consider the following three points in my lecture:

1) Why have the Cyclodextrin (CD)-inclusion compounds the greatest pharmaceutical significance in comparison to other inclusion compounds?

2) I'll give a survey about the most important previous results of pharmaceutical applications of CD-inclusion compounds.

3) What could be the future development of the pharmaceutical use of CD-inclusion compounds?

Urea, thiourea, deoxycholeic acid, the cyclodextrins, and amylose are the most important chemical compounds which can act as host molecules for the formation of inclusion compounds for pharmaceutical purposes.

Why have the CD complexes among these five host molecules the greatest pharmaceutical interest?

There are three reasons to consider:

1) The size of the channel diameter
2) Good stability
3) Absence of toxicological side effects

Regarding the size of the channel diameter.

The diameter of the ring molecule or of the channel determines which compounds or compound classes can be included. The guest molecule neither must be too big, nor too small. The approximate cavity dimension (internal diameter) have the following values:

Urea	5,1 Å
Thiourea	6,1 Å
Deoxycholeic acid	6,0 Å
α-Cyclodextrin	4,5 Å
β-Cyclodextrin	7,0 Å
γ-Cyclodextrin	8,5 Å
Amylose	5,0 Å

Diameters of approximately 5 Å result in a too great limitation of the interesting pharmaceutical compounds to be included. Only long-chain, unbranched molecules like fatty acids may be accomodated within the cavity of the urea channel.

This is not the case with the β- and γ-CD.

Regarding the stability problems.

Urea, thiourea, deoxycholeic acid and CD can form stable solid complexes. This allows us to use them generally in solid drug formulations. But the stability problems after mixing with other additives are different.

As pure complexes the urea adducts with fatty acids can be stored up to a relative humidity of 70 to 75 percent without deterioration. Mixtures with water containing solid pharmaceutical additives like starch, can have a reduced stability. Examples are the addition of starches. The manufacture of the urea complexes is only possible with water-free excipients (1).

This is not necessary with the CD-complexes; CD inclusion compounds are quite stable. They can be recrystallized from water. Higher humidity and higher water contents of added additives in solid formulations have no negative influences.

Heating of salicylic acid containing tablets up to 160° during 15 hours didn't result in a deterioration of the complex with following sublimation of salicylic acid compared to the physical mixture of salicylic acid and CD (2).

CD complexes are not deteriorated in dissolved form. There exists an equilibrium between the dissolved inclusion compound and the dissociated form. The complexes can be pressed in a tablet machine without deterioration.

The third point. CD are nontoxic. Thiourea for example cannot be applied to man because of undesired side effects. Higher doses of deoxycholeic acid produce diarrheal effects.

These three points are the main reasons why CD inclusion compounds have the greatest significance compared to other host molecules.

In which pharmaceutical areas CD-inclusion compounds can be applied?

Generally, we can consider inclusion compounds as a coated form. The drug molecules are surrounded by other molecules. In pharmacy, we have long since other coated forms. Let me mention coated tablets, coated granules, coated pellets or coated isolated crystals, e.g. microcapsules. In these formulations a mixture of drug particles, or a mixture of a drug and one or more additives are surrounded by the coat. These coats can be composed of sugar or of a thin macromolecular film.

In all these cases it is not the single molecule which is surrounded by the coating material. Here seems to be the greatest difference to the inclusion compounds. In the inclusion compounds we have some kind of solid solution. Single molecules are embedded in a solid matrix.

Normally, the industrial drug manufacturer has to guarantee a stability of 3 to 5 years for his drug product. Especially liquid drug products are candidates for stability problems. But, there are a lot of drugs, where we have to apply stability precautions during the manufacture to solid formulations, too.

During the last years, there have been some publications regarding the possible stabilization of prostaglandines with CD. In the early Seventies, Japanese companies reported about this possibility for the first time (3). Japanese authors from Teijin Co. reported the possible stabilization of vitamin D_3 (4). The complex is very stable against heat, light and oxygen. The preparation retained 94 percent of its original activity after storage in an uncovered glass container at 60°C for 7 days.

Very interesting is the stability behaviour of dissolved CD complexes. These adducts can exert an inhibitory or accelarating effect on the hydrolysis of the included guest compound. LACH and CHIN reported such an inhibitory effect on the hydrolysis of benzocaine (5). They attributed this to the complete inclusion of the ester in the ß-CD cavity, thus protecting the ester linkage from attack. Other compounds which can be stabilized by CD are procaine, atropine, or acetylsalicylic acid. On the other hand, CD inclusion compounds can accelerate different reactions ranging from acceleration factors 0.3 to 300 (6).

This is a very important point. If drugs are formulated

with CD to liquid formulations, we have to determine the influence of the CD on the stability of the drug or of another dissolved constituent.

Volatile drugs can be lost from drug formulations or from medicinal tea mixtures during storage through evaporation. The volatile and hydrolyzable nitroglycerine can be stabilized by complexation with CD (7).

Another patent describes the tablet manufacture of a nonvolatile l-menthol-CD inclusion compound which was stable for 6 months (8). Essential oils or their effective parts can be added to tea mixtures as microcapsules or as CD complexes.

Another patent from the field of cosmetic recommends CD inclusion compounds with perfumes for the preparation of bath tablets (9).

Unpleasant smell and taste of drugs can be reduced or covered by inclusion compounds. A patent describes the elimination of bitterness of protein hydrolysates by treating the product with CD (10).

An iodine ß-CD clathrate can be used as deodorant or antiseptic. It has no iodine odor, but slowly releases iodine by heating (11).

Formulation problems occur when smaller amounts of a liquid drug have to be formulated into a solid oral form. One possibility is the manufacturing of microcapsules. This is a complicated, expensive process. Another way could be the adsorption of the liquid drug at the surface of an additive with a great surface area, like colloidal silicic acid. The danger of autoxidation processes of an unstable drug increases with this last process.

Liquid compounds can be manufactured as solid CD complexes into solid drug formulations, e.g. vitamin A, unsaturated fatty acids, prostaglandines, clofibrate. We prepared solid inclusion compounds with linoleic acid, ascaridol and chaulmoogric acid ethylester (2). These three very easily oxidable substances were absolutely stable in the complex.

In a tablet or in a capsule we have to avoid undesirable incompatibilities between two or more substances. The result of such incompatibilities can be a further increase of the chemical instability. Well known examples are the problems with the manufacture of multivitamin preparations. Incompatible compounds can be protected by inclusion formation of one of the incompatible substances.

There is no possibility of an agglomeration among guest molecules when they are in such adduct. This is important during the tabletting process.

Some complexes can be used for the isolation of substances from plant extracts or for analytical determinations of certain substances from mixtures.

Until 10 to 15 years ago we had no knowledge about the behaviour of inclusion compounds in the body. We didn't know anything about the biopharmaceutical properties of these adducts. Biopharmaceutics is a part of the pharmaceutical and medical sciences which consider the change of the drug behaviour in the body depending on formulation factors. Two examples:

A drug from a solution is normally a little faster absorbed than the same drug from a tablet or a coated tablet. If a tablet has been compressed with too great forces, it can be that it passes the GI-tract without disintegration.

The first compounds which we tested in this respect were the deoxycholeic acid inclusion compounds (12). We had the following hypothesis: The biopharmaceutical behaviour of an included compound should be influenced in the first case by the dissolution behaviour of the host molecules.

Deoxycholeic acid is rather insoluble in stronger acid medium; above pH 5,5 it becomes increasingly soluble. In experiments with humans we could prove that the included guest molecules are not released in the acid medium of the stomach but in the less acid medium of the intestine. We have an enteric coated form.

We assumed that CD adducts would have a different behaviour. ß-Cyclodextrin is relatively slightly soluble between pH 1 and 8 without great differences.

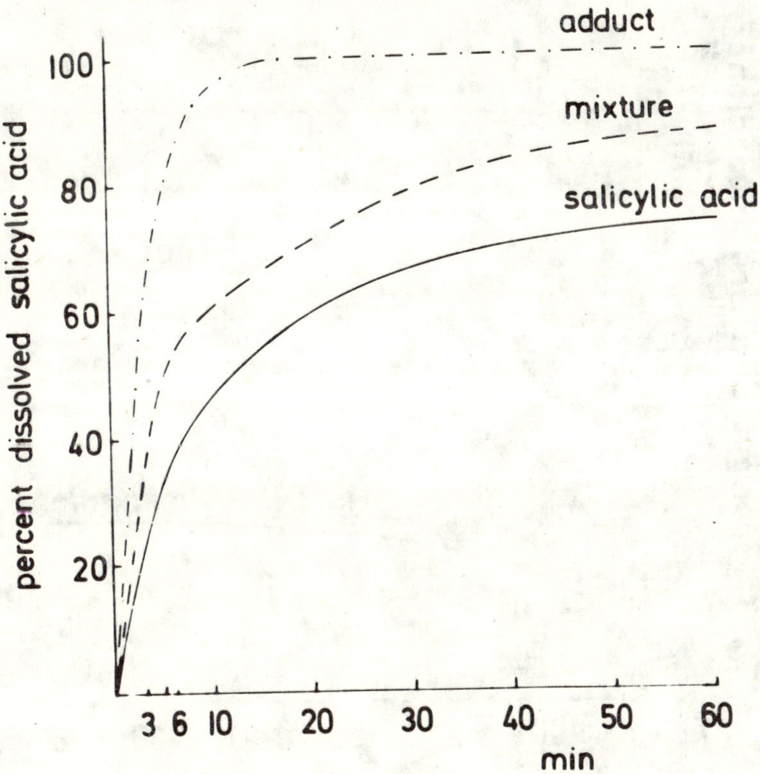

Fig. 1: Dissolution test (beaker method) ß-CD-salicylic acid inclusion compound; 320 mg salicylic acid; pH 1,0; 37°C

If the host molecules are soluble in gastric fluids, it should be expected that the molecularly included drug molecules are released after dissolution of the host molecules. We assumed that there should be an increase in solubility and in dissolution behaviour.

Whether this hypothesis is right can be determined by a simple in vitro dissolution test for drug release. There is a faster drug release into 0.1 N hydrochloric acid from the inclusion compound, the physical mixture shows a higher rate than the pure drug (fig. 1) (13).

These differences are quite impressive with another complex, the menadione-inclusion compound (fig. 2).

Generally, dissolution and absorption enhancing effects are the more pronounced the less soluble the drug substance is.

Identical doses of a ß-CD/salicylic acid adduct, a physical mixture and pure salicylic acid in capsules were administered to each of 10 human volunteers. The fastest and highest urinary excretion of total salicylates is to observe after administration of the inclusion compound (fig. 3).

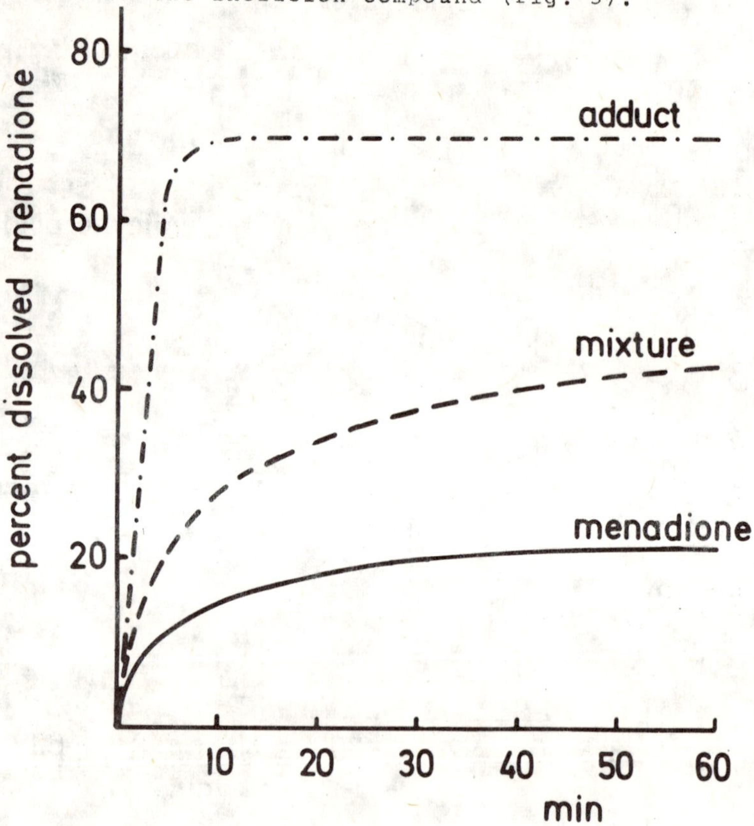

Fig. 2. Dissolution test (beaker method) of ß-CD-menadione inclusion compound; 170 mg menadione; pH 1,0; 37°C

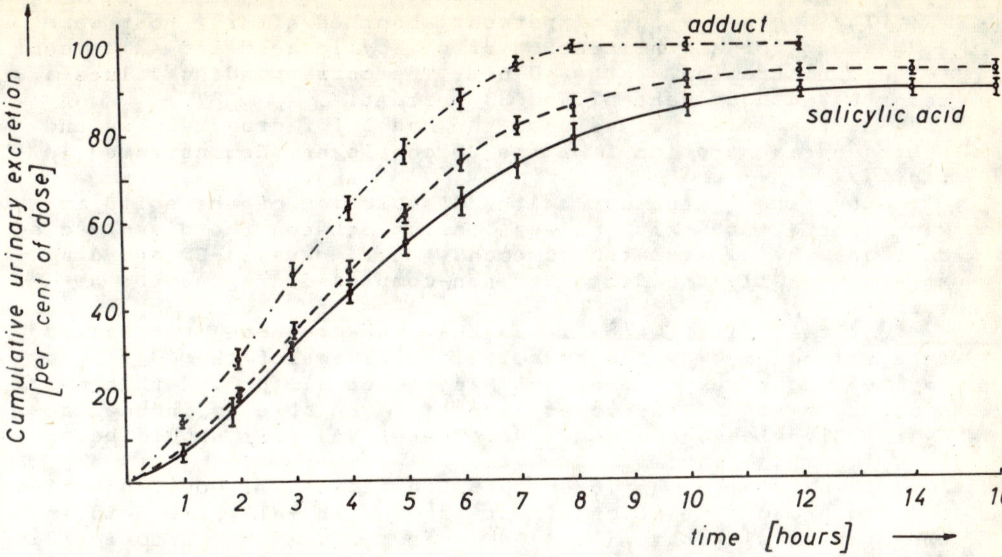

Fig. 3: Cumulative urinary exretion of total salicylates after oral administration (10 volunteers)

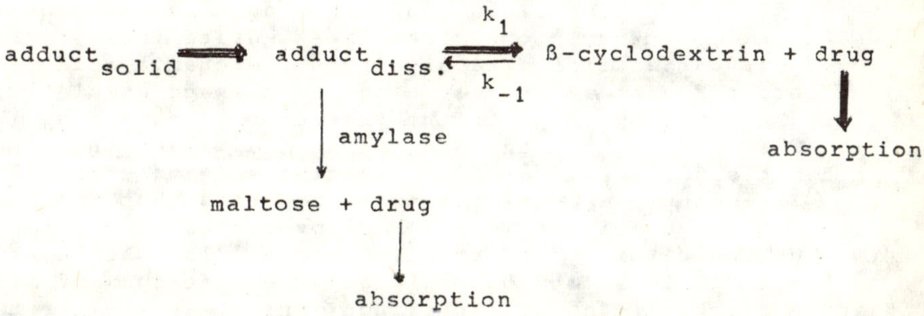

Fig. 4: Model for in vivo drug release from CD-inclusion compounds

The average value of percent absorbed after 1 hour were 25 percent for administration of salicylic acid and 43 percent for administration of the adduct. The corresponding values after the 2nd hour were 57 and 83 percent.

At that time, we assumed the model for drug release and subsequent absorption from the CD complexes, demonstrated in fig. 4.

The first step must be the dissolution of the solid adduct. There must exist an equilibrium between the dissolved adduct and the dissociated compound, that means, ß-CD and drug molecules. Only the dissolved non-complexed drug can be absorbed.

At that time, we believed that the drug can be released in a second way, by the hydrolytic cleavage of the CD ring by amylose. Newer experiments of Szejtli et al. proved that only microorganisms in the lower intestine are able to induce the hydrolytic cleavage of CD (14). Therefore, this should be a less important way.

It should be mentioned that Szejtli et al. could not find this enhanced absorption effect of CD with salicylic acid in experiments with rats (15). The reason should be anatomical and physiological differences between rats and man. But he found no stomach irritating effects of salicylic acid. This is another example for avoiding side-effects by the administration of inclusion compounds.

Our first observations about a possible change of biopharmaceutical properties after application of inclusion compounds were followed by experiments of other authors using other drugs. NAGAI and coworkers proved that freeze-dried inclusion compounds of the antiinflammatory drugs flufenamic acid, ibuprofen and ketoprofen showed higher blood levels and cumulative urinary excretion compared with the freeze-dried pure drug (16).

The same authors and Szejtli et al. proved that the indomethacin complex is free of ulcerating side-effects (17).

In another Japanese patent is reported that the toxicity of a 5-fluorouracil derivative is lower after inclusion than that of the parent fluorouracil compound (18). Administration of the inclusion compound prolonged the life-span of mice bearing Ehrlich carcinoma. The effectiveness was similar to that of the parent compound.

KOIZUMI et al. reported that oral administration of five barbiturate-ß-cyclodextrin complexes to mice gave smaller 50 percent effective doses (ED_{50}) than the intact drug. The authors found a relation between the stability constants of the complexes, the solubilities and the ED_{50} (19). The most remarkable reduction of the ED_{50} is to observe after application of the phenobarbital complex. Phenobarbital forms the most stable complex and shows the greatest enhancement of solubility. With the exception of barbital, complexation significantly increased the sleeping time, too.

Let me summarize and let me give an outlook for the use of CD adducts in pharmaceutical industry.

I don't believe that the general future development of new

drug formulation will yield such spectacular inventions as we had it in the invention of tablets or capsules or ointments. Almost all important drug substances can be formulated to these three established formulations.

The future development of formulation design will take much more into consideration the requirements of the corresponding drug. Today we have very complicated drug products which can only be used for one or two drugs up to now. An example is the transdermal systemic effect for the scopolamin.

Under such aspects we cannot put the question: Are the CD complexes suitable to be used for the formulation of all drugs?

Generally, inclusion compounds are only interesting for low dosed drugs because of the two following reasons:

1) Up to now, the price of the CD is still too high for the manufacture of greater amounts. This point could be changed in the future.

2) Normally, the amount of included guest compound in the CD inclusion compounds is less than 20 to 25 percent. That means that drugs with a single dose of more than 50 mg drug will hardly be candidates for inclusion.

As an advantage for the future use, I see the small toxicological problems. CD are metabolized to natural occuring glucose units.

A very important point seems to be the future development of the economical production of γ-CD. If we had the possibility to produce the β- and γ-isomers to an acceptable price, we could include much more interesting drugs. The production of the γ-CD seems to be more important than that of the α-isomer. In this case we could include larger molecules than with the β-CD.

We should consider that the only addition of pure CD to a drug product can help to produce a better formulation. The biopharmaceutical properties of heavily soluble drugs can be improved in this way, too.

I see the main advantage for the pharmaceutical use of CD-inclusion compounds in solid formulations in the possibility to improve the biopharmaceutical properties of heavily soluble drugs and to increase their stability.

Literature Cited

1. Frömming, K.-H.: Arch.Pharm. 301 548 (1968)

2. Frömming, K.-H., Sandmann, R., Weyermann, I.: Dtsch.Apoth. Ztg. 112 707 (1972)

3. Hayashi, M., Takatsuki, I.: Ger.Offen. 2 128, 674, Dec. 16, 1971; cit. CA 76 59978

4. Ikura, H.: Japan,Kokai 77, 130, 904, Nov. 2, 1977; cit. CA 38 79103

5. Lach, J.L., Chin, T.-F.: J.Pharm.Sci. 53 924 (1964)

6. Bender, M.L., Komiyama, M.: Cyclodextrin Chemistry, Springer Verlag, Berlin-Heidelberg-New York 1978; p.29

7. Akito, E., Nakajima, Y., Horioka, M.: Japan.Kokai 75, 129, 520, Oct. 13, 1975; cit. CA 84 58617

8. Akiyama, Y.: Japan.Kokai 77, 144, 014, Sept. 24, 1977; cit. CA 88 27824

9. Hirano, S., Tsumura, J., Izeki, I., Kawamura, H., Ohhara, M.: Japan.Kokai 75 63, 126, May 29, 1975; cit. CA 120672

10. Suzuki, Y.: Japan.Kokai 75 69, 100, June 9, 1975; cit. CA 84 120071

11. Hirose, T.: Japan.Kokai 76 88, 625, Aug. 3, 1976; cit. CA 85 149145

12. Frömming, K.-H., Sandmann, R.: Arch.Pharm. 303 371 (1970)

13. Frömming, K.-H., Weyermann, I.: Arch.Pharm. 305 290 (1972); Arzneim.-Forsch. (Drug Res.) 23 424 (1973)

14. Szejtli, J., Gerlóczy, A., Fónagy, A.: Arzneim.-Forsch. (Drug Res.) 30(I) 808 (1980)

15. Szejtli, J., Gerlóczy, A., Sebestyén, G., Fónagy, A.: Pharmazie 36 283 (1981)

16. Nambu, N., Shimoda, M., Takahashi, Y., Ueda, H., Nagai, T.: Chem.Pharm.Bull. 26 2952 (1978)

17. Szejtli, J., Szente, L., Dávid, A., Virág, S., Sebestyén, G., Mándi, A.: Hung.Patent Appl. Cl-1679 (1976); cit. Chinoin Budapest: ß-Cyclodextrin

18. Asano, O., Kato, B.: Japan.Kokai 78, 109, 953, Sept. 26, 1978; cit. CA 90 43812

19. Koizumi, K., Miki, H., Kubota, Y.: Chem.Pharm.Bull. 28 319 (1980)

I.Int.Symp. on Cyclodextrins
Budapest, 1981 5.2.

A FORECAST FOR APPLICATION OF CYCLODEXTRINS IN THE PHARMA-INDUSTRY

Á. STADLER-SZŐKE, J. SZEJTLI

CHINOIN Pharmaceutical-Chemical Works, Tó u. 1-5
H-1045 Budapest, Hungary

SUMMARY

During the last decade a rapidly increasing activity was reflected in the number of papers and patents dedicated to the pharmaceutical application of CD-s.

The main results of these investigations should be summarised as follows:

a./ on account of solubility and toxicity reasons the β-CD - until now the only industrially produced CD - can be utilized in oral and rectal preparations

b./ the main features of complexation of active ingradients by β-CD are the stabilization of labile compounds and enhancement of bioavailability

c./ the accumulated knowledge on this field allow to set limits with acceptable certainty to the group of drugs, which can be and which are worth of transformation into CD-complex.

Shape, size and hydrophobicity of molecules are the main factors in determining the complexability.

About 10 % of orally applied drugs - their active ingredient - seems to be complexable; about half of these cases one or other physical or chemical properties can be modified up to such an extent, which give reason of CD complexation.

Taking into account all factors, in at least 2 % of all actually produced tablets the presence of CD as complexing agent /or auxiliary substance/ is favourable. It means, that in a not too far future a several thousand ton fraction of the β-CD market might be consumed by the pharmaceutical industry for production of oral preparations.

During the last decade a rapidly increasing activity was reflected in the number of papers and patents dedicated to the pharmaceutical application of cyclodextrins (1). The published data demonstrate unambiguously, that there are two important fields in the application of cyclodextrins in the pharmaceutical industry (2).
These are:

- enhancement of bioavailability by improving the solubility of the water insoluble or slightly soluble drugs:
- stabilization against oxygen, decomposition, hydrolysis, etc.

In the pharmaceutical application of CD-complexation, there are two fundamental limiting factors: the required dose and the structure. With drugs of high molecular weight the active ingredient content of the inclusion complex is in the range of 15-25 %, with guests of low molecular weight this figure is under 15 %, sometimes even less than 5 %. If not more than 25 mg is required as a single dose of the drug, then even a complex of 5 % active substance content, may carry the necessary dose in a single tablet of 500 mg weight. Thus, with drugs capable to form complexes, the relationship of the required dose and the molecular weight determines the feasibility of oral administration of a cyclodextrin complex in tablets.

The other mentioned factor is the structure, which determines whether the guest molecule is able to form complex at all. The correlation between the drug content and the amount of the complex required for a 25 mg dosis is illustrated in Table I.

Table I.

MW OF DRUG	ACT.INGRED. CONTENT IN THE COMPLEX %	DOSE OF DRUG FREE mg	COMPLEXED mg
378	25	25	100
284	20	25	125
200	15	25	166
126	10	25	250
60	5	25	500

(Molar ratio of drug: β-CD = 1:1)

The most important parameters that determine the inclusion complexation of a given molecule are the hydrophobicity and relative size and geometry in relation to the CD cavity (3). The formation of an inclusion complex with CD can alter some of the physical and/or chemical properties of the guest molecule. The inclusion complex formation can be detected and monitored by a variety of instrumental methods, e.g. UV-visible absorption, ORD, NMR, ESR, MS, fluorescence, Raman X-ray analysis, thermoanalytical methods etc. Equilibrium constant for binding of various guest molecules to the different CD-s are determined and published (4-8).

Within certain families of compounds some correlation exists between the strength of binding and some measures of the guest molecule's hydrophobicity (e.g. polarizability, molar volume, molar refraction and partition coefficient of the guest in an apolar water system (3,9,10). Therefore well-established forecast for the application of CDs in the pharmaceutical industry is beyond our actual possibilities. However no obvious correlation has been found between any physical or chemical properties of different families of guest molecules and the inclusion complex formation equilibrium constants.

Thus we selected arbitrarily the groups according to the amount of data available. These are illustrated on Table II.

Table II.

ANTIPHLOGISTICS	salicylic acid-, anilin-,anthranilic acid-, phenylacetic acid derivs. etc.
BARBITURATES	
STEROIDS	
LOCALANAESTHETICS	
PROSTANOIDS	
PHENOTIAZINES	
SULFONYLUREAS	
CHEMOTERAPEUTICS	sulfonamides alkylparabens etc.
ANTITUMOR AGENTS	

A great variety of drugs have been studied, but these studies were focused on the chemistry of complexes. Only recently were published data on enhanced biological effects of CD-complexed drugs (E.g.11). With constantly increasing amount of informations, and experimental data, such a survey can be broadened, but we do not think that substantially new points may emerge, which fundamentally change the main points, namely the enhancement of bioavailability and improvement of stability (moreover in several cases the reduction of side effects).

Table III. illustrates these points with the group of antiphlogistics. Frömming (12) published the first human experiments with CD--inclusion complex in 1973. He observed an enhanced bioavailability with complexed-salicylic acid - as he just showed in his lecture. In the following several experimental observation will further demonstrate the advantages of CD-complexation.

Table III.

ANTIPHLOGISTICS

Drug	Advantage of CD-complexation			Author
	A	B	C	
SALICYLIC ACID	+			FRÖMMING 1973(12)
ASPIRIN		+		FUJIMOTO 1975(13)
PHENACETIN		+		NAKAI 1979(14)
FLUFENAMIC ACID	+			KUROZUMI 1975(15)
IBUPROFEN	+			NAGAI 1977(16)
KETOPROFEN	+			NAMBU 1978(17)
INDOMETHACIN	+		+	SZEJTLI 1978(18)
PHENYLBUTAZONE			+	NAMBU 1978(19)
MEFENAMIC ACID	+			SANKYO 1976(20)
FLURBIPROFEN	+			OHTA 1979(21)

A: enhanced bioavailability
B: stabilization
C: reduced side effect

Figure 1. illustrates the enhancement of the bioavailability of indomethacin by β-CD complexation, where the molar ratio of the components is 1:2 (18). An unsuccesful attempt to prepare the complex in molar ratio 1:1 was published by Kurozumi et al. (22). In crystalline state there is no unambiguous evidence for the inclusion of indomethacin, however in solution circular dichroism as well as diffusion tests provide direct evidence for inclusion complexation (18). Following oral administration indomethacin-CD complex resulted in 25 % higher blood level as compared with equivalent amount of indomethacin.

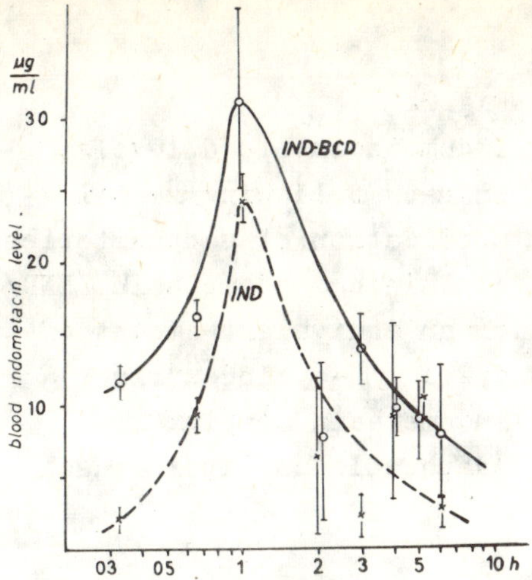

Fig. 1.
Indometacin blood levels in rats treated orally with indometacin /IND/ or with an equivalent amount of 2:1 β-cyclodextrin indometacin complex /IND-BCD/

An unexpectedly high blood level, about two times higher was observed in rats after rectal application of suppositories containing indomethacin-CD complex compared to equivalent amount of free indomethacin, in Adeps solidus lipoid base (Fig.2.). Increased blood level must be manifested itself in an improved antiinflammatory effect too. The experiment was carried out with carragheenane induced oedema in rats. Leg weight increase of the control group (injected with carragheenane, but not treated with indomethacin) was taken as 100 % inflammation. The weight differences in the treated groups were compared to this, and expressed in %. Figure 3. demonstrates a dose-response relation with free (IND) and complexed (IND-BCD) indomethacin providing a direct evidence for the improved bioavailability.

By cyclodextrin complexation the indomethacin induced ulcerogenic effect is reduced. Male and female rats were treated with indomethacin and indomethacin-cyclodextrin complex. 70 % of the males and 60 % of the females treated with a 5 mg/kg dose of indomethacin developed serious ulcer, whereas in the animals treated with the complex containing the same indomethacin dose, these figures were reduced to 20 %

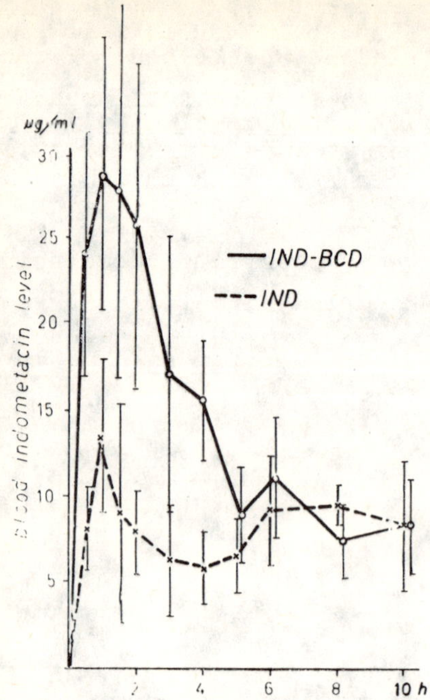

Fig. 2.
Indometacin blood levels in rats following rectal application of suppositories containing indometacin /IND/ or an equivalent amount of a 2:1 β-cyclodextrin indometacin complex in Adeps solidus lipoid base.

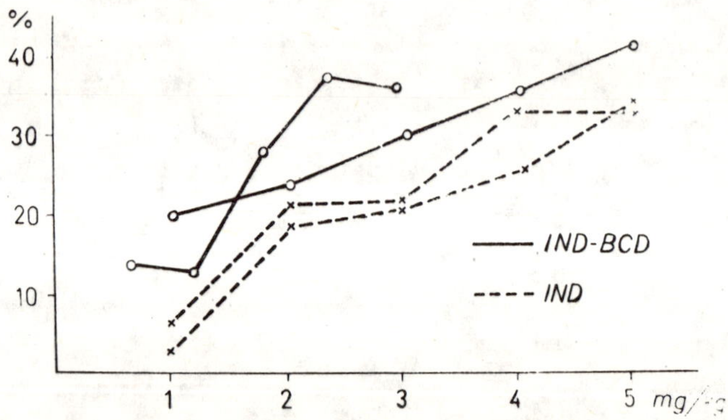

Fig. 3. Potentiation of antiinflammatory effect of indometacin by cyclodextrin complexation: dose-response relation with free /IND/ or complexed /IND-BCD/ indometacin. /Two independent experiments./

and 10 %, respectively (Table IV.). Further results on this topic will be publishsd on the poster session by T. Neumark.

Table IV.

	Ulcus		Erosion	
	male	female	male	female
Control	0/10	0/10	0/10	0/10
40.3 mg/kg IND-CD (= 5 mg IND)	2/10	1/10	4/10	4/10
5 mg/kg IND	7/10	6/10	9/10	7/10

Another interesting and thoroughly studied group is that of the barbiturates. Koizumi (23) compared the hypnotic potency of different barbiturates to those of their β-cyclodextrin complexes on oral administration to mice. The improvement of effectivity - expressed as the ratio of ED_{50} values of the complexed and that of the free drug - is plotted against the improvement of the solubility - expressed as the ratio of the solubility of the complexed and that of the free drug (Fig.4.).

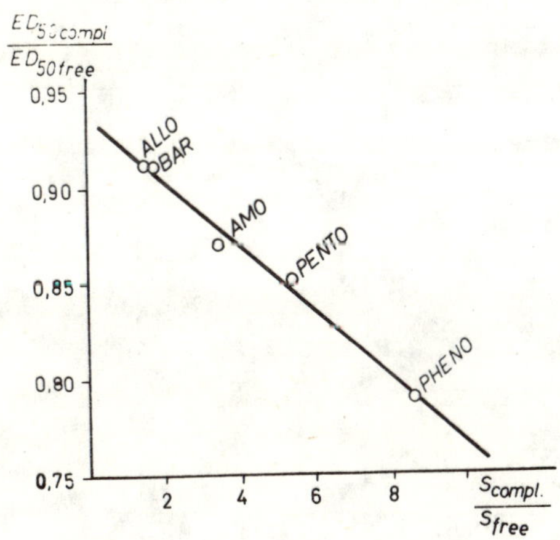

Fig.4.
Correlation between enhancement of solubility $/S_{compl}/S_{free}/$ and enhancement of hypnotic activity /=reduction of ED_{50}/ by β-cyclodextrin complexation of barbiturates. /Based on data published by Koizumi et al., 23/

There is an unambiguous correlation between the enhancement of solubility and enhancement of effectivity. This observation was further substantiated by studying the sleeping lag and sleeping time. Figure 5. illustrates the shortening of the sleeping lag time, and the elongation of the sleeping time by cyclodextrin complexation of the barbiturates. The figure demonstrates the very strict correlation between the stability constants and the pharmacological effect.

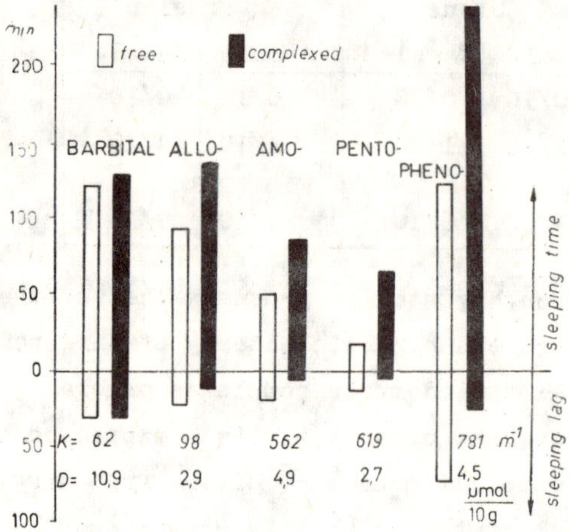

Fig. 5.
Reduction of sleeping lag and elongation of sleeping time in mice by β-cyclodextrin complexation of barbiturates. /K=stability constant of complex, D=orally administered dose per 10 g./ /Based on data published by Koizumi et al., 23/

With steroids considerably improved solubility was observed in dimethyl-β-cyclodextrin solution (Table V.). By complexation of the cholecalciferol an enhanced solubility, biological effect and stability was attained (24).

Table V.

STEROID	SOLUBILITY IN WATER S_1 g/100 ml	SOLUBILITY IN 10 % di-Me-β-CD SOLUTION S_2 g/100 ml	S_2/S_1
NORTESTOSTERONE	0,31	14,7	47,4
HYDROCORTISONE	0,36	20,3	56,4
METHYLSECODIONE	0,06	4,5	79,0
ANDROST-4-EN-3,17-DION	0,82	13,0	158,0
OESTRONE	0,03	4,75	158,0
METHYL-TESTOSTERONE	0,07	13,7	193,0
PROGESTERONE	0,02	13,0	813,0

In the case of the localanaesthetics by cyclodextrin complexation the duration of analgesia is elongated (see in the lecture of Dr.Ferenczy). Another example for such an effect was observed with marcaine. Injections containing identical doses of marcaine-HCl and marcaine base dissolved in dimethyl-β-cyclodextrin solution, respectively, were administrated subcutaneously to human volunteers and analgesia was detected. 5 persons at 9 intervals gave 45 answers (Fig.6.).

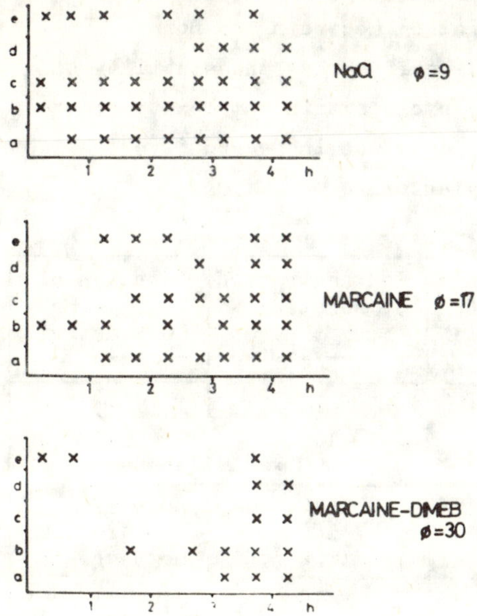

Fig.6. Response of five (a - e) volunteers to pinpricking following the subcutaneous administration of physiological saline, marcaine.HCl, or equivalent marcaine base dissolved in dimethyl-β-CD solution.

The number of analgesia after physiological saline administration were 9, after marcaine-HCl administration 17, after marcaine-dimethyl-β-cyclodextrin administration 30. According to these results the complexed form elongated the duration of analgesia about twice as compared to the marcain-HCl in an interval of 270 minutes. Prostaglandins are very effective even in extremely small quantities, however the most active ones are very unstable. Cyclodextrin complexation showed to be a succesful method for their stabilizations as well as for enhancement of their solubility. There are numerous patents dedicated to this stabilization effect (e.g. 25,26,27). PGE_2-β-cyclodextrin complex was the first marketed cyclodextrin complexed drug, named PROSTARMON-E, produced by ONO, Japan. So far only the pharmacological effect of PGE-type prostaglandin-cyclodextrin complex have been published. Kawasaki (28) and Nishibori (29) concluded that the complex has the same pharmacological effect as the free PGE_2.

Recently the interest is centred within this group of compounds on the prostacyclin. The PGI_2 is the most effective antiplatelet agent of natural origin. The spreading application was limited by its instability, and this substantial hindrance could be by-passed with CD-complexation (30). The stability of the PGI_2-ethylester-β-cyclodextrin complex was studied in aqueous solution at room temperature by inhibition of rabbit platelets aggregation according to Born.

Results presented on Table VI. verify, that the stability of PGI_2-ethylester can be increased by cyclodextrin complexation; the activity half-life of the complexed drug was increased five to eight fold referred to the free PGI_2-ethylester.

Table VI.

	Content of PGI_2 %	Contrentration of solutions µg/ml	Duration of losing the 50 % activity min
PGI_2-Et-β-CD complex crystallized	6,0	10 100 1000	75 270 no decrease[x]
PGI_2-Et-β-CD complex freeze-dried	4,2	10 100 1000	87 420 no decrease[x]
PGI_2-Et		10 100 1000	11 45 -

[x]during the measuring period (5h)

It is supposed, that about 10 % of the orally applied drugs can be complexed with β-cyclodextrin; in about half of these cases one or more physical or chemical properties can be modified to an extent that gives reason of CD-complexation.

Taking into account the pros and contras, in at least 2 % of all actually produced tablets, the presence of CD as complexing agent (or auxiliary substance) is favourable. It means that in the not-too-far future a several thousand ton fraction of the β-CD market might be consumed by the pharmaceutical industry for production of oral preparations.

Acknowledgement: Thanks are due for the antiphlogistic investigations to Dr.Gy.Sebestyén and V.Gergely in the Toxicological Lab. of

Chinoin, for literatury search to Mrs.M.Vikmon, for the steroid solubility studies Miss I.Habon.

Literature Cited

1. Szejtli, J.: Cyclodextrins and their Inclusion Complexes. Akadémiai Kiadó, Budapest, 1981.
2. Szejtli, J.: Die Stärke, 33 in press (1981)
3. Bender, M.L. and Komiyama, M.: Cyclodextrin Chemistry, Springer Verlag, Berlin-Heidelberg-New York, 1978.
4. Uekama, K., Hirayama,F. and Irie, T.: Chem.Lett. (Japan) 661 (1978)
5. Uekama, K., Hirayama,F., Ikeda, K. and Inaba, K.: J.Pharm.Sci., 66 706 (1977)
6. Uekama, K., Hirayama,F., Wakuda, T. and Otagiri, M.: Chem.Pharm. Bull. 29 213 (1981)
7. Uekama, K., Hirayama, F., Otagiri, M., Ikeda, K.: Chem.Pharm. Bull. 26 1162 (1978)
8. Ikeda, K., Uekama, K., Otagiri, M., Hatano, M.: J.Pharm.Sci. 63 1168 (1974)
9. Van Etten, R.L., Sebastian, J.F., Clowes,G.A. and Bender, M.L.: J.Amer.Chem.Soc., 89 3242 (1967)
10. Griffiths, D.W. and Bender, M.L.: Adv. Cat., 23 209 (1973)
11. Uekama, K., Matsuo, N., Hirayama, F., Yamaguchi, T., Imamura, Y., Ichibagase, H.: Chem.Pharm.Bull. 27 398 (1979)
12. Frömming, K.H., Weyermann, I.: Arzneim. Forschung 23 424 (1973)
13. Fujimoto, S.: Japan Kokai, 78,124,607 /1978/
14. Nakai, Y.: Japan Kokai, 79 86 607 (1979)
15. Kurozumi, M., Nambu, N., Nagai, T.: Chem. Pharm.Bull. 23 3062 (1975)
16. Nagai, T., Nambu, N., Shimoda, M., Takahashi, Y.: Expo.Congr. Int. Technol. Pharm., 1st 4 204 (1977)
17. Nambu, N., Shimoda, M., Takahashi, Y., Ueda, H., Nagai, T.: Chem.Pharm.Bull. 26 2952 (1978)
18. Szejtli, J., Szente, L.: Pharmazie 36 694 (1981)
19. Nambu, N., Shimoda, M., Takahashi, Y., Ueda, H., Nagai, T.: Chem.Pharm.Bull. 26 2952 (1978)

20. Sankyo KK: Japan Patent 119818 (1976)
21. Ohta Pharm. Co., Ltd.: Japan Kokai 81 34 618 (1981)
22. Kurozumi, M., Nambu, N., Nagai, T.: Chem.Pharm.Bull. 23 3062 (1975)
23. Koizumi, K., Hiroko, M., Kubota, Y.: Chem.Pharm.Bull. 28 319 (1980)
24. Szejtli, J., Bolla-Pusztai, É., Szabó, P. and Ferenczy, T.: Pharmazie, 35 779 (1980)
25. Hayashi, M., Ishihara, T.: Japan Patent 50 119 70 (1971)
26. Hayashi, M., Nishibori, T.: Japan Patent 4 739 057 (1972)
27. Hatachie, Y., Inaba, K.: Japan Kokai 74 26 416 (1974)
28. Kawasaki, A., Nakai, K., Ishii, A., Akimoto et al.: Oyo Yakuri 8 61 (1974)
29. Nishibori, T., Matsuoka, Y., Matsumoto, T.: Oyo Yakuri 8 795 (1974)
30. Szejtli, J., Szejtli, M., Cseh, Gy., Stadler, I.: Hung.Patent Appl. CI-1770 (1977)

I.Int.Symp. on Cyclodextrins
Budapest, 1981

EFFECTS OF CYCLODEXTRINS ON THE HEMOLYSIS INDUCED WITH PHENOTHIAZINE NEUROLEPTICS

M. OTAGIRI, K. UEKAMA, T. IRIE, M. SUNADA, T. MIYATA, Y. KASÉ

Faculty of Pharmaceutical Sciences, Kumamoto University
Kumamoto 862, Japan

SUMMARY

Cyclodextrins (CyDs) protected the human erythrocytes from hemolysis and shape changes induced with phenothiazines (PZ) in isotonic solution, depending upon the magnitude of the stability constant of PZ-CyD complexes (β-> γ-> α-CyD). A good correlation between the stability constant of inclusion complexes and the inhibitory effect on drug uptake into erythrocytes was found. CyD itself did not show any effects on the discocytic form of erythrocytes and any remarkable changes in osmotic pressure and viscosity of PZ solution under these experimental conditions. From these observations, the protective effects of CyDs *in vitro* appeared to be due to the decrease in the effective hemolytic concentration of PZ through inclusion complex formation rather than the direct interaction of CyDs with the erythrocyte membrane. The effect of β-CyD on some central nervous system (C.N.S.) actions of chlorpromazine (CPZ) in rats was also investigated to see if there are any advantages in the use of β-CyD complexes given by injection. The results suggest that β-CyD does not alter the time-course or magnitude of the effects of CPZ on the C.N.S..

Drug-induced hemolysis may limit clinical trial of compounds having excellent pharmacological activity, yet the protootion of human erythrocytes against drug-induced hemolysis has received little attention (1). Cyclodextrins (CyDs), cyclic oligosaccharides, have been successfully applied to pharmaceutical fields because of improved physical and chemical properties of the drug molecules through inclusion complex formation (2,3). Recently, we reported that β-CyD significantly protects human erythrocytes from hemolysis and shape changes induced by chlorpromazine (CPZ) in isotonic solution (4,5). The present paper deals with the effects of CyDs on the hemolysis induced with ten phenothiazine (PZ) neuroleptics and discusses the protective mechanism of CyDs on the basis of inclusion complexation. Furthermore, the effect of β-CyD on some central nervous system (C.N.S.) actions of CPZ was examined to see if CyD complex would use as an injecting agent.

Fig. 1 shows the hemolytic effects of CPZ and prochlorperazine (PCZ) on human erythrocytes in the absence and presence of three CyDs in isotonic solution, as a typical example. The hemolytic activities of CPZ and PCZ were significantly decreased by addition of β- and γ-CyDs, but no appreciable effects were observed with α-CyD. This dependency on the molecular size implies complementarity between host and guest molecules and suggests that to prevent the hemolysis CPZ and PCZ are preferably included in the β-CyD cavity. Fig. 2 shows the effects of CyD concentrations on the hemolysis

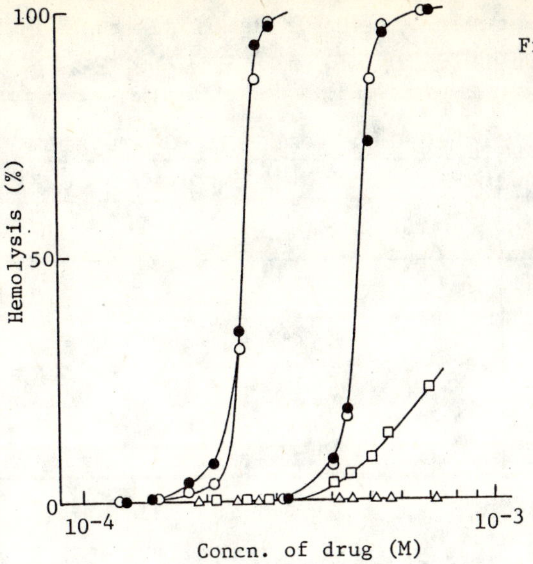

Fig. 1. Hemolytic Effects of CPZ and PCZ on Human Erythrocytes (1%) in the Absence and Presence of CyDs (1×10^{-3} M) in 10 mM Isotonic Phosphate Buffer (pH 7.4) at 37°

Percent hemolysis is expressed by the ratio of the absorbance at 543 nm of hemoglobin released from erythrocytes after incubation(5 min) with drug to the absorbance after the complete hemolysis of erythrocytes in water.

○: drug alone ●: drug + α-CyD △: drug + β-CyD □: drug + γ-CyD.

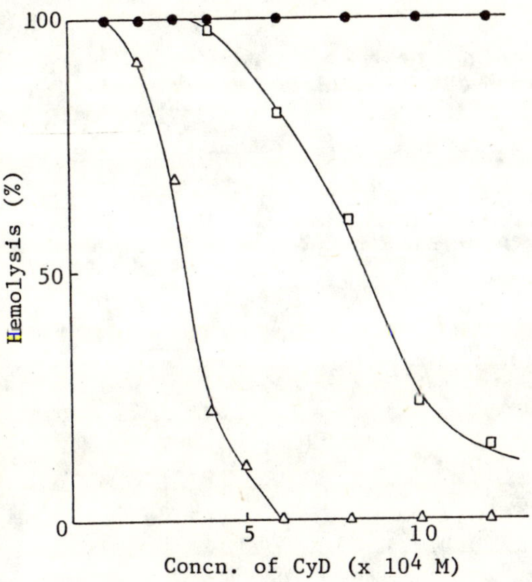

Fig. 2. Effects of CyD Concentrations on the Hemolysis Induced with CPZ at 7.0×10^{-4} M in 10 mM Isotonic Phosphate Buffer (pH 7.4) at 37°

●: CPZ + α-CyD △: CPZ + β-CyD □: CPZ + γ-CyD.

induced by 7.0×10^{-4} M CPZ. It suggests that β-CyD is the most effective in protecting against hemolysis. Since the hemolysis was reduced as a function of CyD concentration, attention was directed toward the stability constant, K_c of inclusion complex. Taking into account the free and complexed species in the CPZ-CyD systems, the free CPZ concentration, $(CPZ)_f$ was calculated from

Table I. Relationship between Extent of Hemolysis and Free CPZ Concentration in CPZ-CyD Systems[a)]

Initial concn. of CyD ($\times 10^4$ M)	α-CyD system[b)]		β-CyD system[b)]		γ-CyD system[b)]	
	Hemolysis (%)	$(CPZ)_f$ ($\times 10^4$ M)	Hemolysis (%)	$(CPZ)_f$ ($\times 10^4$ M)	Hemolysis (%)	$(CPZ)_f$ ($\times 10^4$ M)
0.0	100	7.0	100	7.0	100	7.0
2.0	100	6.7	92	5.3	100	6.2
4.0	100	6.5	22	3.5	97	5.6
6.0	100	6.3	0	2.5	81	5.0
8.0	100	6.1	0	1.7	60	4.5
10.0	100	5.9	0	1.2	23	4.1
12.0	100	5.8	0	0.9	15	3.7

a) Initial concentration of CPZ, 7.0×10^{-4} M.
b) Stability constants: 200 M^{-1} for α-CyD complex, 12000 M^{-1} for β-CyD complex, 1000 M^{-1} for γ-CyD complex.

Table II. Relationship between Hemolytic Data and Stability Constants

Compound	Hemolytic concentration ($\times 10^4$ M)				K_c ($\times 10^{-3}$ M^{-1})
	(PZ) 0%	PZ-β-CyD system			
		$(PZ)_t$ 100%	$(\beta\text{-CyD})_t$	$(PZ)_f$	
1) Promazine	12.0	24.0	15.0	11.1	5.6
2) Methoxypromazine	13.0	32.0	20.0	13.1	13.0
3) Chlorpromazine	3.0	7.0	6.0	2.5	12.0
4) Promethazine	10.0	18.0	12.0	9.9	2.1
5) Trimeprazine	7.0	15.0	10.0	6.7	7.0
6) Perazine	11.0	22.0	15.5	10.1	9.0
7) Prochlorperazine	1.2	3.0	3.5	1.2	34.0
8) Trifluoperazine	1.1	3.0	3.8	1.3	14.0
9) Perphenazine	2.2	4.0	1.5	2.7	22.0
10) Fluphenazine	1.4	3.5	3.5	1.4	11.0

equation 2, using previously reported K_c values (6), where (complex) and $(CyD)_f$ are concentrations of complex and free CyD, and $(CPZ)_t$ and $(CyD)_t$ are total concentrations of CPZ and CyD, respectively. It is evident from Table I that the

$$CPZ \quad + \quad CyD \overset{K_c}{\rightleftharpoons} complex \tag{1}$$

$$K_c = \frac{(complex)}{(CPZ)_f (CyD)_f} = \frac{(complex)}{[(CPZ)_t - (complex)][(CyD)_t - (complex)]} \tag{2}$$

extent of hemolysis is well correlated with $(CPZ)_f$, indicating that the complexed CPZ may be essentially inactive and thus unable to induce the hemolysis. Complete inhibition was achieved by β-CyD at concentration above 6.0×10^{-4} M, where $(CPZ)_f$ values were smaller than the effective hemolytic concentration of CPZ (3.0×10^{-4} M). A similar relationship between the extent of hemolysis and the free concentration of PZ, $(PZ)_f$ was observed for all other PZ-β-CyD systems. The initial hemolytic concentrations, $(PZ)_{0\%}$, are in good agreement with $(PZ)_f$ obtained from K_c values, as illustrated in Table II. These data clearly indicate that PZ may be essentially inactive to induce the hemolysis. Therefore, the protective effects of CyDs appeared to be due to the decrease in effective hemolytic concentration of PZ through inclusion complexation.

Fig. 3 shows some scanning electron micrographs of human erythrocytes treated with CPZ in the absence and presence of β-CyD. The untreated erythrocytes have a biconcave disk shape with smooth featureless surface (Fig. 3-A). At the concentration of 6.0×10^{-5} M, CPZ caused the erythrocytes to become cup-shaped (Fig. 3-C). With higher CPZ concentrations, the erythrocytes become spherical (Fig. 3-E). When β-CyD was added to the erythrocyte suspension, the shape changes induced with CPZ were significantly prevented, as shown in Fig. 3-D and 3-F, respectively. β-CyD itself had no effect on the discocytic form of erythrocytes under this experimental condition. Similar results were obtained for other PZ-CyD systems. The order of protective effect was β-> γ-> α-CyD, consistent with that of reduction in drug-induced hemolysis. In our preliminary studies, however, it has been recently found that CyDs themselves caused the partial

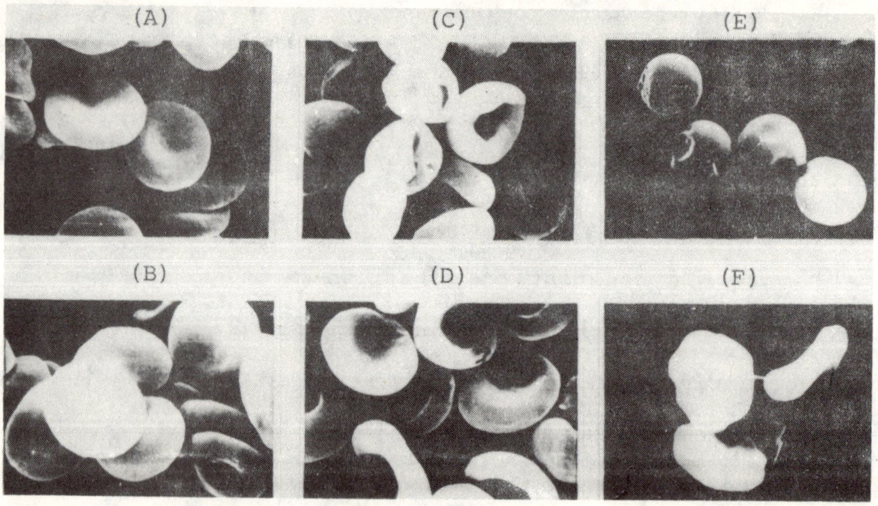

Fig. 3. Scanning Electron Micrographs of Erythrocytes Treated with CPZ in the Absence and Presence of β-CyD (1×10^{-3} M)

Magnification of the photographs is 7000 x.
(A): control (B): β-CyD (C): CPZ(6×10^{-5} M) (D): CPZ(6×10^{-5} M) + β-CyD (E): CPZ(2×10^{-4} M) (F): CPZ(2×10^{-4} M) + β-CyD.

swelling of cells at the relatively high concentrations of CyDs.

To gain insight into the protective mechanism of CyD, some factors responsible for PZ-induced hemolysis (7) were investigated. A good correlation between the stability constants and the inhibitory effects on the drug uptake into erythrocytes was found, as shown in Fig. 4, indicating that the binding activities of the complexed PZ are much smaller than those of PZ. The surface activities of PZ were also reduced by CyDs in the order of β-> γ-> α-CyD, consistent with those of reduction in PZ-induced hemolysis. These may be due to the reduction of hydrophobicities and the membrane permeability of PZ molecule by sheaths of hydrophilic CyDs. By the addition of CyDs, no remarkable changes in osmotic pressure and viscosity of the drug solutions were observed under these experimental conditions. Moreover, CyD showed a little protective or accelerative effect against the osmotic hemolysis. These results clearly indicate that the protective effect of CyD was probably due to the

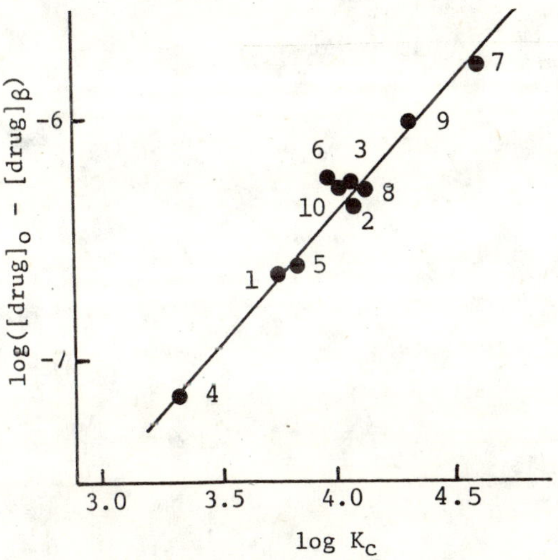

Fig. 4. Relationship between ($[drug]_o$ - $[drug]_\beta$) and K_c

Concentration: drugs, 1.0×10^{-4} M; β-CyD, 1.0×10^{-3} M.
$[drug]_o$: uptake of drug in the absence of β-CyD; $[drug]_\beta$: uptake of drug in the presence of β-CyD.
Numbers refer to compounds in Table II.

decrease in effective hemolytic concentration of PZ through inclusion complex formation rather than the direct interaction of CyD with erythrocyte membrane.

The therapeutic effects of PZ including CPZ are known to be proportionate to the extent of hemolysis as well as the interactive abilities of the drugs on the erythrocyte membrane (8). Since β-CyD was found to significantly inhibit the interaction of human erythrocytes with PZ, it seemed interesting to study the effect of β-CyD on the pharmacological activities of PZ *in vivo*. Some pharmacological

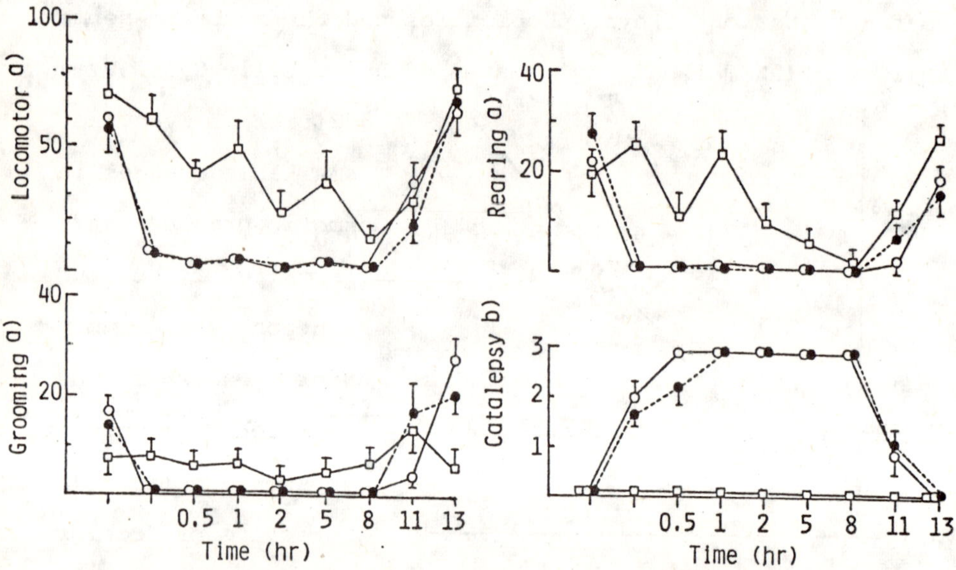

Fig. 5. Effect of β-CyD on Some Pharmacological Activities of CPZ in Rats

Mean ± S.E. obtained from 6 animals. In the abscissa, the time is spaced ununiformly.
a) Psychomotor activities (counts/10 min/animal)
b) (score/animal)
○: CPZ (10 mg/kg) ●: CPZ (10 mg/kg) + β-CyD (32 mg/kg)
□: saline.

activities of CPZ and its β-CyD complex on C.N.S. were thus compared (Fig. 5). Locomotor activity of the rats in the control group after saline administration somewhat decreased and fluctuated when measured at different times of the day. The fluctuation was likely to depend upon biological rhythm known to occur in the spontaneous motor activity of rats. Locomotor activities in the groups administered CPZ and its complex were almost completely suppressed for 8-11 h. The suppression was initiated at 15 min after the administration of both samples. Similarly to locomotor activity, grooming and rearing markedly decreased by 15 min after administration of CPZ and its complex. The behaviors were completely suppressed for at least 8 h and recovered 11-13 h after administration of the samples. Thus, psychosedative actions of CPZ and its complex appeared to be closely similar and neither qualitative nor quantitative difference was observed. There found a slight difference between CPZ and its complex in the time-course of their cataleptogenic actions. Pronounced catalepsy (score 3) appeared in all the animals by 30 min after administration of CPZ, while the complex induced the same degree of catalepsy by 1 h after administration. The cataleptogenic actions of both samples diminished and disappeared 11 h and 13 h after administration, respectively.

Aforementioned results apparently suggest that β-CyD does not seem to alter the time-course or magnitude of the studied effects of CPZ on C.N.S.. This might be explained on the basis of dissociation equilibrium of the complex. Soon after the end of injection, the complexed CPZ may be quickly dissociated by the circulating blood. If such was the case, no appreciable difference in the pharmacological activities of CPZ and its β-CyD complex could be observed. Although detailed mechanism should be investigated further, these limited data suggest that β-CyD is particularly useful in alleviating an injury to the membrane at the injection site.

ACKNOWLEDGEMENT

The authors are grateful to Dr. K. Tsubaki, Mr. Y. Okano, and Mr. K. Iwasaki of Kumamoto University for their helpful discussions throughout this work.

"Literature Cited"

1) Ogiso, T., Masuda, H., Oue, S.: Biochem. Pharmacol., 27, 1263 (1978).
2) Uekama, K.: Jpn. J. Antibiot., 32, s-103 (1979).
3) Saenger, W.: Angew. Chem. Int. Ed. Eng., 19 344 (1980).
4) Uekama, K., Irie, T., Sunada, M., Otagiri, M., Tsubaki, K.: J. Pharm. Dyn., 4, 142 (1981).
5) Uekama, K., Irie, T., Sunada, M., Otagiri, M., Iwasaki, K., Okano, Y., Miyata, T., Kasé, Y.: J. Pharm. Pharmacol., in press.
6) Uekama, K., Irie, T., Hirayama, F.: Chem. Lett., 1978, 1109.
7) Ahtee, L.: Ann. Med. Exp. Fenn., 44, 453 (1966).
8) Seeman, P.: Pharmacol. Rev., 24, 583 (1972).

I.Int.Symp. on Cyclodextrins
Budapest, 1981 5.4.

ENHANCED BIOAVAILABILITY OF DIGITALIS GLYCOSIDE BY CYCLODEXTRIN COMPLEXATION

K. UEKAMA,[1] T. FUJINAGA,[1] F. HIRAYAMA,[1] M. OTAGIRI,[1] H. SEO,[2] M. TSURUOKA[2]

[1] Faculty of Pharmaceutical Sciences, Kumamoto University
Kumamoto 862, Japan
[2] Department of Pharmacy, Miyazaki Medical College Hospital
Miyazaki 889-16, Japan

SUMMARY

Inclusion complex formations of some digitalis glycosides (digoxin, digitoxin, methyl digoxin) with three cyclodextrins (α-, β-, γ-CyDs) in aqueous solution and in solid phase were studied by solubility method, spectroscopies (UV, CD, IR, NMR) and X-ray diffractometry, and their mode of interactions were assessed. Acid hydrolysis of digoxin was found to be retarded by CyDs (in the order of β-> γ-> α-CyD), depending upon the fitness of the drug to CyD cavity. The solid complexes of digitalis glycosides with γ-CyD in a molar ratio of 1:4 were obtained, and their dissolution behavior, chemical stability, and oral bioavailability were examined. The rapid dissolving form of digoxin—γ-CyD complex, for example, increased the plasma levels of digoxin (about 5.4 fold) after oral administration to dogs. The enhanced bioavailability of the drug by inclusion complexation suggests the possibility of smaller doses and fewer side effects in oral digitalis glycoside therapy.

The bioavailability of digitalis glycoside preparation is known to vary significantly among brands and batches. The main cause of this variability seems to be related to the unsatisfactory factors of the drug such as extremely low water solubility and chemical instability in acidic medium (1-3). Cyclodextrins (CyDs) have been extensively applied to improve the physicochemical properties of various drug molecules (4-6). One of the important characteristics of CyDs is the formation of inclusion complex in solid phase and in solution, in which the drug molecule is included in the relatively hydrophobic cavity of CyDs (7). The present study dealt with inclusion complexations of digitalis glycosides (digitoxin, digoxin, methyl digoxin) with three CyDs (α-, β-, γ-CyDs) in anticipation of improved dissolution and chemical stability of the drugs. In addition the study on bioavailability of the inclusion complex was conducted by oral administration to dogs, measuring the plasma levels of the drug (8).

RESULTS AND DISCUSSION

Evidence of Inclusion Complex Formation

Complex formation in aqueous solution was studied by solubility method (9) and spectroscopies. Figure 1 shows the phase solubility diagrams obtained for digitoxin with three CyDs in water, as a typical example. The solubility of digitoxin increased by the addition of CyDs, showing a B_S type of solubility curve (9). In the higher concentration range of β- and γ-CyDs, microcrystalline complexes were deposited.

In sharp contrast, no precipitation was observed for α-CyD system. This might de due to the smaller cavity size of α-CyD, therefore allowing very little penetration of the bulky guest molecule. Table I summarizes the apparent stability constants and stoichiometries of the complexes estimated from solubility data. The cavity size dependency of inclusion complexation was further confirmed by CD (circular dichroism), UV, and IR spectroscopies.

^{1}H-NMR technique (200 MHz) was employed to gain insight into the inclusion mode of the digitalis glycoside-CyD complexes in aqueous solution. Table II summarizes the effects of CyDs on some ^{1}H-chemical shifts (18-methyl and 19-methyl) of digoxin (10). Unfortunately, the other proton signals were too weak to be quantitatively analyzed under the experimental conditions used. In the presence of CyDs, both signals moved downfield probably due to the steric perturbation through inclusion complexation. Interestingly, the cavity size of β-CyD appeared to be most appropriate to include the A-ring moiety of digoxin molecule, since

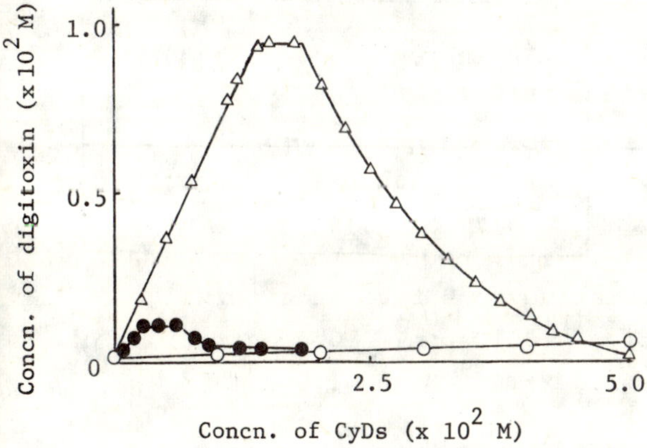

Fig. 1. Phase Solubility Diagrams of Digitoxin-CyD Systems in Water at 25°

Key: o; α-CyD system, •; β-CyD system, Δ; γ-CyD system.

Table I. Apparent Stability Constants (M^{-1}) of Digitalis Glycoside-CyD Complexes Determined by Solubility Method in Water at 25°

guest molecule	host molecule		
	α-CyD	β-CyD	γ-CyD
Digoxin	180	11200	12200 (4:1)
Digitoxin	290	17000	63600 (4:1)
Methyl digoxin	400	11400	13600 (4:1)

Values in parenthesis are molar ratio (CyD : guest) of solid complexes, isolated from the descending curvature of B_s type phase solubility diagrams.

Table II. Effects of CyDs on ^{1}H-Chemical Shifts (ppm) of Some Digoxin Protons in D_2O[a)]

proton[b)]	without CyDs	^{1}H increment,[c)] with CyDs		
		α-CyD	β-CyD	γ-CyD
18-methyl	0.772	0.038	0.055	0.040
19-methyl	0.932	0.017	0.198	0.081

a) Measured at 200 MHz relative to DSS as an external reference, accuracy of ± 0.0012 ppm. Concentrations of digoxin and CyDs were 1.0×10^{-4} M and 1.0×10^{-3} M, respectively.
b) Assigned according to ref. 10.
c) The values are positive if any upfield shift of the resonance occurs.

the signal for 19-methyl significantly moved downfield upon binding to β-CyD. The magnitude of the downfield shifts of both methyl protons decreased in the order of β- > γ- > α-CyD, suggesting that the digoxin molecule is located at the entrance of the α-CyD cavity, it could penetrate further into the β-CyD cavity, and is loosely bound to γ-CyD.

Effects of CyDs on Acid Hydrolysis of Digoxin

It is known that the digitalis glycosides are susceptible to hydrolysis in acidic media and the hydrolysis products exhibit different therapeutic efficacy: e.g. the cardioactivity of digoxigenin is one-tenth of that of digoxin (11). Digoxin hydrolysis is a complex combination of parallel reaction (2) that can proceed initially by three alternative pathways to yield digoxigenin, which may undergo further hydrolysis (scheme 1). In the present study, the effects of three CyDs on the hydrolysis of digoxin were investigated at 37° at pH 1.66. From the high

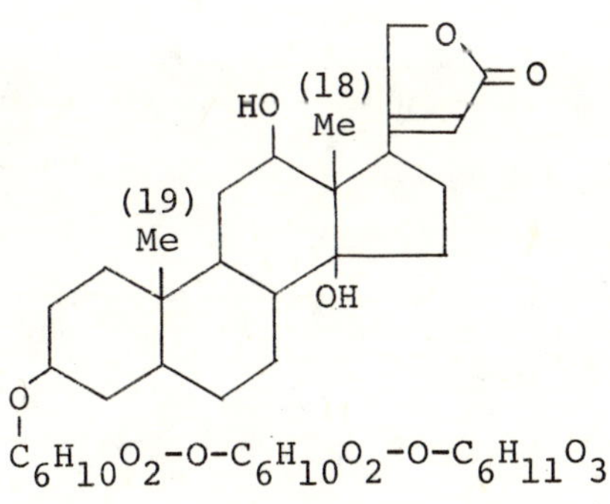

Digoxin

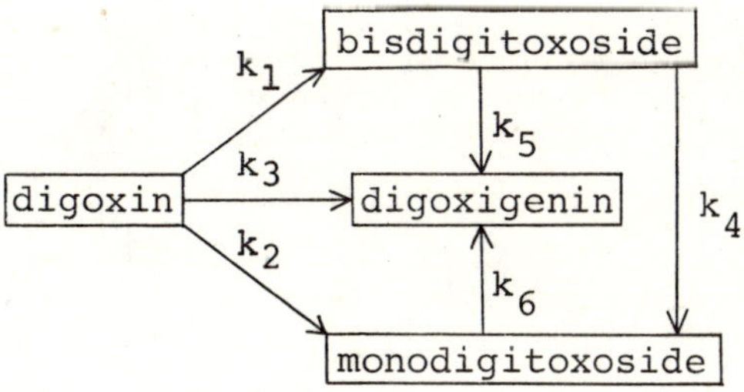

Scheme 1

performance liquid chromatography (HPLC) tracing of each of the four components in reaction mixtures, the individual hydrolysis rate constants (k_1-k_6) were determined by analog computer fitting procedure. As shown in Table III, CyDs retarded the hydrolysis in the order of β-> γ- > α-CyD, where β-CyD inhibited the conversion of monoglycoside to digoxigenin (k_3, k_5, and k_6) almost completely. This indicates that the smaller or larger CyD cavity than β-CyD is less effective to protect the breakdown of monoglycoside bond, as would be expected from NMR study. Above results suggest that the variation in oral bioavailability of digitalis glycosides owing to acid hydrolysis in stomach may be improved by inclusion complexation.

Table III. Effects of CyDs on the Rate Constants (hr^{-1}) of Digoxin Hydrolysis[a)] in HCl-KCl Buffer (pH 1.66) at 37°

	without CyDs	with α-CyD	with β-CyD	with γ-CyD
k_1	0.375	0.340	0.338	0.345
k_2	0.145	0.130	0.101	0.118
k_3	0.171	0.108	0.002	0.025
k_4	0.325	0.288	0.255	0.295
k_5	0.189	0.070	0.005	0.020
k_6	0.285	0.128	0.002	0.056

a) Concentrations of CyDs and digoxin were 1×10^{-2} M and 1×10^{-4} M, respectively.

Dissolution Behavior of γ-CyD Complex

Interaction of digitalis glycosides with γ-CyD in solid phase were confirmed by IR spectroscopy and X-ray diffractometry. Since the X-ray data for γ-CyD complex showed rather amorphous patterns than that of digitalis glycoside (see Fig. 2), the effect of crystalline modification on the dissolution rate of the drug was examined in acidic medium (pH 1.52). Figure 3 shows the release profiles of digoxin and its γ-CyD complex from the rotating disk with constant surface area, where intact digoxin was quantitatively determined by HPLC. It is evident that the complexed form of digoxin dissolved much more rapidly (about 100

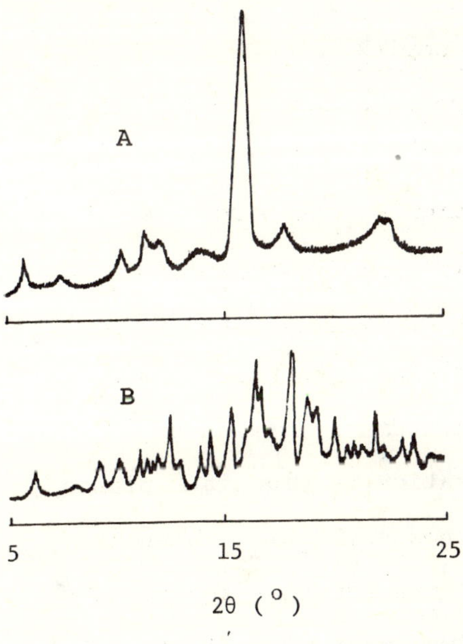

Fig. 2. Powder X-Ray Diffraction Patterns of Digoxin—γ-CyD System
Key: A; complex of digoxin with γ-CyD,
B; physical mixture of digoxin and γ-CyD.

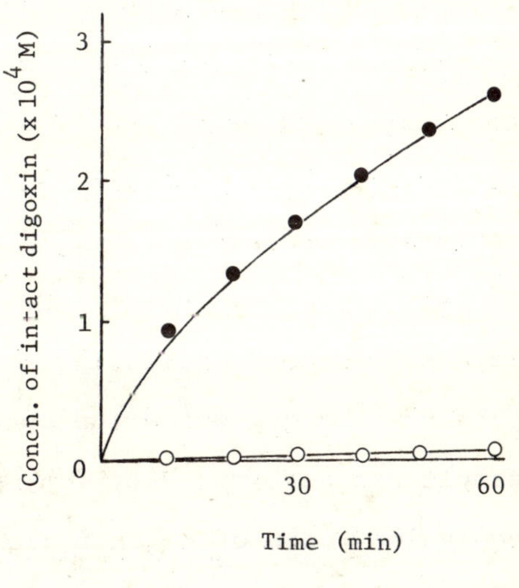

Fig. 3. Dissolution Behaviors of Digoxin (o) and Its γ-CyD Complex (•) in Acidic Medium (pH 1.52) at 37°, Measured by Rotating Disk Method

fold) than digoxin itself, which may be due to the increase in solubility and/or the decrease in crystallinity of the drug by γ-CyD complexation. During the dissolution experiments, the simultaneous conversions of digoxin to hydrolysis products were also followed by HPLC. The rapid dissolving form of γ-CyD complex showed the reduced decomposition compared to that of digoxin alone. Similar results were obtained for other digitalis glycoside—γ-CyD systems. The increased dissolution rate together with improved chemical stability in acidic medium suggest that the γ-CyD complex of digitalis glycoside may have a good oral bioavailability.

Bioavailability of Digoxin—γ-CyD Complex

In vivo absorption studies were undertaken to find if the *in vitro* dissolution enhancement of digoxin from its γ-CyD complex increases in the GI absorption of the drug. The fine powder of digoxin and its γ-CyD complex were compressed to tablets in the average weight of 50 mg to give final digoxin content of 0.1 or 0.05 %. Figure 4 shows the mean plasma levels of digoxin following the oral administration of the tablets to dogs, where the concentration of digoxin in plasma sample was determined by enzymeimmunoassay (12). When the equivalent dose of digoxin (100 μg) was administered to dogs, the γ-CyD complex attained the maximum plasma level of 0.90 ± 0.14 mg/liter at 45 minutes, which was about 3 times higher than that of digoxin alone. The area under plasma concentration time curve (AUC) of the complex up to 24 hr was found to be 5.4

times as much as that of digoxin alone. The AUC of γ-CyD complex containing 50 μg of digoxin was found to be superior to that of digoxin alone (100 μg). Thus, the improved bioavailability of digoxin by γ-CyD complexation suggests the decrease in doses and fewer side effects in oral digitalis glycoside therapy. It should be also noted that the 1 : 4 complexation of digitalis glycosides with γ-CyD results in approximately 8 fold increase in molecular weight of the drug, which may facilitate the pharmaceutical preparation, particularly from the viewpoint of content uniformity test.

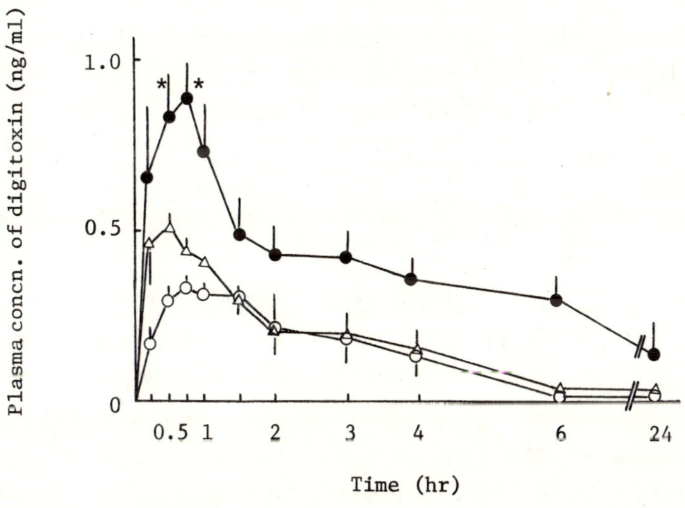

Fig. 4. Plasma Levels of Digoxin Following the Oral Administrations of Tablets Containing Digoxin or Its γ-CyD Complex to Dogs

Key: o; digoxin tablet containing 100 μg digoxin,

•; γ-CyD complex tablet containing 100 μg digoxin,

Δ; γ-CyD complex tablet containing 50 μg digoxin,

Each point represents the means ± S.E. of six dogs.

* $P < 0.01$ in (•) versus (o).

ACKNOWLEDGEMENT

The authors are grateful to Dr. M. Yamasaki, Faculty of Medical Sciences, Kumamoto University, for the analysis of NMR spectra. Thanks are also due to Dr. Y. Kurono, Faculty of Pharmaceutical Sciences, Nagoya City University, for the computer analysis of kinetic data.

"Literature Cited"

1 Reddy, R.K., Khalil, A., Gouda, M.W.: J. Pharm. Sci., 65, 1753 (1976).

2 Sternson, L.A., Shaffer, R.D.: J. Pharm. Sci., 67, 327 (1978).

3 Sonobe, T., Hasumi, S., Yoshino, T., Kobayashi, Y., Kawata, H., Nagai, T.: J. Pharm. Sci., 69, 410 (1980).

4 Frömming, K., Weyerman, I.: Arzneim-Forsh., 23, 424 (1973).

5 Uekama, K., Hirayama, F., Yamada, Y., Inaba, K., Ikeda, K.: J. Pharm. Sci., 68, 1059 (1979).

6 Saenger, W.: Angew. Chem. Int. Ed. Engl., 19, 344 (1980).

7 Bender, M.L., Komiyama, M.: Cyclodextrin Chemistry, Springer-Verlag, Berlin-Heidelberg-New York, 1978.

8 Uekama, K., Fujinaga, T., Otagiri, M., Seo, H., Tsuruoka, M.: J. Pharm. Dyn., 4, 726 (1981).

9 Higuchi, T., Connors, K.A.: Adv. Anal. Chem. Instr., 4, 117 (1965).

10 Voigtlander, H.W., Balsam, G.: Arch. Pharmaz., 310, 208 (1968).

11 Kroneberg, G.: Arch. Exp. Pathol. Pharmakol., 237, 222 (1965).

12 Drost, R.H., Plomp, T.A., Teunissen, A.J., Maas, A.H.J., Maes, R.A.A.: Clin. Chim. Acta, 79, 557 (1977).

INFLUENCING DRUG ABSORPTION BY β-CYCLODEXTRIN COMPLEXATION

ANNA FÓNAGY,[1] ANDREA GERLÓCZY,[1] PÉTER KERESZTES,[1]
JÓZSEF SZEJTLI[2]

[1] "Frédéric Joliot-Curie" National Research Institute
for Radiobiology and Radiohygiene, Pentz K. u. 5
H-1221 Budapest, Hungary

[2] CHINOIN Pharmaceutical-Chemical Works, Tó u. 1-5
H-1045 Budapest, Hungary

By cyclodextrin complexation the solubility of poorly water-soluble compounds of various drugs can be increased, i.e. higher bioavailability can be achieved following oral or rectal administration.

Absorption of two practically insoluble drugs and their beta-cyclodextrin inclusion complexes were investigated.

Rats were treated orally with ^{3}H-Vitamin D_3 (cholecalciferol) or ^{3}H-Vitamin D_3-beta-cyclodextrin complex. The radioactivity of blood was measured by the liquid scintillation method. The radioactivity of blood was higher, i.e. absorption from the intestine was larger after treatment with ^{3}H-Vitamin D_3-beta-cyclodextrin complex. The radioactivity was approximately twice as much in the case of complex as in case of free Vitamin D_3 during the first 3 hours.

Indomethacin and indomethacin-beta-cyclodextrin complex was administered orally to rats. Indomethacin level of blood was measured by high pressure liquid chromatography. The blood indomethacin maximum was found between 1 and 2 hours after treatment. It was approximately 25% higher in the case of complex.

By cyclodextrin complexation the solubility of poorly water-soluble compounds of various drugs can be increased, i.e. higher bioavailability can be achieved following oral or rectal administration. Absorption of two practically insoluble drugs and their β-cyclodextrin /βCD/ complexes were investigated.

A considerable difficulty in the application of vitamin D_3 /cholecalciferol/ in human and veterinary medicine is its extreme instability. It is namely rapidly decomposed by oxidation, heat, light and metal ions. Inclusion of vitamin D_3 into βCD significantly improves its resistance towards these effects /1,2/. The improved bioavailability was described by Szejtli and co-workers /1/, who induced rachitis in rats by feeding them a vitamin D_3 deficient diet, followed by treatment by uncomplexed and complexed vitamin D_3, respectively. Oral administration of the complex over a period of 28 days proved to be more effective in normalizing P and Ca serum levels as well as the calcification of the bones than the uncomplexed vitamin.

It was assumed that complexed vitamin D_3 exerts its higher biological effect by ensuring higher blood levels of vitamin D_3 than treatment with the uncomplexed vitamin. Rats were treated orally with ^{3}H-vitamin D_3 /107 μg/kg/ and its βCD complex /81 μg/kg ^{3}H-vitamin D_3; percentage of ^{3}H-vitamin D_3 of the complex was 2.9/ respectively, and the change of radioactivity as a function of time was determined /Fig.1/.

Oral administration of complexed ^{3}H-vitamin D_3 resulted in a higher level of relative radioactivity in blood than the uncomplexed ^{3}H-vitamin D_3. This result is in accordance with the enhanced biological effect of complex described by Szejtli and co-workers /1/.

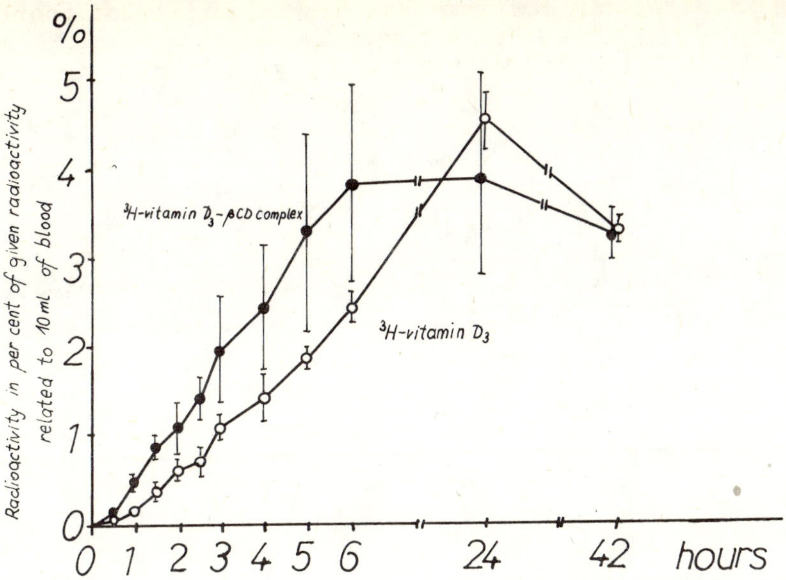

Fig.1

Absorption of radioactivity after oral administration of ^{3}H-vitamin D_3 and its βCD complex

One of the harmful side effects of the well-known non-steroidal antiinflammatory agent, indomethacin is its ulcus-inducing effect. This ulcus-inducing effect of orally administered indomethacin is much lower if the drug is applied in form of βCD inclusion complex /molecular ratio: indomethacin : βCD = 1:2/ than added as a "free" molecule /3/. At the same time the effect of complex proved to be better in diminishing the carrageen-induced inflammation in rats than indomethacin alone /3/.

It was supposed that the improvement of bioavailability of complexed form is due to a higher blood level of indomethacin. Rats were treated orally with 30 mg/kg indomethacin and its βCD complex /30 mg/kg indomethacin, percentage

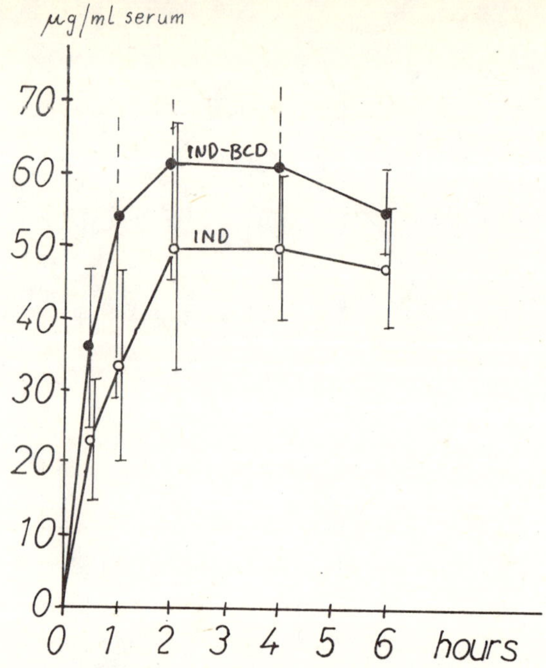

Fig.2

Absorption of indomethacin and its βCD complex after oral administration

of indomethacin of the complex was 14/, respectively. Change of blood level of indomethacin was detected by high pressure liquid chromatography method /Fig.2/.

Oral administration of complexed indomethacin resulted in about 25 % higher blood level of indomethacin.

The conclusion may be drawn that complexation of poorly water-soluble drugs with βCD results an increased bioavailability following oral administration.

Literature Cited

1. Szejtli, J., Bolla-Pusztai, É., Szabó, P., Ferenczy, T.: Pharmazie 35 779 /1980/

2. Szejtli, J., Bolla, É.: Stärke 32 386 /1980/

3. Szejtli, J., Szente, L., Dávid, Á., Virág, S., Sebestyén, Gy., Mándi, A.: Hung.Patent Appl. CI-1679 /1976/

I.Int.Symp. on Cyclodextrins
Budapest, 1981

COMPLEX EQUILIBRIUM AND BIOAVAILABILITY

I. HABON, J. SZEJTLI

CHINOIN Pharmaceutical-Chemical Works, Tó u. 1-5
H-1045 Budapest, Hungary

SUMMARY

Pharmacokinetic properties and bioavailability of drugs can be modified by complexing them with cyclodextrins.

The majority of these complexes can be prepared in solid, crystalline form. Complexation means enhanced molecular dispersity and consequently the dissolution rate of drugs of poor water solubility increases generally by about 1 order of magnitude. This rapid dissolution makes possible to reach complex equilibria in the gastrointestinal tract.

Complex dissociation equilibrium of a definite drug in cyclodextrin containing system is governed primarily by temperature, dilution and concentration ratios. At physiological temperature the equilibrium can be modified either by dilution or by varying the concentration of the complexing agent according to the law of mass action.

A simplified theoretical model based on complex chemical considerations could help to examine under what conditions enhanced, or on the contrary, retarded biological effect of a drug can be reached by cyclodextrin complexation.

Results of biological experiments indicate that the use of cyclodextrins and their complexes in pharmaceutical products often modify the pharmacokinetic properties and the bioavailability of drugs. In the present paper we should like to call the attention to some complex chemical equilibria which must have an influence on the biological effect of cyclodextrin-drug complexes.

The considerations presented here are simplified to rather chemical than biological systems, are based on calculations and remain hypothetical until they are proved by experimental chemical and biological tests which are awaited in the near future.

The fate of an orally administered solid drug can be described by the following simplified model /Fig.1./.

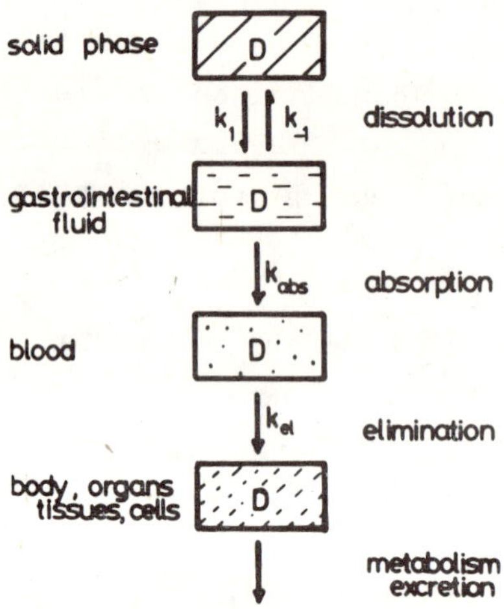

$$\frac{d[D]_{aq}}{dt} = k_1 - k_{-1}[D]_{aq} - k_{abs}[D]_{aq} \qquad \text{I.}$$

$$\frac{d[D]_{blood}}{dt} = k_{abs}[D]_{aq} - k_{el}[D]_{blood} \qquad \text{II.}$$

Fig. 1. Fate of an orally administered solid drug in the organism.

The solid drug dissolves in the gastrointestinal fluid and only the dissolved drug molecules are able to enter the blood circulation, get to the organs, tissues and cells where they are metabolized and finally excreted. All these processes can be characterized by various rate constants. In this model only the dissolution rate constants and its reverse as well as the absorption and elimination rate constants are considered. The concentration of the dissolved drug in the gastrointestinal fluid can be described with good approximation by means of the equation N° I., while equation N° II. represents the blood level curve. Rather unusually in the pharmacokinetics the concentration of the dissolved drug in the gastrointestinal fluid has been chosen in this model and regarded as a variable determining the rate of all the following steps and the drug levels in the blood or in other parts of the body. It is this variable that can be modified by the use of cyclodextrins or their complexes in the pharmaceutical products.

It is known that complex formation usually has two main effects on the properties of the drugs. The dissolution rate of the complex is often higher than that of the free solid drug by about one order of magnitude, and the solubility of the complex is also increased compared to the free drug. It means that after the administration of the complexed drug the total amount of the drug in the aqueous phase is much higher compared to the free drug. It is also known that complex dissociation equilibria in aqueous phase can be reached in fractions of a second - so the dissolution process of the solid complex leads to a complex chemical equilibrium state in a rather short time /3-5 minutes/, during this time the absorption is almost negligible /Fig.2./.

The partition of the drug in its different forms, such as dissolved free drug molecules, dissolved complex, solid free drug and solid complex, depends on the solubility of the free drug and the complexed drug in the given medium, the stability constant of the cyclodextrin-drug complex, on the molar ratio of the components and on the volume of the liquid phase.

Assuming, that a drug, the water solubility of which is 10^{-4} M, forms a cyclodextrin complex with the stochiometry 1:1 and with the watersolubility 10^{-3} M, some equilibrium calculations were carried out, the results of which are shown in Table I. Having dissolved 10^{-4} mole of the complexed drug in different volumes of water and assuming hypothetical stability constants for the cyclodextrin- drug complex, well

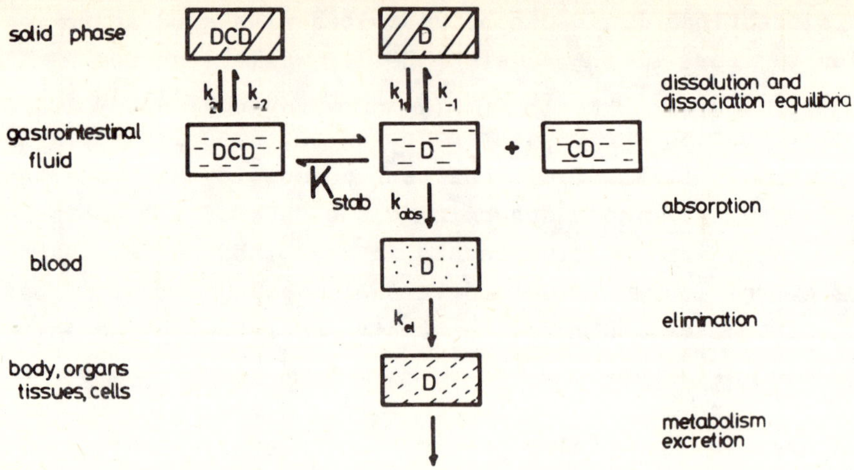

Fig. 2. Fate of an orally administered solid cyclodextrin-drug complex in the organism

$$K_{stab} = \frac{DCD}{D\ \ CD}$$

defined equilibrium states are reached that can be characterised by the partition of the drug in its different forms. These values are given in the table as percents of the total dose. Increasing the molar ratio of the components to 2:1 by additional amount of cyclodextrins the equilibrium states can be modified.

It is often the free drug that forms the solid phase on account of its lower water solubility, but with a well chosen dissolution volume and cyclodextrin excess an equilibrium state can be reached with the complex in the solid phase. This fact must have practical importance with drugs which have e.g. stomach wall irritating effect in their free form but exert considerably diminished local side effect in their complexed form. It can also be seen, that by stability constants high enough the whole dose can be transferred into the aqueous phase with additional cyclodextrin amounts.

It means that knowing the solubility of the drug and its complex as well as the stability constant, - at least theoretically - the desired equilibrium in any of the fol-

Table I.

Partition of the drug in its different forms, after the dissolution equilibrium is reached							
D = drug, CD = cyclodextrin DCD = complex		Partition in % of the dose /10^{-4} mole/					
		Dissolved in 10 ml		Dissolved in 100 ml		Dissolved in 1000 ml	
$K_{stability}$	Cyclodextrin mole/ drug mole ratio =	1:1	2:1	1:1	2:1	1:1	2:1
$K = 10^2$	Aqueous phase D_{aq}	1	1	10	10	99	98,1
	Aqueous phase DCD_{aq}	1	2	0,99	2	1	1,9
	Solid phase D_{Solid}	98	97	89	88	-	-
	Solid phase DCD_{Solid}	-	-	-	-	-	-
$K = 10^3$	Aqueous phase D_{aq}	1	0,99	10	10	91,6	84,4
	Aqueous phase DCD_{aq}	9,09	10	9,1	18,2	8,4	15,6
	Solid phase D_{Solid}	89,9	-	80,9	71,8	-	-
	Solid phase DCD_{Solid}	-	89,01	-	-	-	-
$K = 10^4$	Aqueous phase D_{aq}	1	0,1	10	8,44	61,8	41,4
	Aqueous phase DCD_{aq}	10	10	50	91,56	38,2	58,6
	Solid phase D_{Solid}	9	-	40	-	-	-
	Solid phase DCD_{Solid}	80	89,9	-	-	-	-

lowing four cases - as shown in Fig.3. - can be reached simply by varying the dilution volume and the cyclodextrin excess in the stomach after the administration of the same dose of the drug.

In case 1. the whole dose is to be found in the aqueous phase in form of free and complexed drug molecules; in case 2. the solid free drug precipitates in a fine microcrystalline form; in case 3. a part of the complex remains in the solid phase; in case 4. the solid phase consists of a mixture of the free drug and its complex.

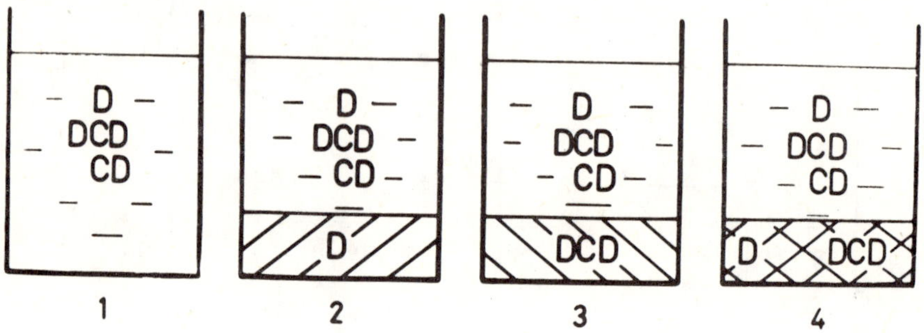

Fig.3. Dissolution and dissociation equilibria in aqueous cyclodextrin-drug complex systems.

Generally only dissolved drug molecules are ready to enter the circulation. Now let us consider, how the intestinal concentration of free dissolved drug changes during the absorption process.

Figure 4. shows dissolved drug concentration curves in the intestine after administration of the free drug. By a given absorption rate constant and dissolution rate the free drug level, shows a maximum, the height of which decreases with the increase of the diluting volume. It means, that the maximum temporary rate of the absorption proportional with the free drug concentration also decreases.

Enhancement in the dissolution rate, resulting in a higher concentration, might lead to higher blood level as

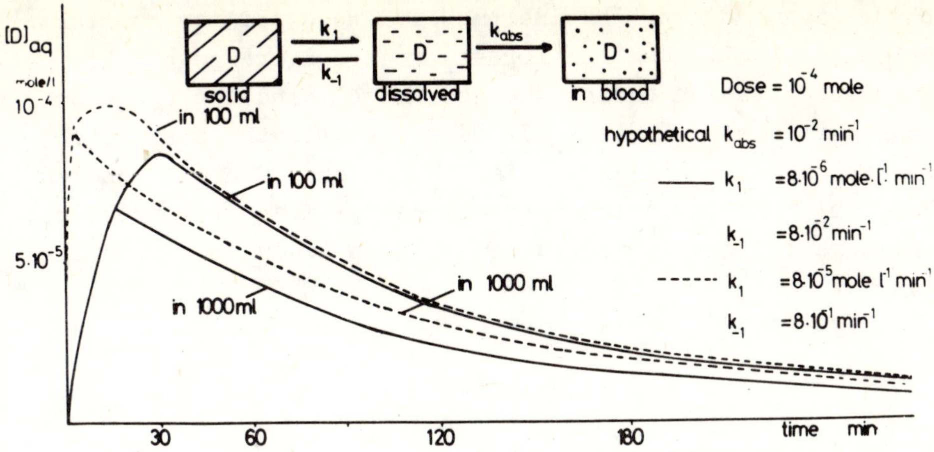

Fig.4. Time-dependent concentration of the dissolved drug in the gastrointestinal fluid

well. However it is not the only consequence of the complexation of a drug by cyclodextrin.

During the absorption only the free drug molecules can enter the circulation and the increasing excess of the left free cyclodextrin will push the complex equilibria toward the complex formation. This effect may bring dramatical decrease in the concentration of the free drug in the gastrointestinal fluid, which will be manifested in the absorption rate.

To show this trend in the case of the above mentioned four starting equilibria the free drug concentration in the intestine is represented as a function of the absorbed part of the dose in Fig.5-8. In the calculations we used the approximation that the system is closed - the volume of the aqueous phase remains constant, the total amount of the given cyclodextrin remains in the intestine - only the absorption of the drug is permitted.

Figure 5. shows the results of these calculations in the case 1. /see Fig.3./, where the absorption starts from the equilibrium state in which the total dose is dissolved in the given volume. The increase in the stability constant of the complex leads to a more expressed decrease of the free drug concentration during the absorption, and the cyclodextrin excess has similar effect. It means that if the stability constant of the complex is high enough and if a well chosen excess of the complexing agent is used, a very low free drug concentration can be maintained - which leads to a kind of retardation.

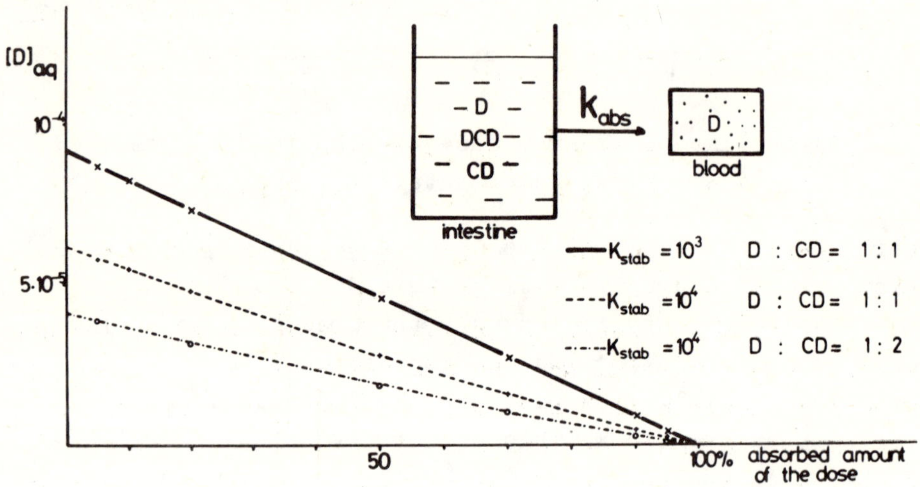

Fig. 5. Free dissolved drug concentration in the gastrointestinal fluid during the absorption, for case 1. in Fig. 3.

Starting from the case 2. /Fig. 3./ type equilibrium state the hypothetical dissolved drug concentrations are illustrated in Fig. 6. We have to emphasize that in this case the solid free drug is to be found in a fine microcrystalline form, which means a higher dissolution rate as compared to the normally administered free drug. It was observed that the dissolution rate of the drug is enhanced in cyclodextrin solutions - consequently it keeps increasing during the absorption with the continuously increasing free cyclodextrin concentration. The shape of those curves depends on the ratio

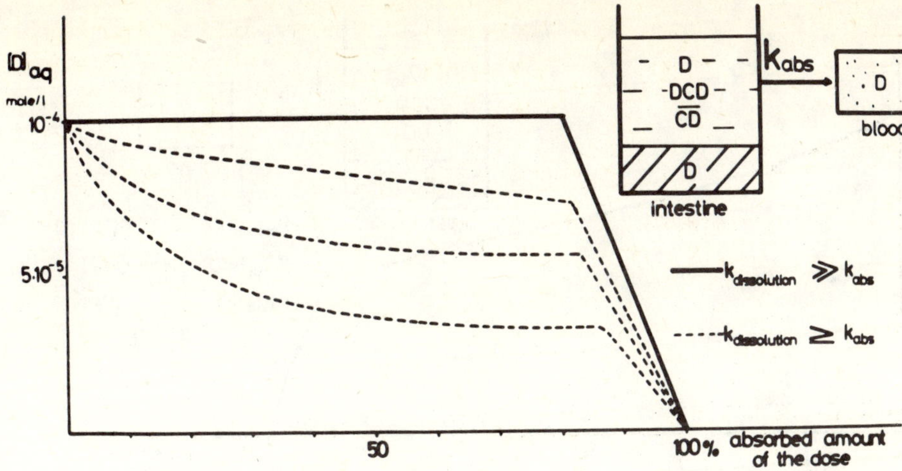

Fig. 6. Free dissolved drug concentration in the gastrointestinal fluid during the absorption, for case 2. in Fig. 3.

of the dissolution and absorption rate. Nevertheless the observation, that starting from this equilibrium an enhanced biological effect can be reached, seems to be of general character.

The case 3. type system /Fig. 7./ implies a model for retardation. While the complex, being in equilibrium with its solid form, has a constant concentration in the aqueous phase, the free drug concentration drops to low values.

The case 4. type system /Fig. 8./ is a special mixture of the two latter ones. First a high absorption rate can be reached and with the disappearance of the free solid drug it drops again as shown in the previous case.

Summarized, we wanted to show in a very simplified approximation that complex chemical considerations are indispensable in the interpretations of the biological effects of the cyclodextrin - drug complexes. The same complex may provoke different - enhanced or decreased - biological responses depending on the additional amount of cyclodextrin or even on the volume of water consumed.

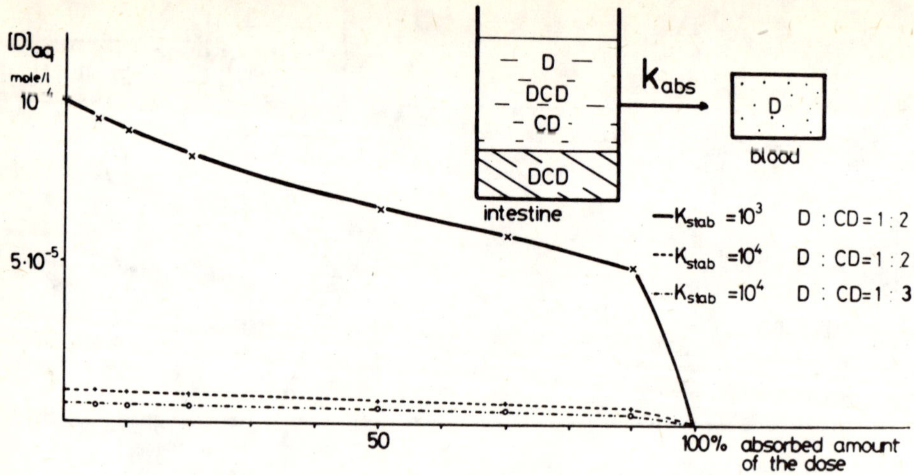

Fig. 7. Free dissolved drug concentration in the gastrointestinal fluid during the absorption, for case 3. in Fig. 3.

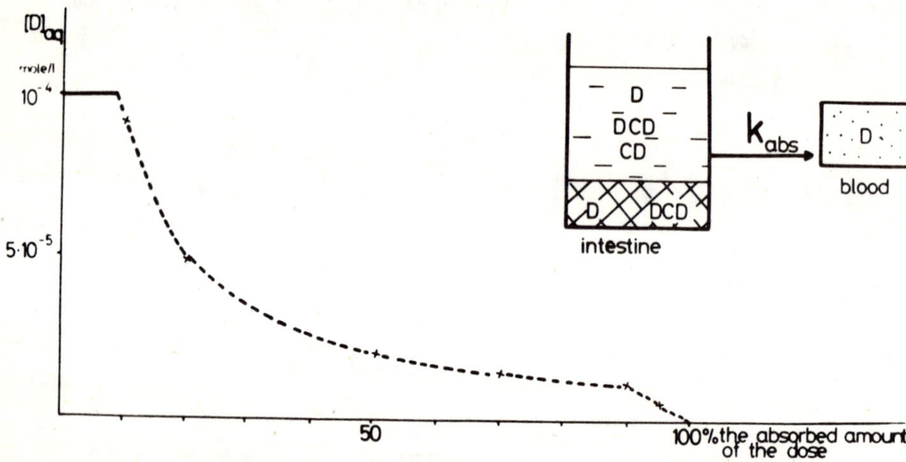

Fig. 8. Free dissolved drug concentration in the gastrointestinal fluid during the absorption, for case 4. in Fig. 3.

I.Int.Symp. on Cyclodextrins
Budapest, 1981 5.7.

MENADIONE-β-CYCLODEXTRIN COMPLEX: CHEMICAL PROPERTIES AND BIOLOGICAL EFFECTS

M.T. LENGYEL,[1] J. SZEJTLI,[1] P. SZABÓ,[2] T. FERENCZY[2]

[1] CHINOIN Pharmaceutical-Chemical Works, Tó u. 1-5
H-1045 Budapest, Hungary

[2] Department of Zoology and Anthropology, Kossuth Lajos University, Egyetem tér 1
H-4010 Debrecen, Hungary

SUMMARY

Properties of menadione-β-cyclodextrin complex in solutions were studied by UV spectrometry, circular dichroism and ^{1}H-NMR spectroscopy proving the existance of 1:1 molar complex in solutions. Stability constant K_c = 200 (20 °C).

X-ray diffraction and thermoanalytical studies as well as vacuum-sublimation experiments confirmed complex formation in crystalline state with 1:3 menadione: β-cyclodextrin molar ratio.

Recalcification, prothrombine and blood coagulation times determined in two strains of baby chickens show increased biological activity of menadione-β-cyclodextrin complex. Improved stability to heat and sublimation make the complex applicable for veterinary use as feed additive.

INTRODUCTION

Menadione (Vitamin K_3, 2-methyl-1,4-naphtoquinone) is an important feed additive in stock-breeding, especially in poultry raising. Baby chickens are extremely sensitive to vitamine K deficiency, leading up to 40 % mortality in the first few weeks of their life. But menadione easily undergoing sublimation rapidly disappeares from feeds and premixes. Thus a stable, biologically active vitamin K preparation has long been desired.

These expectations might be fulfilled by Menadione-β-cyclodextrin inclusion complex first prepared by Frömming (1). Physical, chemical properties and biological effects of the complex are reported here.

RESULTS and DISCUSSION

Menadione-β-cyclodextrin complex in solution

Results of circular dichroism studies prove that the molar ratio of menadione to β-cyclodextrin is 1:1 in aqueous solutions (Fig.1.).

Stability constant calculated: ~ 200 at 20 °C.

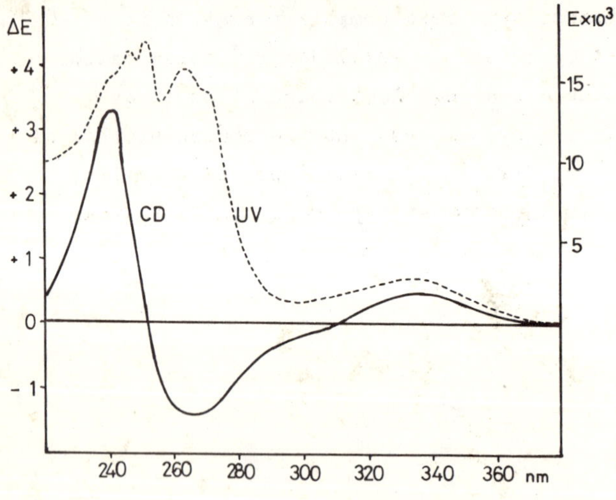

Fig.1.
Circular dichroism spectrum of menadione (dotted line represents the UV spectrum)

Shifts in the UV spectrum of the complex make it likely that the apolar benzene ring is located inside the cyclodextrin cavity, while the quinone part of the molecule is protruded, immersed into the more polar aqueous media.

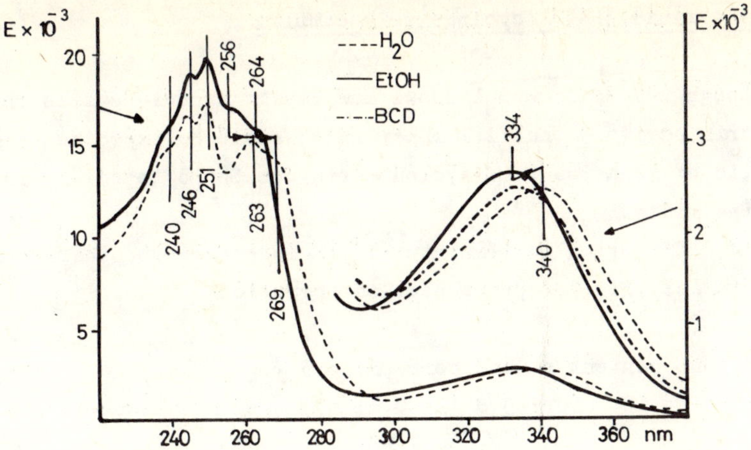

Fig.2. UV spectrum of menadione in Water, in ethanol and aqueous β-cyclodextrin solution

^{1}H-NMR studies also seem to support the above structure and prove that no dimer formation could be detected in β-cyclodextrin containing solutions.

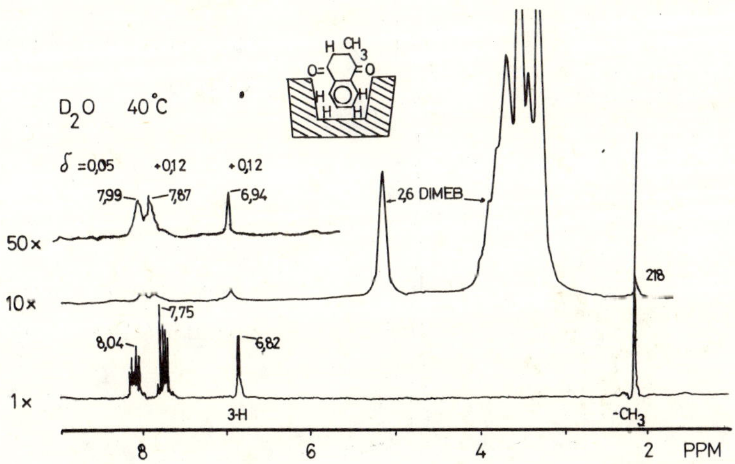

Fig.3. ^{1}H-NMR (100 MHz) spectrum of menadione in deutero-chloroform and in D_2O-2,6-dimethyl-β-cyclodextrin (2,6-DIMEB) solution.

Details of the above measurements are published in die Pharmazie (2).

Crystalline menadione-β-cyclodextrin complex

Although the scale model shows how easily menadione fits the cyclodextrin cavity - thus 1:1 molar ratio could be expected - the molar ratio of menadione: β-cyclodextrin was found to be 1:3 in crystalline state.

X-ray diffraction pattern, turbidity measurements, thermoanalytical studies (Fig.4.) prove complex formation.

Menadione content of the complex: 4,5 %
Solubility in water: 3,8 g/100 ml = 170 mg menadione/100 ml
37 °C, 3 hrs
Solubility of menadione: 16,5 mg/100 ml

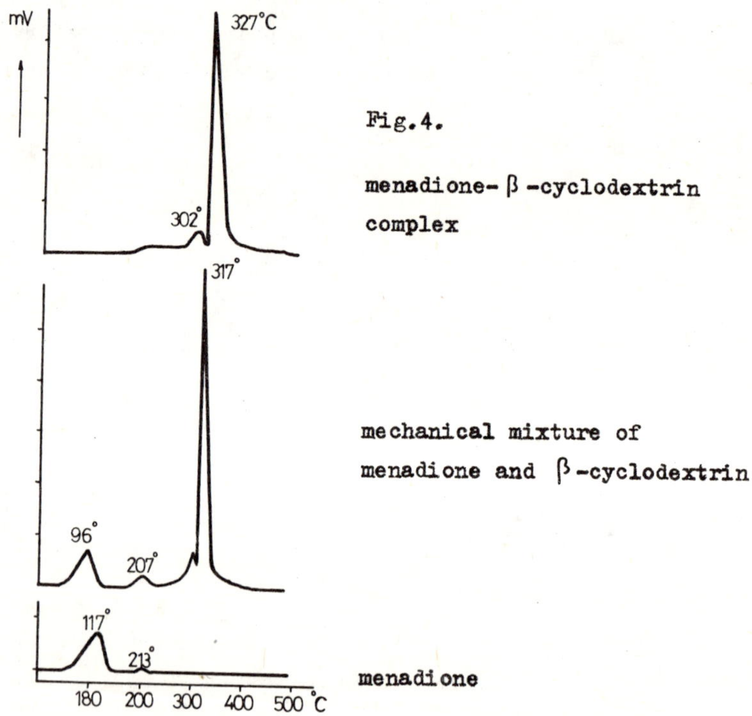

Fig.4. EGA (Evolved Gas Analysis) pattern of menadione, menadione-β-cyclodextrin complex and the mechanical mixture of menadione and β-cyclodextrin.

Menadione seems to have been stabilized by complex formation. Enhanced stability to heat and sublimation (Fig.5.) make possible the application of menadione-β-cyclodextrin complex in feeds and premixes for veterinary use.

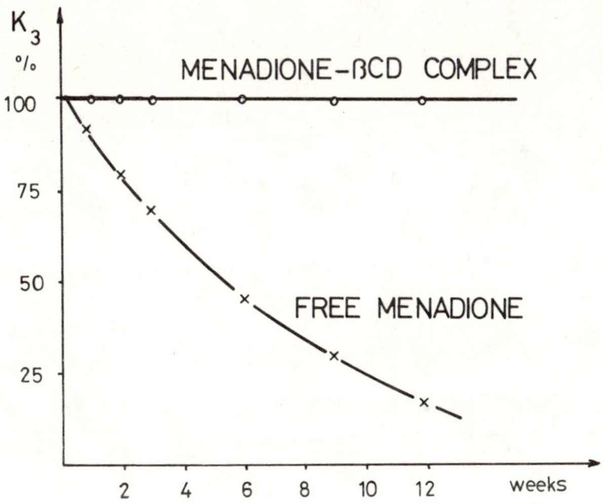

Fig.5. Loss of menadione in dry mixtures with amino acids at 60 °C

Biological effects:

Baby chickens on vitamin K free diet develop Dam syndrome (3) that means that blood coagulation time increases while prothrombine concentration drops (4,5). The effects of menadione and menadione-β-cyclodextrin complex treatments were studied in hypovitaminotic Leghorn and Rhode Island baby chickens monitoring clinical parameters.

Treatment pattern is represented in Table I.

As Fig.6. shows, high blood coagulation times of hypovitaminotic birds are restored by both menadione (1,5 µg/day) and its β-cyclodextrin complex (33 ug/day = 1,5 µg menadione/day) treatment.

Table I.

a.) One day old Leghorn chickens fed with corn grits mixed with hard boiled eggs for 7 days (to drink camomile tea was given)

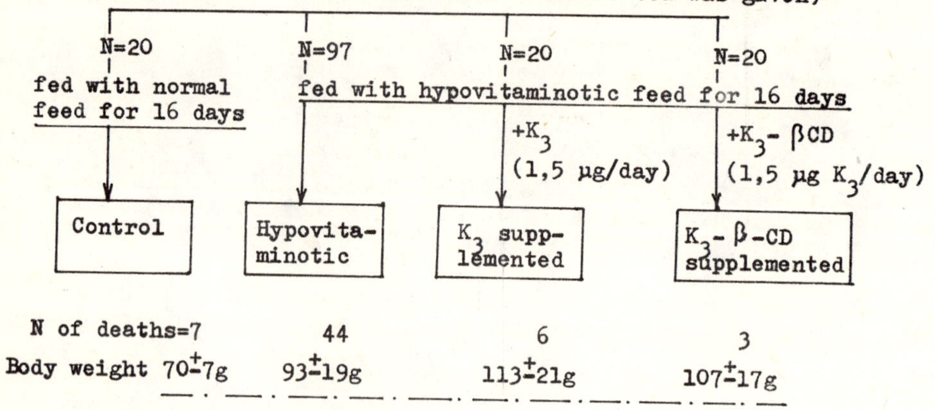

b.) One day old Rhode Island chickens (1 % benzoic acid solution was given to drink to prevent formation of intestinal bacterium flora)

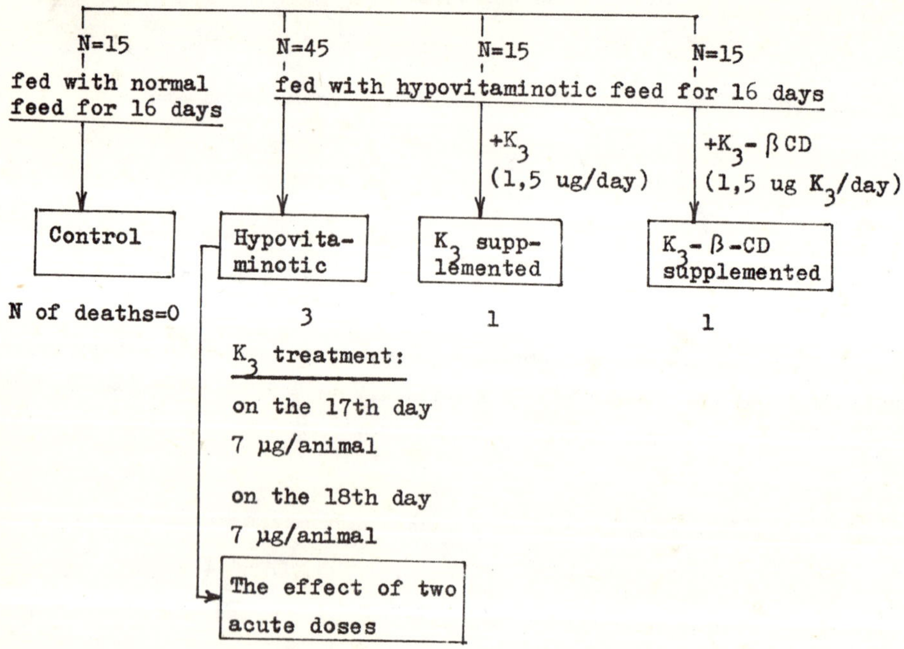

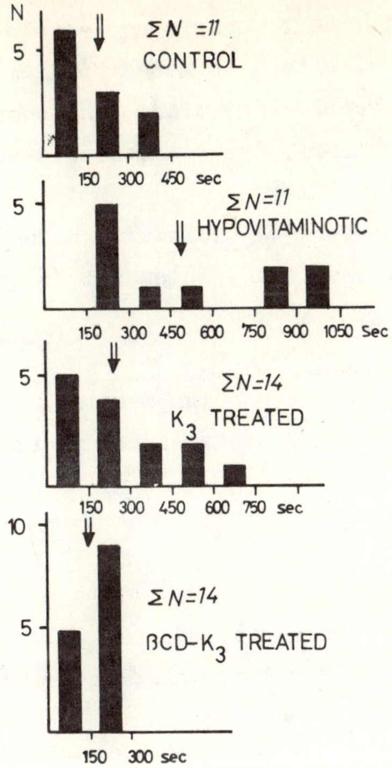

Fig.6.
Distribution of blood coagulation times of Leghorn chickens

Fig.7.
Recalcification times of Rhode Island chickens

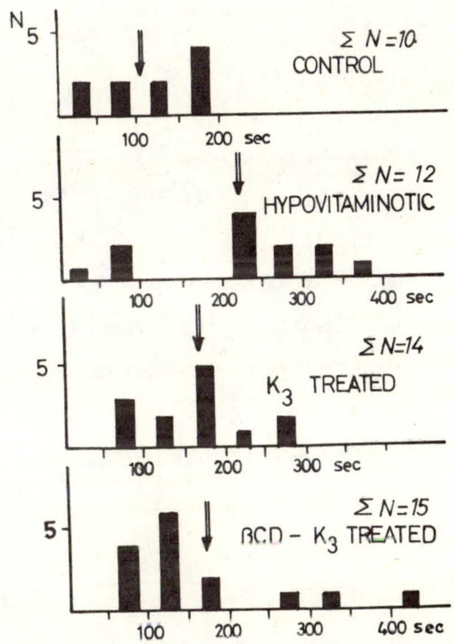

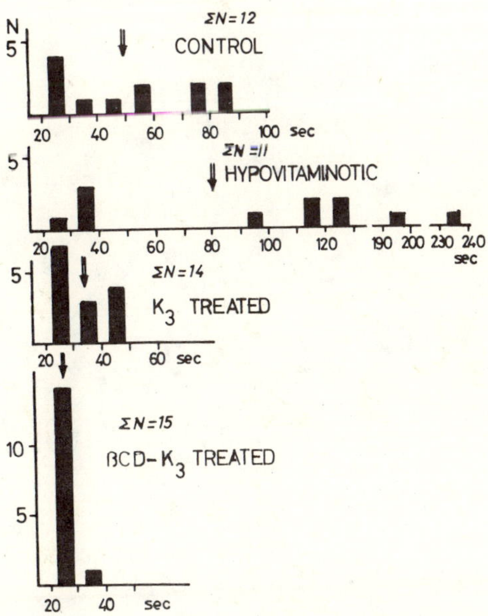

Fig.8.
Protrhrombine times of Rhode Island chickens

Summarizing the results shown in Fig. 6-8. menadione-β-cyclodextrin complex exerts the same biological effect as menadione itself. Concerning blood coagulation times and prothrombine times, the complex seems to have even enhanced biological activity due to increased solubility and consequently improved bioavailability.

Detailed studies on crystalline menadione-β-cyclodextrin complex and its biological effects will be published in die Pharmazie (6).

Acknowledgements:

Thank is due to Dr.M.Kajtár for circular dichroism studies, to Dr.A.Neszmélyi for NMR measurements and to Dr.J.Sztatisz for thermal analysis of the samples.

Literature Cited

1. Frömming, K.M., Weyermann, I.: Archiv.Pharmazie, 305 290 (1972)
2. Szejtli, J., Bolla-Pusztai, É., Kajtár, M.: Pharmazie, in press (1981)
3. Dam, H.: Biochem.Z. 215 475 (1929)
4. McFarlane, W.D., Graham, W.R., Hall, G.E.: J.Nutrit. 4 331 (1931)
5. Schonheyder, F.: Nature 135 653 (1933)
6. Szejtli, J., Bolla-Pusztai, É., Lengyel, M.T., Szabó, P., Ferenczy,T.: Pharmazie, in press .

I.Int.Symp. on Cyclodextrins
Budapest, 1981

STABILIZATION OF CAMOMILE OIL WITH β-CYCLODEXTRIN

L. SZENTE, M. GÁL-FÜZY, J. SZEJTLI

CHINOIN Pharmaceutical-Chemical Works, Tó u. 1-5
H-1045 Budapest, Hungary

SUMMARY

Stability of camomile oil can be significantly enhanced by inclusion complexation with cyclodextrin. In complexed state heat and oxygen sensitive components of camomile oil do not suffer significant decomposition even at 150 °C in vacuum or at 37 °C in an athmosphere of pure oxygen. The complex is a fine powder which seems to be applicable with advantage in various pharmaceutical preparations.

Introduction

Camomile is among the few medicinal plants the popularity and extended application of which has not been impaired by the multitude of newly developed synthetic or biosynthetic preparations of high efficiency. There is even an ever increasing demand for camomile (1).

Camomile is marketed either as the dried plant (flower) from which an aqueous decoction is prepared or as an extract, first of all the so called blue oil, which is prepared by steam distillation and which is then used for various preparations such as alcoholic solutions, ointments, tablets, suppositories, etc.

Ways of application are rather varied: inhalation, bath, stupe, gargling, painting, etc.

It is characteristic for the biologically active ingredients of camomile that their beneficial pharmacological effects have rather broad spectrum. The effects are not very intensive, but are not associated with side effects or toxic symptoms (2,3,4).

The results of pharmacological and biochemical studies on the individual components and the whole extract have been reviewed in 1980 by ISAAC and KRISTEN (2). It has been proved by convincing data that the antiphlogistic, spasmolytic, further antibacterial and antimycotic effects of the active ingredients of camomile can be utilized with almost no risk in dermatology, stomatology, pulmonology, pediatrics, otolaryngology and radiation therapy. Thus the camomile still might be of vital importance for a long period to come.

In the aqueous decoction of dried camomile some sesquiterpenoids, but first of all flavonoids, coumarins and characteristic essential oils are present. The blue coloured essential oil of camomile prepared by steam distillation contains the pharmacologically most important components in higher concentration consisting of sesquiterpenoid hydrocarbons, oxygen containing derivatives thereof, as well as non-terpenoid spiroethers or en-in-dicycloethers (Table I.).

The quantity and concentration of the active ingredient in commercial camomile preparations are usually small. Thus Kamillosan (Chemie Werke Homburg, Frankfurt a.M.) an alcoholic solution applied for the treatment of the inflammations on skin and mucosa contains 150 mg camomile oil in 100 g, a minimum of 3 mg of chamazulene and 50 mg of (-)-alphabisabolol. 100 g of Kamillosan Salbe contains 400 mg of dry camomile extract a minimum of 0.4 mg of chamazulene and 7 mg of (-)-alpha-bisabolol.

There are two basic problems of camomile oil containing preparations: standardization of composition and stability of ingredients. The ratio of components in preparations of different origin shows large variations (1) and under the effect of light, heat, oxygen or acid all pharmacologically important components of camomile oil decompose rapidly. Thus e.g. guaiazulene decomposes to 96 % in 0.16 N hydrochloric acid at 37 ^{o}C within 10 min . The same compound decomposes to 74 % under a lamp of 10 000 lux simulating sunshine over a period of 160 h (5). We could not identify any of the components of the original camomile oil in a product prepared by spray drying of camomile decoction after 2 months. From a commercial alcoholic camomile preparation only 15 % of the original chamazulene content along with traces of bisabolol and a great variety of non identified decomposition products could be detected 8 months after its production. Significant portion of the liquid and unstable components of camomile can be transformed with cyclodextrin to a stable, crystalline in-

Table I.

MATRICIN
/Prochamazulene/

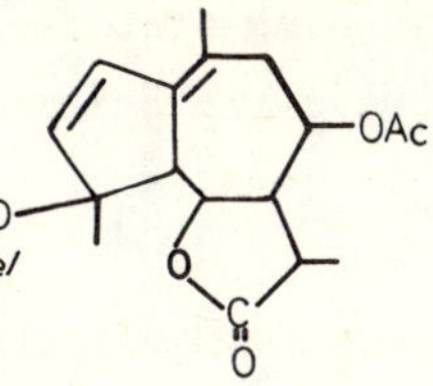

colourless

CHAMAZULENE
CARBONIC ACID

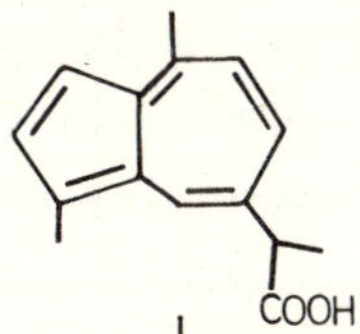

CHAMAZULENE

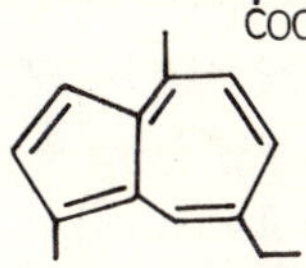

GUAIAZULENE

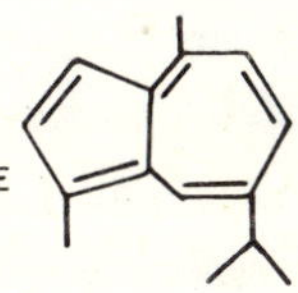

intensive blue colour

these are formed during steam distillation

antiphlogistic effect

(-)-α-BISABOLOL

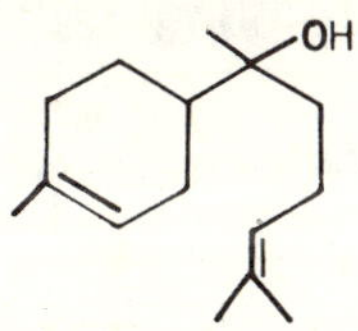

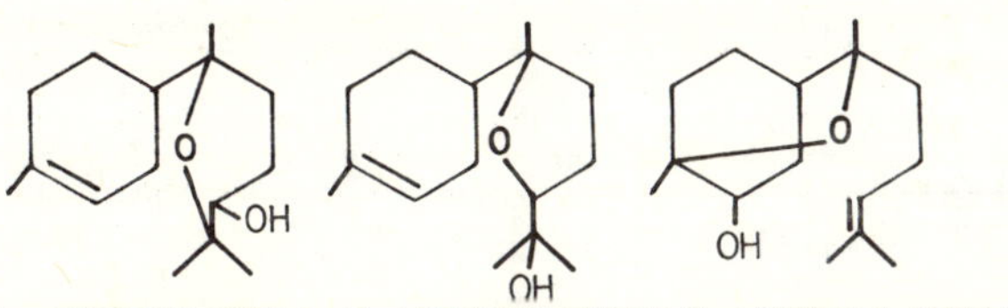

(-)-α-BISABOLOL OXIDE A; (-)-α-BISABOLOL OXIDE B; (-)-α-BISABOLOL OXIDE C

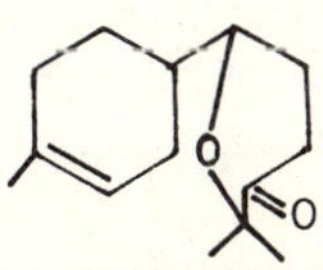

(-)-α-BISABOLON OXIDE A

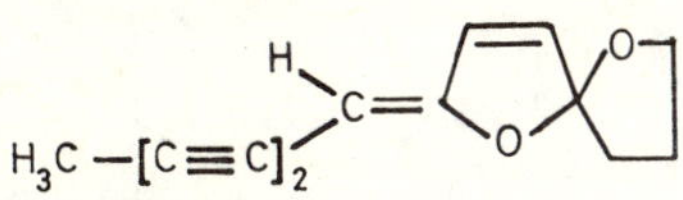

CIS-SPIROETHER

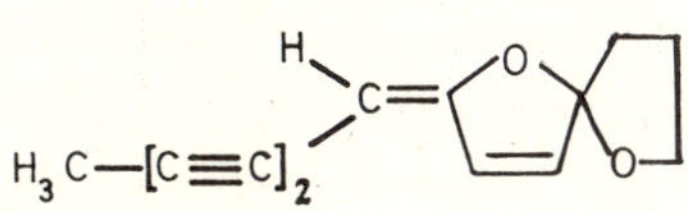

TRANS-SPIROETHER

spasmolytic effect

clusion complex which can be stored for a long time (6,7). In recent years cyclodextrin complexation of several aroma substances (8) and drugs has been reported, but it deserves mentioning that already the first patent covering the preparation of cyclodextrin complexes described the preparation of the β-cyclodextrin complex of isopropylazulene in 1953 (9).

Since β-cyclodextrin has become nowadays a compound produced on industrial scale and it proved to be non-toxic (10,11) the preparation of its inclusion complexes opens new ways in the pharmaceutical technology. The inclusion complexes are microcrystalline substances in which the active ingredient is present in a molecularly encapsulated state. In the apolar cavity of the cyclodextrin ring one single molecule of the active ingredient is included and because of the dry state it is even less exposed to decomposition than in a microencapsulated product. In aqueous media the complex dissociates immediately, the complexed compound is released and is available for resorption or to exert its effect.

The technology of preparation of cyclodextrin inclusion complex is much more simple than microcapsulation. Lipophilic components are thus brought into a molecularly dispersed form for utilization and can be resorbed quickly, without needing emulgeation. The molecules of the individual components are physically separated from each other which excludes the possibility of intermolecular reactions (e.g. disproportionation, polymerization or oxidation (7).

Characterization of the cyclodextrin complex of camomile oil

The β-cyclodextrin complex of camomile oil is a pale blue coloured fine microcrystalline powder, has faint camomile odour. The content of active ingredient (as expressed in the total content of volatiles) is 8.1-8.5 %.

Under microscope the product appears to be amorphous. The X-ray diffraction pattern of the complex shows sharp peaks and is significantly different from that of the β-cyclodextrin. This result proves indirectly the fact of complex formation and the crystalline structure of the product (camomile oil having no crystalline structure does not give any diffraction pattern) (12) (Fig.1.).

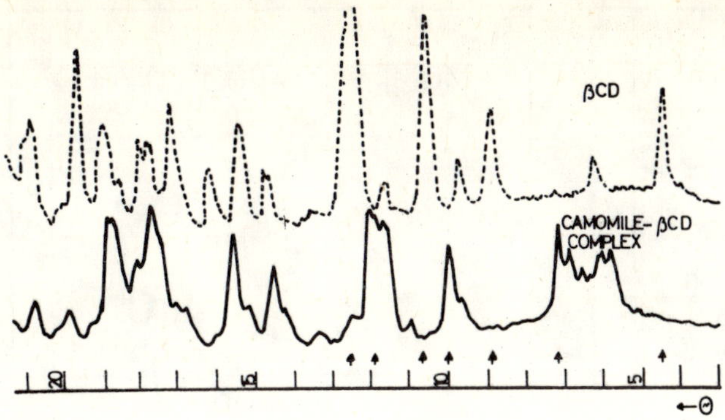

Fig.1.
X-ray diffraction pattern of β-cyclodextrin and its camomile oil complex

Qualitative and quantitative thin layer and gas chromatographic investigations showed that camomile oil recovered from the complex by extraction is identical with the original material (Fig.2. and 3.).

Thermal stability of the complex was measured by thermoanalytical methods (pyrolytic gas chromatography, TAS chromatography, differential scanning calorimetry and thermoderivatography).

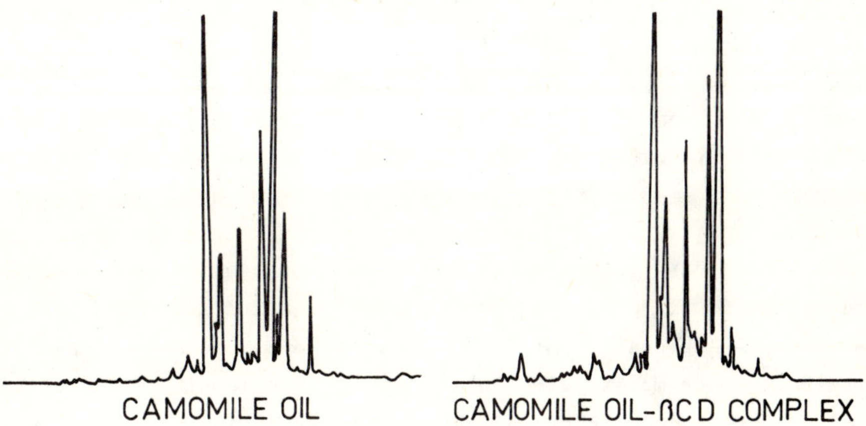

Fig.2. GLC of original and from complex re-extracted camomile oil

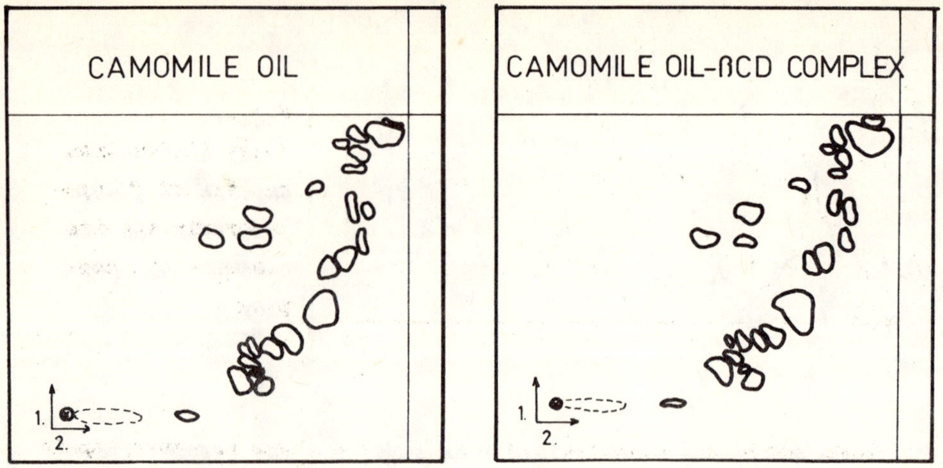

Fig.3. Two-dimensional TLC of original and from complex re-extracted camomile oil (silica gel layer, developed first in benzene-ethylacetate 95:5, second in cyclohexane-glacial acetic acid-ethylacetate 79:1:20)

Only 0.02 w% of the sample of the complex was detectable as camomile oil by pyrolytic gas chromatography at 200 °C and this value remained less than 0.1 w% even at 270 °C while the quantitative composition of this oil was different from that of the extract of the complex and component distribution was also dependent on the temperature of pyrolysis. At about 300 °C the decomposition of β-cyclodextrin becomes significant, but due to the decomposition of included camomile components the release of larger quantities of the essential oil cannot be observed any longer at such a high temperature.

The stabilizing effect of complex formation could also be demonstrated by means of "pyrolytic" thin layer chromatography (TAS technique). The TAS chromatograms of camomile oil adsorbed on lactose and of the inclusion complex both containing 10 % camomile oil show that the release of the components of the mechanical mixture of camomile oil-lactose is significant already at 120 °C while the same effect with the complex sets in only at 180-200 °C (Fig.4.).

Figure 5. demonstrates the results of thermoanalytical studies of camomile oil and its inclusion complex. According to DTG curves free camomile oil gets released at 140 °C while the complexed oil is released only with the decomposition of the β-cyclodextrin. Before the decomposition temperature of cyclodextrin only the loss of water of crystallization can be observed. In accordance with the above facts, TG curves also support the significantly enhanced thermal stability of the complexed camomile oil. The apparent differences of the DSC curves provide additional proof for the fact that the complex does not suffer significant changes up to about 200 °C.

Accelerated long term stability tests of pure camomile oil and of the complex were performed at 150 °C in vacuum. Although such extreme conditions never occur during normal storage, in this way the expiry time at normal temperature can be estimated. For rapid qualitative evaluation thin layer chromatography was applied. It could be observed that by prolonged exposure to heat first of all the en-in-dicycloethers and chamazulene were affected. Some unidentified components having Rf values near that of chamazulene were also rather unstable. After 40 min exposure there was no significant qualitative difference between the composition of uncomplexed oil and the sample extracted from the complex, but the content of poly-ines and chamazulene decreased in the uncomplexed oil. After 80 min exposure decomposed products appeared in the uncomplexed sample and the quantity of the more unstable components decreased significantly while in the complexed sample no change could be found. After 24 hours only one component with the Rf value of farnesene could be detected in the uncomplexed oil. The complexed oil showed significant changes as well, namely a decrease of poly-ine and chamazulene content, whereas the quantity of the other components remained unchanged.

After 48 hours uncomplexed camomile oil showed only traces of detectable components, whereas in the sample extracted from the complex two unidentified components were found, along with a decrease of poly--ines, chamazulenes and (-)-alpha-bisabolol; the rest of the components were practically unchanged.

After 72 hours of heat treatment no components could be detected. The uncomplexed sample was transformed to a sparingly soluble, brownish, ill-smelling resin. From the complexed oil most of chamazulene and spiroethers disappeared, the content of (-)-alpha-bisabolol dicreased significantly, while the quantity of the other components (herniarin, farnesene, bisabolol oxides) remained nearly unchanged.

Oxygen uptake was studied in Warburg apparatus in pure oxygen.

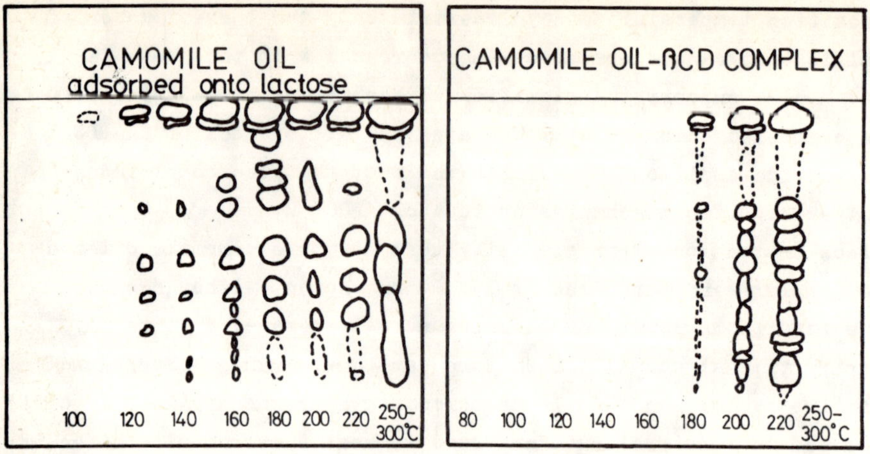

Fig.4. TAS chromatograms of camomile oil and of its β-cyclodextrin complex

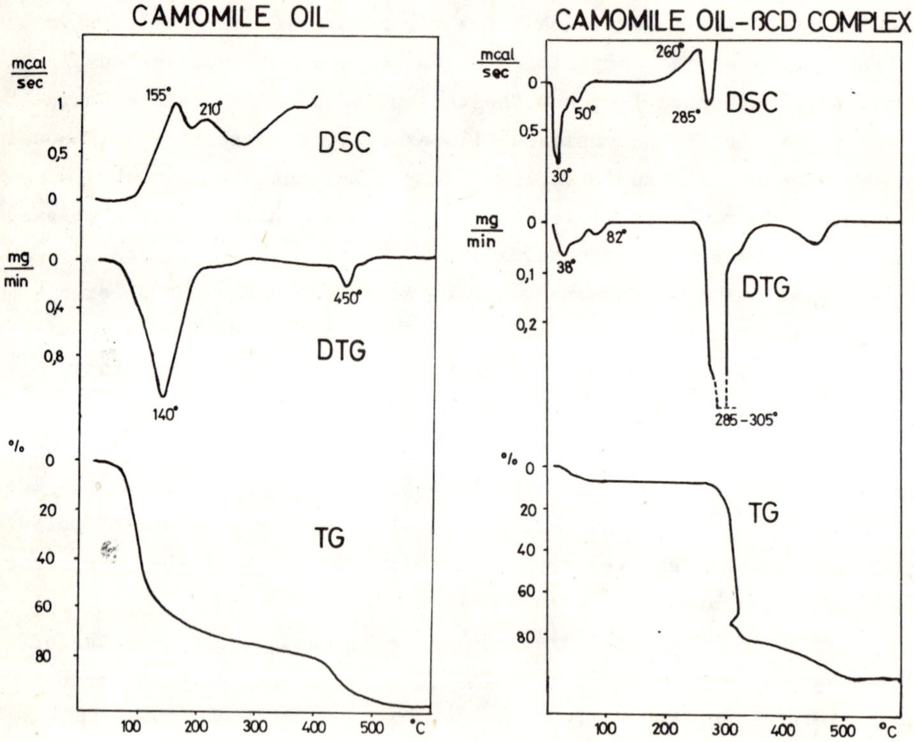

Fig. 5. Thermoanalytical curves of camomile oil and of its β-cyclodextrin complex

It was established (Fig.6.) that the complex is far more resistant to oxidation than the uncomplexed oil.

Samples recovered from the Warburg oxidation experiment were examined by thin layer chromatography in order to establish the changes caused by oxidation. Peroxides were detected with iron(II) sulphate - rhodanide reagent, olefins with bromofluoresceine reagent(13). Results are summarized in Table II. In the original camomile oil only one peroxide-type component could be detected, while after **exposure** to oxygen their number grew to six. In the sample recovered from the inclusion complex befor and after of oxygen treatment, only one peroxidic component was found. The original camomile oil contained eight olefinic components, the number of which decreased to three on oxygen exposure while in the complexed oil no changes were found.

On oxidation first of all chamazulene, farnesene and the poly-ines were affected while none of the components of camomile oil extracted from the oxygen treated sample suffered any change.

Table II. TLC of original and from β-cyclodextrin complex re-extracted camomile oil both after exposure to oxygen.

R_f values	Peroxides (ferrosulfate-rhodanide reagent)		
	camomile oil original	re-estracted from complex after exposure to oxygen	camomile oil oxidized
0,90-0,92			+
0,75			+
0,66			+
0,35			+
0,10-0,15	+	+	+
0,0			+
	unsaturated components (fluoresceine-bromine-reagent)		
0,90-0,92	+	+	
0,80-0,82	+	+	
0,70-0,72	+	+	
0,58-0,60	+	+	
0,45-0,48	+	+	+
0,40			+
0,28-0,30	+	+	
0,25		+	
0,20	+	+	+
0,10-0,12	+		

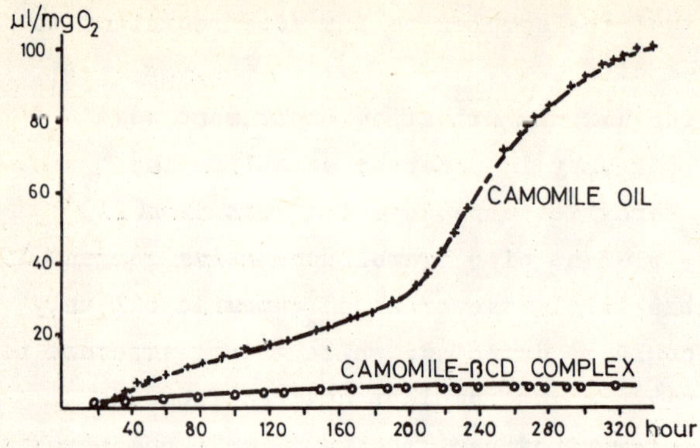

Fig.6. Oxygen consumption of froo and complexed camomile oil

These results prove that complexation with β-cyclodextrin can efficiently protect all of the pharmacologically active components of camomile oil against decomposition induced by heat and oxygen. Yonezawa (5) investigated the decomposition of guaiazulene under the effect of acid and light and established that this compound can also be effectively stabilized by β-cyclodextrin; e.g. in 0.16 N HCl at 37 °C after 20 min only 1,2% of guaiazulene remained intact, while 95% of that of the sample complexed with β-cyclodextrin.

All this suggests that by the aid of β-cyclodextrin camomile oil containing pharmaceutical preparations of higher stability can be manufactured than with the conventional methods.

Experimental, Materials and Methods

β-cyclodextrin product of Chinoin, Budapest, loss on drying 14%. Camomile oil (prepared by Research Institute for Medicinal Plants, Budakalász, Hungary) from Matricaria chamomillae, a deep blue oil of characteristic odour, density 0.925 g/cm^3 at 25 °C, complying with the specifications of Ph.Hg. VI. The complex was prepared according to Szejtli at al. (14).

X-ray diffraction diagrams were recorded on a Phillips PW 1060 diffractometer.

Camomile oil content of the complex was determined by the method of Clevenger (15): the sample is subjected to steam distillation. The volatile oil is collected in a calibrated flask where it separates from water.

For gaschromatography Hewlett-Packard 5830 instrument was used under the following conditions:

column: capillary column of Carbowax 20M (20m x 0.25mm)

injector temperature: 250 °C

FID detector temperature: 300 °C

carrier gas: nitrogen at 30 ml/min

Chromatography was performed with temperature program. The chromatograms were evaluated by attached computer unit.

For pyrolytic chromatographic experiments micropyrolysis cell of closed volume was attached to the gaschromatograph.

TAS chromatography was carried out according to Stahl (16) modified for the study of cyclodextrin-essential oil complexes (17).

Thin layer chromatography was applied using the method of Tyihák and Vágujfalvi (19). Silica gel layers were developed in benzene - ethyl acetate 95:5 and 98:2 mixtures. If necessary two dimensional separation was applied using cyclohexane - ethyl acetate - acetic acid 79:20:1 mixture for the second development. As a TLC reagent 10% ethanolic phosphomolibdenic acid was used.

Stability to oxygen was measured in a Warburg apparatus in oxygen atmosphere at 37°C, using Brodie solution (density 1.033 g/cm^3). Equivalent quantities of pure chamomille oil and its β-cyclodextrin complex were weighed into Warburg vessels. Samples were kept in oxygen athmosphere for 320 h. Oxygen consumption was registered every 40 hours.

Acknowledgement: X-ray diffraction studies are greatfully acknowledged to dr. Kálmán Simon, gaschromatography to dr. János Harangi and thermo-analytical measurements to dr. Sándor Gál.

References:

1./ Stettler, J.: Chimia 35,143 (1981)

2./ Isaac, O., G.Kristen: Med. Welt 31, 1145 (1980)

3./ Isaac, O.: Planta Medica 35, 118 (1979)

4./ Jakovlev, V., O.Isaac, K.Thiemer, R.Kunde: Planta Medica 35, 125 (1979)

5./ Yonezawa, Y., S.Maruyama, K.Takagi: Agric. Biol. Chem. 45, 505 (1981)

6./ Bender, M.L., M.Komiyama: Cyclodextrin Chemistry, Springer-Verlag, Berlin-Heidelberg-New York, 1978

7./ Szejtli, J.: Cyclodextrins and their inclusion complexes, Akadémiai Kiadó, Budapest, 1981

8./ Szejtli, J., L.Szente, E.Bánki-Előd: Acta Chim. Acad. Sci. Hung. 101, 27 (1979)

9./ Freudenberg, K., F.Cramer, H.Plieniger: Deut. Patent 895,769 (1953)

10./ Makita T., M.Ojima, Y.Hashimoto, H.Ide, M.Tsuji, Y.Fujisaki: Oyo Yakuri 10, 449 (1975)

11./ Szejtli, J., J.Sebestyén: Die Stärke 31, 385 (1979)

12./ Szejtli, J.: Stärke 30, 427 (1978)

13./ Fryklöf, L.E.: Farmacevtisk Rewy 56, 233 (1957)

14./ Szejtli, J., L.Szente, J.Körmöczy, P.Tétényi, Á.Dávid, T.Zilahy, A.Kelemen, J.Hárshegyi: Hung. Patent 176217 (1978); Ger.Offen. 2,944,350, 29.May 1980.

15./ Gildemeister, E., Fr.Hoffmann: Die Ätherishen Öle, Akademie Verlag, Berlin, 1960. Vol. II. p.22.

16./ Stahl, E., J.Fuchs: Dtsch. Apoth. Ztg. 108, 1227 (1968)

17./ Kernóczy, L., P.Tétényi, E.Mincsovics, J.Szejtli: Quart. J. Crude Drug Res. 16, 153 (1978)

18./ Sztatisz, J., S.Gál, J.Kőmüves, Á.Stadler-Szőke, J.Szejtli: Thermal Analysis, Birkhäuser Verlag, Basel-Boston-Stuttgart, 1980. Bd. 2. p. 487.

19./ Tyihák, E., D.Vágujfalvi: Herba Hung. 1, 213 (1963)

I.Int.Symp. on Cyclodextrins
Budapest, 1981

PHARMACOLOGICAL EFFECT OF LIDOCAINE DIMETHYL-β-CYCLODEXTRIN INCLUSION COMPLEX

T. FERENCZY,[1] P. SZABÓ,[1] J. SERFŐZŐ,[1] Á. STADLER-SZŐKE,[2] J. SZEJTLI[2]

[1] Department of Zoology and Anthropology, Kossuth Lajos University, Egyetem tér 1
H-4010 Debrecen, Hungary

[2] CHINOIN Pharmaceutical-Chemical Works, Tó u. 1-5
H-1045 Budapest, Hungary

SUMMARY

Protraction of the anaesthetic effect of well-known local anaesthetics is possible by depot-formation. A new possibility for depot formation is coupling the anaesthetics to a big, slowly diffusing moiety, assuming that the anaesthetic molecule will be released at a satisfactory rate.

Lidocaine base is soluble in an aqueous solution of heptakis(2,6-dimethyl)-β-cyclodextrin, forming a stable, injectable solution. No covalent bond is established between the two components, the dissociation equilibrium is governed by concentrations of the components at a given temperature. Toxicity and metabolism of lidocaine is not modified, however its diffusion is considerably retarded by the relatively big "anchor" /the mol. weight of dimethyl-β-cyclodextrin is 1288/.

The duration of anaesthetic effect of lidocaine basis dissolved in dimethyl-β-cyclodextrin solution was compared with that of conventional lidocaine -HCl injection on corneal reflex of rabbit, intracutaneous reaction of guinea pig and conduction anaesthesia test in rat tail. The effect of the inclusion complex proved to be dose-dependent, while the time of the anaesthesia was elongated by 35 to 109 % as compared to the conventional lidocaine -HCl injections at equivalent doses.

The aqueous solution of the heptakis (2,6-di-O-methyl)-β-cyclodextrin (= DM-β-CD) is a good solvent for a number of slightly soluble drugs, e.g. cholecalciferol, various steroids, lidocaine base, etc. (1).

The formation of a localanaesthetic lidocaine depot was aimed by applying the DM-β-CD complex lidocaine base. It was assumed, that the bulky host molecule (DM-β-CD) reduces the migration of the relatively small guest molecule (= lidocaine) which results in prolonged localanaesthetic effect.

MATERIALS and METHODS

1. The preparation of lidocaine-heptakis (2,6-di-O-methyl)-β-cyclodextrin /further on: DM-β-CD-LID/: To 95 ml of physiological salt solution, 15 g of DM-β-CD then - under vigorous stirring - 1,5 g of pulverized lidocaine base were added at 30 °C. After the complete dissolution, the pH was adjusted to 6,4 to 6,7, and the volume was brought up to 100 ml with physiological NaCl-solution. A stable and pure solution was obtained that may be stored at room temperature for months and could be diluted with distilled water or physiological saline without precipitation of lidocaine.

As control the adrenalin-free 2 % lidocaine hydrochloride injection /a product of EGYT, Budapest/ was diluted with 0,9 % NaCl-solution until the lidocaine content became identical to that of the solution of DM-β-CD-LID.

2. The corneal reflex method: this is the most appropriate method for the examination of surface anaesthesia, the main point of which is that the irritation of the cornea elicits the reflex closure of the eye lid.

In case of dropping locally effective anaesthetics in the eye, the reflex may not be elicited by irritations of the intensity applied previously. The method of Soehring, Frahm and Mletzko (2) was applied, as follows.

Two drops /50 mg/ of the solution were dropped upon the upper margin of the cornea of the rabbit /Orychtolagus cuniculus/. The solutions of lidocaine HCl /further on: LID/ and the DM-β-CD-LID contained in the drops 0,125; 0,250 or

0,375 mg lidocaine, respectively. Meanwhile the eye lid was kept open, thus the solution might be spread evenly. The loose end of the irritating bristle /of wild boar/ was 1,3 cm long, the interface between bristle and cornea was 0,0804 mm^2 the pressing force at the starting point of its bend was 3,99 g, i.e. the applied pressure had the calculated value of 49,62 g/mm^2.

The schema suited for the quantitative evaluation of surface anaesthesia in case of rabbit cornea makes it possible to draw substantial conclusions.

The examination began two minutes after the local anaesthetics had been dropped, and it was carried out repeatedly in 16 times every two minutes, thus altogether during 32 minutes. A single stimulation lasted about 20 sec. A case was regarded as total anaesthesia when there was not a single point at all where the reflex could be elicited. In each one of the experimental groups lidocaine was dropped in the one eye and DM-β-CD-LID in the other one. In a double logarithmic chart the percentual values of the effect /= no reflex elicited/ was plotted against the time. The relationship between effect and time proved to be linear, thus the half--effect time ($t_{eff\ 50}$) was read from the straight line obtained.

3. Wheal test /intracutaneous reaction/ was carried out on guinea pig /Cavia porcellus/, according to Bülbring and Wajda (3). This method was used also for investigating the infiltration anaesthesia. Since the sensibility of the skin is the most intense in the vertebral line, furthermore since the anterior areas are slightly more sensible than the posterior ones, the procedure applied involved the examination of sensibility in both areae. In this case, too, selfcontrol was applied by injecting on the left side lidocaine and on the right side DM-β-CD-LID intracutaneously, ahead as well as at the back beside the spine, in doses of 0,1 ml. After the injection the border of the wheal caused by it was marked. 5 minutes after the injection the skin was irritated by standard pinpricks. The normal reaction /scream/ was checked by pricking outside the area of the

wheal. The irritation was carried out repeatedly every 5 minutes by pricking 6 times in the area of the wheal and counting the negative reactions. The numbers obtained in each one of the 10 animals at the same points of time were summarized, and the sums were divided with the number of the animals. From the results obtained partly by examinations in the anterior juxtaspinal areas and partly by those in the posterior ones, mean values were calculated. In this experiment, too, the quantitative evaluation was meant by giving the value of $t_{eff\ 50}$.

4. The examination of conduction anaesthesia by rat tail test was made according to Therr (4) in mouse and according to Herr (5) in rat. The solutions were injected 1 cm from the tail-base after having verified that the reflex might be elicited by electric excitation. In each one of the animals 0,3 ml of LID or LID-DM-β-CD solutions were injected in two doses of 0,15 ml by giving one of them close to the left--sided nerve trunk and the other close to the right-sided one. The injected doses were spread evenly by a light massage. Before the excitation began, the tail was degreased at full length with alcohol, then gauze layers of about 1,0 cm^2 surface wetted by 0,9 % NaCl-solution were put on the excitatory electrodes. The distance of the two electrodes from each other was 11 mm. The electric impulses were excited by a four-channel myograph of typ MG-42 produced by Medicor Works, Budapest. The parameters of the excitation were as follows:

frequency.......100 c/s	tension.................90 V
pulse lenght......2 ms	stimulation intervall...10 min

Based upon data measured in several control animals (6) the tension threshold value was found to be 28 V. The reactions were considered positive of the animal's response appeared in pulling away the tail vigorously and in screaming. After summarizing the intervals between the injection and the reflex response, the results were evaluated statistically by Weber's method (7).

RESULTS

1. Results of the study of corneal reflex: The results are summarized in Table I., which presents the $t_{eff\ 50}$ values of lidocaine and DM-β-CD-LID at various concentrations.

Table I.

The $t_{eff\ 50}$ values of LID and DM-β-CD-LID in rabbit cornea			
concentration in %	$t_{eff\ 50}$ value in minutes		changes of $t_{eff\ 50}$ in %
	LID	DM-β-CD-LID	
0,25	4'54''	6'36''	+ 34,69
0,50	11'45''	18'40''	+ 58,86
0,75	12'10''	24'40''	+102,73

The anaesthetic effect of both lidocaine and DM-β-CD-LID seems to be dose-dependent.

By calculating the precentage of the differences between the $t_{eff\ 50}$ values presented, it may be established that the complex at all the three concentrations show a more active anaesthetic effect than the lidocaine.

For diminishing the errors arising from random reactions, at calculating the $t_{eff\ 50}$ values the experimental values over 90 % and under 15 % anaesthesia were neglected.

2. Results of the wheal test carried out in guinea pigs: The $t_{eff\ 50}$ values obtained by applying various concentrations of the anaesthetics are summarized in Table II.

The pharmacological effect of both preparations seems to be dose-dependent in this case, too.

The effect expressed in $t_{eff\ 50}$ values of the DM-β-CD--LID proves to be superior with about 60 % to that of the lidocaine. An increase to such an extent in the length of time of the anaesthesia may be of importance in the surface anaesthesia applied in the clinical practice.

Results of conduction anaesthesia are seen in Table III. In accordance with the former experimental results, the

Table II.

The $t_{eff\ 50}$ values of wheal test			
mg/ml	$t_{eff\ 50}$ value in minutes		changes of $t_{eff\ 50}$ in %
	LID	DM-β-CD-LID	
5,0	19'55''	31'00''	+ 55,64
7,5	28'10''	46'45''	+ 64,28
10,0	38'00''	60'00''	+ 58,79

Table III.

Duration of conduction anaesthesia in rat tail test			
mg/ml	Duration of anaesthesia in minutes		change in %
	LID	LID-DM-β-CD	
5,0	76 ±15,75	146 ±40,38	+ 92,10
7,5	117 ±30,49	215 ±56,51	+ 83,76
10,0	222 ±39,66	464 ±42,73	+109,00

dose-dependence as well as the better anaesthetic effect of the complexed lidocaine is verified by this test, too.

Acknowledgement

Thanks are due to Mrs. Zádor, Mr. Tibor Tóth and Miss Gabriella Molnár for their contribution in the experimental work.

Literature Cited

1. Szejtli, J., Bolla-Pusztai, É., Stadler-Szőke, Á.: Hung. Patent Appl. 1141/80, (1980)

2. Soehring, K.L., Frahm, M. and Mletzko, K.: Arch. Int. de Pharmacodyn. 91 112 (1952)

3. Bülbring, E. and Wajda, I.: J. Pharmacol. 85 78 1945

4. Therr, L.: Arzneimittel-Forsch. 3 345 (1953)

5. Herr, F., Nádor, K., Pataky, Gy. and Borsi, J.: Arch. exp. Path. u Pharmacol. 217 447 (1953)

6. Carsuto, J., Jodal, M., Ruttle, R. and Lundgreen, O.: Experientia 35 1467 (1979)

7. Weber, E.: Grundriss der biologischen Statistik, Springer Verlag, Jena (1964)

I.Int.Symp. on Cyclodextrins
Budapest, 1981

INVESTIGATIONS OF DIGITALIS GLYCOSIDES AND THEIR CYCLODEXTRIN COMPLEXES

B. LENKEY

Institute of Biochemistry, Kossuth Lajos University
Egyetem tér 1
H-4010 Debrecen, Hungary

SUMMARY

It is known that the absorption of digitalis glycosides of various origin - used in the medicine - is different. The discrepancy in the absorption is caused by the polarity of the compounds and the excretion of glycosides also depends on the polarity. Digoxin used in the largest extent belongs to the more polar glycoxides.

Digo in, similarly to steroids, is practically insoluble in water. Cyclodextrin and its derivatives being able to form complexes may cause a significant increase in the solubility of glycosides in water. The absorption features of such watersoluble complexes of digitalis glycosides were investigated by per os administration on rabbits. The absorption of glycosides and their cyclodextrin complexes were different.

Acknowledgement: The author wishes to express his thank to T.Forgách and P.Szabó their contribution.

I.Int.Symp. on Cyclodextrins 5.11.
Budapest, 1981

EFECT OF SEVERAL DRUGS AND THEIR CD-COMPLEXES ON HEART-RHYTHM OF BLABERA CRANIIFER

L. PAP

CHINOIN Pharmaceutical-Chemical Works, Tó u. 1-5
H-1045 Budapest, Hungary

SUMMARY

The effect of the phenobarbital and its CD-complex on heartbeat of an insect, Blabera craniifer was studied. The heartbeat was recorded with two different methods: 1. semiisolated heart preparation, and 2. intact animals with implanted electrodes.

The phenobarbital and its CD-complex significantly decreased the frequency of heartbeat in the semiisolated preparations. In the long--term tests similar results were obtained when the drug was administered in the pericardial sinus, however on the perivisceral application only phenobarbital CD-complex decreased the frequency.

Correspondingly only the CD-complexed phenobarbital could reduced the noradrenalin-induced increase in frequency.

A possible explanation for this observation is the improvement of bioavailability, probably on shifting the ionization equilibrium of the phenobarbital by CD-complexation.

I.Int.Symp. on Cyclodextrins
Budapest, 1981

REDUCTION OF INDOMETHACIN SIDE EFFECT WITH β-CYCLODEXTRIN IN GASTRO-INTESTINAL TRACT

T. NEUMARK,[1] J. TAKÁTS,[1] A. GERLÓCZY[2]

[1] National Institute of Rheumatism and Physiotherapy
Frankel L. u. 17-19
H-1027 Budapest, Hungary
[2] CHINOIN Pharmaceutical-Chemical Works, Tó u. 1-5
H-1045 Budapest, Hungary

SUMMARY

In a 3 week comparative study the side effect of indomethacin and indomethacin-β-cyclodextrin complex treatment was compared on the mucous membrane of the gastro-intestinal tract in rats. The drugs were administered orally and the tissue alterations were studied in both groups under light and electron microscope. The investigations indicate that the side effects on the mucous membrane (erosions, haemorrhage, ulceration) could be markedly reduced and retarded in time with indomethacin-β-cyclodextrin complex. The most peculiar protective effect of β-cyclodextrin were observed on the gastric mucous membrane. The side effects on the mucous membrane of small intestine developed several days later in case of indomethacin-β-cyclodextrin treatment as compared with pure indomethacin administration.

I.Int.Symp. on Cyclodextrins 5.13.
Budapest, 1981

CYCLODEXTRINS AND CONGENERS IN PARENTERAL APPLICATIONS

J. PITHA, L. SZENTE

National Institute on Aging
Baltimore, Maryland, U.S.A.

SUMMARY

The natural fat soluble compounds and lipophilic drugs circulate in the organism in the form of complexes with specific carrier proteins. The formation of these protein complexes not only increases the water solubility of lipophilic compounds but also, in many instances, decreases their toxicity through a decrease in their effective concentration (activity). Vitamin A may serve as an example of this phenomenon; its toxic effects are observed only when the concentration of vitamin is higher than the concentration of its specific carrier protein. The role of carrier proteins in the organism may possibly be imitated by synthetic complexing agents. If these agents would have low toxicity, then useful therapies may be developed based solely on strong complexing between the synthetic agent and the natural or xenobiotic compound. Our efforts in the development of such complexing agents have concentrated on cyclodextrins and their alkyl derivatives and on digitonin and compounds derived from this glycoside by derivatization of its saccharide part. Both approaches lead to compounds which form stable, water soluble complexes with representatives of fat soluble vitamins. From these compounds derivatives of vitamin A, retinoids, are of greatest interest since these compounds have beneficial effects on tumor formation.

Introduction

The formation of complexes between organic compounds has been of such interest to chemists that already in the early fifties an authoritative book could be written on the subject (Cramer, 1954), a book which introduced many of us to the subject. In spite of these many studies the biomedical applications of this phenomenon are still in their infancy. It seems that finally the improvements of drugs administered per os by cyclodextrins will find their way into practice and replace less suitable or more complex modes of drug processing, e.g. tableting with the aid of nondegradable polymers or microencapsulation. Parenteral application of cyclodextrins and congeners is still far from being close to the practical stage; nevertheless its potential is far-reaching. The outline of this potential is the subject of this contribution.

An organism has barriers to penetration of either polar or nonpolar compounds. Polar compounds, like cyclodextrins, cannot penetrate any membranes of the organism. This inability results in a lack of absorption into the organism when polar compounds are applied per os; when these compounds are introduced parenterally they stay in circulation and do not penetrate into the cytoplasm of cells or into organs separated from circulation by cellular barriers, e.g. brain. Nonpolar compounds, on the other hand, penetrate the membranes of the organism but accumulate there and in other lipid-rich parts of the organism. This accumulation results in damage to membranes and in the escape of nonpolar compounds from the disposal processes the organism imposes on all of its components.

Cyclodextrins are able to complex and solubilize nonpolar compounds can thus diminish their membrane toxicity and may help to return these compounds to the circulation and consequently to expose them to disposal processes.

To illustrate the problems which arise from membrane toxicity of nonpolar compounds and the problems which arise from the escape of nonpolar compounds from metabolism and excretion two examples will be given--retinoids for the former and iodinated derivative for the latter.

γ-(3-Hydroxy-2,4,6-triiodophenyl)-α-ethylpropionic acid was formerly used in x-ray diagnosis of gallbladder problems but its use was discontinued after it was found that even one year after administration there was still half of this compound in the organism (Goldstein et al., 1974). This iodinated compound is very strongly bound to the nonpolar interior of serum proteins and thus completely avoids the processes of metabolism and excretion. In this case a compound which escaped the excretion is nontoxic; unfortunately there are many toxic nonpolar compounds which bind to nonpolar regions of protein and to lipids and form a depot of poison there. Solubilization of these toxic compounds by cyclodextrins and congeners would assist with their metabolism and excretion.

Vitamin A and structurally similar compounds, collectively called retinoids, at low doses which are required to prevent avitaminosis A (about 80 μg/kg/day) are nontoxic. High doses of retinoids (about 80 mg/kg/day) are toxic, i.e. lead to hypervitaminosis A, which is manifested by pathological changes in epithelial tissues, weight loss, and fragility of bones. High doses of retinoids at the same time have prophylactic and therapeutic effects on carcinogen-induced cancers (Bollag, 1972; Bollag, 1974; Bollag, 1975; Sporn & Newton, 1979). Thus it would be of great value to decrease the toxic effects, i.e. hypervitaminosis symptoms, of retinoids. To understand how cyclodextrins can assist in this particular problem the mechanism of retinoid toxicity should be described. In the organism retinoids circulate bound to proteins. Three such proteins are known (Goodman, 1980; Ross & Goodman, 1979; Smith & Goodman, 1979; Chytil & Ong, 1979); these are retinol-binding protein which is serum protein and cell-contained proteins called cytosol binding protein for retinol, and cytosol binding protein for retinoic acid. These proteins bind retinoids strongly and thus decrease the effective concentration of retinoid under the level which leads to toxic effects. When the concentration of a retinoid surpasses that of its binding proteins, free retinoid appears and its toxic effects are enacted through direct effects on membranes (Meeks et al., 1981). Cyclodextrin and congeners are able to complex retinoids and thus may form, in addition to retinoid binding proteins, another nontoxic depot for these compounds in the organism.

With the idea of creating an artificial, nontoxic depo of retinoid we started several years ago with studies of the chemical aspects of the problem. Apparently such a depo of retinoids should be water soluble. Therefore dextran, a hydrophilic glucose polysaccharide, was used to create this depo. Dextran was derivatized by introduction of a hydrazido group through subsequent condensations with chloroacetic acid in alkaline media and hydrazine in the presence of carbodiimides (Hurwitz et al., 1980). This polysaccharidic hydrazide was then condensed with retinal, which is an aldehyde form of vitamin A. The products were only a little water soluble. Fortunately soon it was found that addition of cyclodextrins to the condensate led to a well soluble depo of retinal. Having a suitable depo preparation, which in further will be called retinal-dextran, suitable tests for its bioactivity were devised (Pitha et al., 1980). In collaboration with Dr. F. Chytil the ability of retinal-dextran to cure avitaminosis A in rats was tested. The first experiment measured the stability of the preparation *in vivo* situation; retinal-dextran or a control (retinol) was administered per os to rats and the serum level of vitamin A was measured two hours thereafter. Administration of retinol led to the expected increase while administration of retinal-dextran was without any effect. Thus retinal-dextran does not dissociate in the gastrointestinal tract in a fast manner. In the second experiment it was tested whether vitamin A activity is released from the retinal-dextran at all, i.e. whether the condensation reaction destroyed the bioactivity of retinal or not. To this purpose retinal-dextran was injected into the peritoneum of avitaminose rats. These animals were cured of symptoms of avitaminosis, while the control animals, i.e. treated only with physiological saline, succumbed to the disease. Thus retinal-dextran is indeed a depo form of vitamin A activity.

Anti-cancer properties of retinal-dextran were tested by Dr. Lotan using cells grown *in vitro*. This system has the advantage of simplicity compared to *in vivo* tests, even when neither of these tests can fully replace the other. Mouse melanoma cells sensitive to retinoid were used; these cells are inhibited in their division by a low concentration of retinoids, while a certain higher concentration of retinoid kills the cells, presumably of membrane damage. The threshold of cell death by retinoid is about the same for both cancer and normal cells; the phenomenon of inhibition is specific to the former. Thus an effective retinoid is able to prevent division of cancer cells before the onset of universal cell death at high doses. The experiments showed that retinal-dextran is less effective than retinal at the same concentration in the inhibition of cancer cell growth. But on the other hand, retinal-dextran is about twenty times less toxic than retinal and thus could be used at much higher concentrations. Consequently maximal inhibition values obtainable with retinal were about 45% while with retinal-dextran inhibitions up to 80% were achieved (Pitha et al., 1980).

Retinal-dextran thus seems to be an effective depo form for both cancer and avitaminose curing effects of retinoids. There is an intrinsic difficulty with this preparation--it contains a macromolecular component, dextran. Dextran is nontoxic; indeed it was used as a blood plasma extender in medical practice. Nevertheless there are deep-rooted fears between physicians of applying a macromolecule parenterally. These fears are based on the long persistence of a fraction of injected macromolecule in the body. During the Second World War polyvinylpyrrolidinone was used as a blood plasma extender and it was later found that this polymer could be detected histochemically in the body even ten years after transfusion. Intuitively it seems healthier not to use compounds which, even when nontoxic, may stay in the organism that long. To avoid the use of dextran, a direct solubilization of retinoids by cyclodextrins was attempted. Unfortunately this approach in all combinations tested leads to only low and unsatisfactory solubilization effects.

As the next step, a chemical modification of cyclodextrins was performed with the hope of obtaining more potent complexing agents (Pitha, 1981). A method for alkylation of cyclodextrins in hydroxyl groups in 2 and 6 positions were well worked out (Casu et al., 1968); consequently we prepared these derivatives. Three cyclodextrins (α, β, and γ) and substitution by methyl, ethyl, and allyl groups were used. Optimal results seem to be obtained with methyl substituted β-cyclodextrin.

This 2,6-O-tetradekakismethylcycloheptaamylose at 5% concentration solubilized retinoids up to concentrations of 0.5-1 mg per ml; these are lower values than achievable with retinal-dextran (up to 2 mg/ml) but definitely in a usable range.

The derivatives of cyclodextrins which are effective solubilizers of retinoids may also be suitable in the removal of nonpolar toxins from the body by solubilizing them in water. 2,6-O-Tetradekakismethylcycloheptaamylose was found to solubilize quite effectively aromatic hydrocarbons (Szente & Pitha, unpublished results); napthalene was solubilized by 5% solution of the agent to 1460 μg/ml compared to less than 0.5 μg/ml in physiological saline alone. The corresponding numbers for anthracene are 26 μg/ml, compared to nondetectable and for tetracene 33 μg/ml, compared to nondetectable. Thus this compound is well suited for further in vivo testing.

The fact that cyclodextrins have negligible toxicity per os has been fully and repeatedly established (Szejtli & Sebestyén, 1979). These findings are in accordance with expectations: (a) cyclodextrins are highly hydrophilic compounds and thus do not penetrate from the gastrointestinal tract into circulation as such, (b) when cyclodextrins are degraded by gastrointestinal bacterial flora only glucose and its oligomers are produced and no toxic effect can be expected from them.

The situation with parenteral applications of cyclodextrins is, on the other hand, quite different. Cyclodextrins cannot be easily excreted as such and their degradation may be a slow process. Cyclodextrin glycosyl transferases, enzymes of cyclodextrin synthesis and degradation, are not present in detectable quantities in mammals and amylases and glycogenases present may not be expected to degrade cyclodextrins effectively. In accordance with these facts cyclodextrins were found rather toxic and the high doses applied parenterally are definitely kidney damaging (Frank et al., 1976).

Careful examination of published results enables some insight into the mechanism of cyclodextrin toxicity. In the kidney, the filtration of blood by glomeruli leaves many useful low molecular weight compounds and proteins in filtrate, i.e. in the nascent urine. These compounds are re-absorbed into the body by kidney tubules. Proteins, in particular, are re-absorbed by vesicular transport, i.e. as a result of the re-adsorption process these proteins are contained in small, membrane enclosed vesicles, in cells lining the kidney tubules. In these vesicles proteins are degraded to amino acids and these are released for re-use in the body. Cyclodextrins, being non-degradable, apparently overload this system to the point of cell death and necrosis of the organ. One way to avoid this overloading would be to lower the spikes of cyclodextrin concentration in blood. The published findings are in accordance with such expectations: repeated doses of cyclodextrins applied parenterally were less toxic than a single cummulative dose (Frank et al., 1976). In our experiments we went one step further. Intravenous application by injection takes about a minute or two and thus leads to a spike in the blood concentration of the injected substance. In our work we applied derivatives of cyclodextrins by infusion, i.e. peristaltic pump was used to inject slowly the solution and thus a "spike" was spread to 6-8 hours (Szente & Pitha, unpublished results). Using this technique toxicities of the effective solubilizer, 2,6-O-tetradekakismethylcycloheptaamylose, found were close to the acceptable range of LD_{50} 1.5g/kg in mouse. Doses of 1g/kg not only did not result in animal deaths, but did not even influence the growth of the young animals compared to controls--a clear sign of lack of toxic effects.

Studies of the effects of parenterally applied cyclodextrins are only few and our knowledge of these effects is admittedly very limited; nevertheless the few facts known indicate that eventually practical accomplishments may result from further work in this direction.

References

Bollag, W., Eur. J. Cancer 8: 689-693 (1972).

Bollag, W., Eur. J. Cancer 10: 731-737 (1974).

Bollag, W., Eur. J. Cancer 11: 721-724 (1975).

Casu, B., Reggiani, M., Gallo, G. G., and Vivegami, A., Tetrahedron 24: 803-821 (1968).

Chytil, F., and Ong, D. E., Fed. Proc. 38: 2510-2514 (1979).

Cramer, F., Einschlussverbindungen, Springer Verlag, Berlin (1954).

Frank, D. W., Grey, J. E., and Weaver, R. N., Amer. J. Pathol. 83: 367-374 (1976).

Goldstein, A., Aranov, L., and Kalman, S. M., Principles of Drug Action, John Wiley & Sons, Inc., New York, p. 163 (1974).

Goodman, D. S., Fed. Proc. 39: 2716-2722 (1980).

Hurwitz, E., Wilchek, M., and Pitha, J., J. Appl. Biochem. 2: 25-35 (1980).

Meeks, R. G., Zaharevitz, D., and Chen, R. F., Arch. Biochem. Biophys. 207: 141-147 (1981).

Pitha, J., Zawadzki, S., Chytil, F., Lotan, D., and Lotan, R., J. Natl. Cancer Inst. 65: 1011-1015 (1980).

Pitha, J., Life Sci. 29: 307-311 (1981).

Ross, A. C., and Goodman, D. S., Fed. Proc. 38: 2515-2518 (1979).

Smith, J. E., and Goodman, D. S., Fed. Proc. 38: 2504-2509 (1979).

Sporn, M. B., and Newton, D. L., Fed. Proc. 38: 2528-2534 (1979).

Szejtli, J., and Sebestyén, Gy., Starch 31: 385-389 (1979) and references therein.

Szente, L., and Pitha, J., unpublished results.

APPLICATION OF CYCLODEXTRINS IN FOODS AGRICULTURE AND OTHER INDUSTRIES

I.Int.Symp. on Cyclodextrins
Budapest, 1981

6.1.

CYCLODEXTRINS IN FOODS, COSMETICS AND TOILETRIES

J. SZEJTLI

CHINOIN Pharmaceutical-Chemical Works, Tó u. 1-5
H-1045 Budapest, Hungary

SUMMARY

Th major advantages of application of CD-complexation in foods, cosmetics and toiletries are as follows:

1. Protection of active ingredients against oxydation, light induced reactions, heat promoted decomposition, self-decomposition, loss by volatility, sublimation;

2. Elimination or reduction of undesired tastes and odours, microbial contaminations, fibres and other undesired components, hygroscopicity;

3. Technological advantages: stable, standardized compositions, simple dosability and handling of dry powders, reduced packing and storage costs, more economical technological processes, saving of energy and manpower;

Stabilization of food flavours and fragrances seems to be the biggest market for cyclodextrins in the 80'-es.

Possible or already realized applications of CDs are reported in:

Stabilization of flavours for food products

Elimination of undesirable tastes and odours

Preparation of antifungal complexes for foods and toiletries

Stabilization of fragrances, dyes and reduction of foaming in cosmetics and toiletries

Improvement of quality, stability storability of foods, etc.

The major advantages of application of CD-complexes in foods and cosmetics are as summarized in Table I. (1).

Table I. ADVANTAGES OF APPLICATION OF CYCLODEXTRINS IN FOODS AND COSMETICS

1. Protection of active ingredient(s) against:
   - oxydation
   - light induced reactions
   - heat promoted decomposition, self-decomposition
   - loss by volatility, sublimation

2. Elimination (or reduction) of
   - undesired tastes/odours
   - microbiological contaminations
   - fibres/other undesired components
   - hygroscopicity

3. Technological advantages
   - stable, standardizable compositions
   - simple dosability and handling of dry powders
   - reduced packing and storage costs
   - more economical technological processes, manpower saving

Stabilization of food flavours and fragrances seems to be the biggest market for cyclodextrins in the 80'-es.

Aroma-containing raw materials of natural origin (overwhelmingly of plant origin) are widely used in human alimentation. The main disadvantages of the application of these are as summarized in Table II.

Table II. DISADVANTAGES OF DIRECT CONSUMPTION OF NATURAL SPICES

- their preparation and processing is usually laborious, e.g. cleaning and chopping of onion;
- raw materials are not of constant composition, their aroma content depends on harvesting, processing, size, sort and growing site of the product;
- natural products may often contain microbiological contamination and sometimes parasitic infestation;
- their aroma content decreases on storage, and may also involve an undesirable alteration of the ratio of certain components;
- storability is limited: the storage of natural vehicles of certain aromas constitute serious storage problems (room, optimal temperature and humidity, etc.)

Liquid aroma extracts are well known for long, and are still marketed. The stabilization of most of them has not been solved so far, moreover the concentrates are often less stable then the original aromas. Their scope is restricted, and they are applied only in certain branches of the food industry (e.g. in tinned foods). Utilization of such extracts in households is rather limited.

The loss (volatility, oxidation, chemical decomposition, etc.) of aromas on solid vehicles is enhanced by their dispersion on a large surface.

Losses due to volatility and oxidation are not negligible even with microcapsulated products. Moreover release of the aroma occurs only after the dissolution of the wall of the microcapsule. Thus, a complete release requires a dissolution in hot water. That is why in foods eaten without cooking the microencapsulated flavours cannot get released completely on chewing. This limits the application of microcapsuled aroma products.

A very promising method for the stabilization of food flavours and fragrances is their complexation with cyclodextrins (2,3,4). The main fields of their applications are summarized in Table III.

The preparation of β-cyclodextrin complexes of food flavours and fragrances is a relatively simple process which is already realized on industrial scale.

The active ingredient content of such aroma-complexes is 6-15 % w/w, most often 8-10 % (3). They are very stable in dry state, their oxygen uptake measured by the Warburg method is less than that of free substances (Fig.1.).

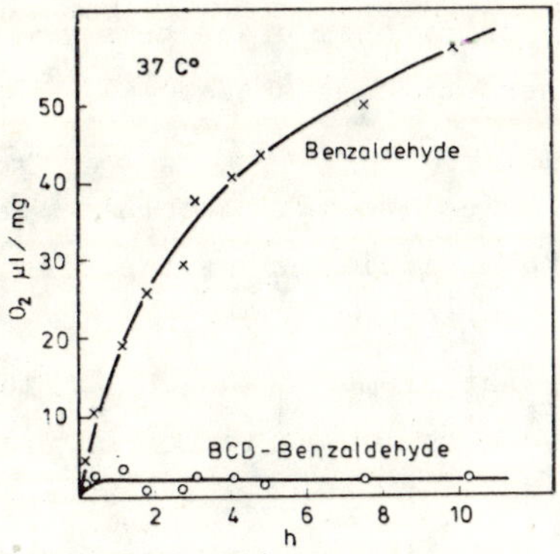

Fig.1.
The oxygen uptake of benzaldehyde and β-cyclodextrin complexed benzaldehyde in pure oxygen atmosphere at 37 °C, measured by Warburg's method

Table III. THE MAIN POTENTIAL FIELDS OF APPLICATIONS OF CYCLODEXTRIN COMPLEXED FLAVOURS

1., In households

It simplifies cooking by providing a wider choice of tastes and aromas. The dispensing units (e.g. tablets) containing various aromas can be stored in a small place for a long time without risking any loss of the active ingredient.

2., In the catering trade

By the elimination of manual processing of raw materials containing aroma substances, work and storage room can be saved, transport is simple, loss on storage can be avoided and a wider variety of tastes and aromas is provided. By using proper recipes identical taste and aroma can be provided independently of the location.

3., In dietetics and hospitals

The consumption of fibrous and seed shaped aroma carrying raw materials irritating the gastro-intestinal tract can be avoided, appetizing aromas can be applied in hospitals, many - otherwise forbidden - aromas can be consumed by persons on diet, the majority of problems with their eating being caused not by the aroma itself, but its vehicle mostly of plant origin.

4., In the tinned food and meat industry

In dehydrated soups or sausages without risking microbiological contamination stable aroma products resisting environmental effects can be used in a constant composition and in a well processable form.

At 150 °C in vacuum only 25-30 % of the active ingredient was lost after 24 hours, whereas none of the components of spice aroma substances could be recovered under these circumstances.

Their water solubility is only 0.1-0.035 g/100 ml at room temperature, however it is always sufficient to make the taste of the released component detectable by the dissociation of the dissolved complex.

Heat treatment studies proved that aroma inclusion complexes obtained in this way, when stored in a sealed container or wrapping at room temperature, lose no more than 5 % of their active ingredient content even after two years. Heat stability of such complexes can be illustrated very convincingly by TAS-chromatography or by thermofractometry (5,6), (Fig.2.).

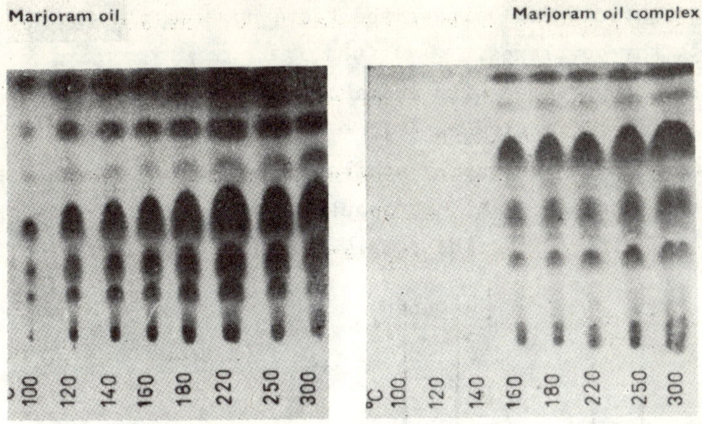

Fig.2. TAS-chromatogram of marjoram oil and
marjoram-β-cyclodextrin inclusion complex

Vanillin-glucose mixture and vanillin-β-cyclodextrin complex stored in open Petri-dishes at room temperature showed that after 240 days practically no detectable aroma substance was left in the glucose mixtures, whereas with the cyclodextrin complexes even after much longer time the loss was only about 20 % (Fig.3.).

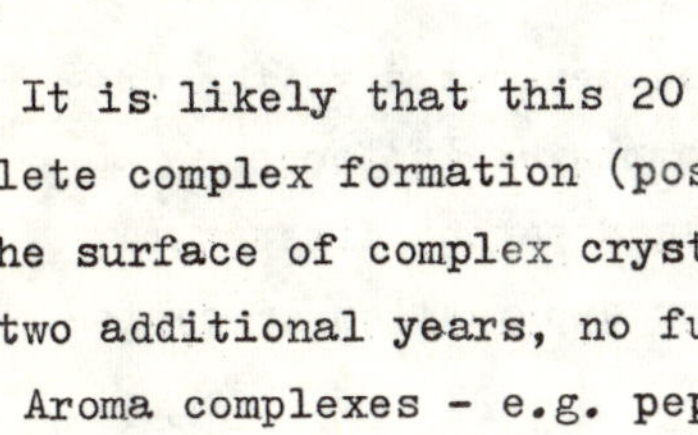

Fig.3.
Loss of vanillin in the case of vanillin-glucose mixture stored at room temperature in an open Petri dish and for the vanillin-β-cyclodextrin complex

It is likely that this 20 % loss of vanilline was due to incomplete complex formation (possibly some molecules were only adsorbed on the surface of complex crystals) because continuing the observation for two additional years, no further loss of vanillin was observed (3).

Aroma complexes - e.g. peppermint-cyclodextrin complex can be blended easier with chewing gum than the liquid aromas. In this form the aroma remains longer in the chewing gum and ensures a longer lasting intensive taste.

The taste of dextrins cannot be detected in the taste of complexes and they effect neither the taste nor the plasticity of food mixed with the complexes.

The meals prepared with aroma complexes were received favourably by a panel of consumers (7).

Aroma substances usually consist of several components and therefore it is important that they should be incorporated with an unchanged composition into the complex. As it is demonstrated in Fig.4.

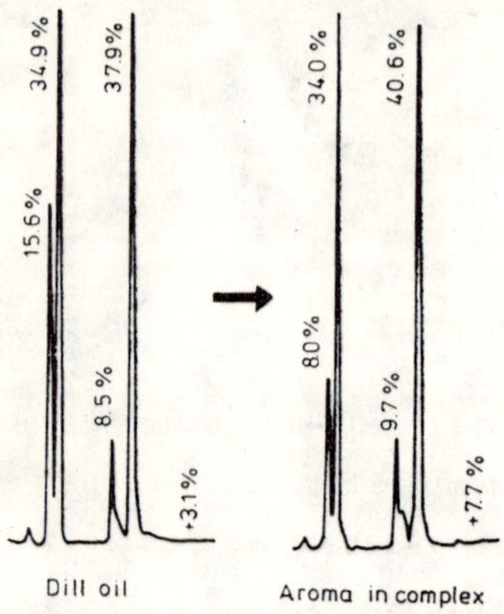

Fig.4.
Gas-liquid chromatogram of free and β-cyclodextrin-complexed dill oil

this requirement can be satisfied to a fair degree and therefore, the aromatizing effect of aroma complexes are not different from that of the original material. Sometimes however it may be definitely of advantage to alter the original composition. Appropriately selecting the host:guest ratio, one or more components can be relatively enriched in the complexed (or in the non-complexed) fraction. E.g. β-cyclodextrin on complexation of fatty acid mixture prefers the more unsaturated molecules (8).

Numerous patents and papers have been dedicated to the preparation and application of various CD-complexes in foods.

Table IV. summarizes the scientific and patent literature, dedicated to investigation, preparation or application of cyclodextrin complexed flavours.

Table IV. LITERATURE REFERENCES ON STABILIZATION OF FLAVOURS BY CYCLODEXTRIN COMPLEXATION

Substances	Reference
- Flavour extracts (e.g. fruit juices, essential oils of spices, meat extracts, etc.)	2,3,4,9
- Farnesane	10
- Limonene	11
- Citral and citronellol	12
- Butter flavour	13,20
- Menthol (e.g. in hard candies)	14
- Aromatizing of tea (with e.g. bergamot oil or jasmin oil complex)	15
- Natural food coloring agents	16
- Aroma powder for instant beverages (e.g. instant martini powder)	17
- Artificial spices (e.g. artificial cinnamon)	18
- Horse raddish and mustard powder	19

Milk casein hydrolysate is a readily digestible protein source, however its bitterness is a heavy problem. Adding 10 % β-cyclodextrin to the protein hydrolysate the bitterness can be eliminated (22,25). Similarly the bitter taste of ginseng extract (23), propylene glycol (24) was eliminated by cyclodextrin.

The characteristic odour of mutton and fish (26), the disagreable odour of bone powder (used as Ca-supplement in animal fodders) (27), of yeast extract (30) sodium caseinate solution (34) etc. can be removed with cyclodextrin. Ethanolic-aqueous solution of β-cyclodextrin (29), or the iodine-β-cyclodextrin complex (33) can be utilized as a deodorant (Table V.).

In many cases adding cyclodextrins or their appropriate complexes to food products important physical properties can be favourably modified, e.g. water retention (57), emulsion stability (42,43,49), egg-white beatability (40), etc. Examples are summarized in Table VI.

Another growing field of cyclodextrin utilization is the cosmetics and toiletry (Table VII.). In laundry cyclodextrin in rinsing water act as a defoaming agent by complexing the rest of detergents (67). Solid perfumes, incenses, fragrant candles (60,61,62,71), emulsified cosmetics bases (63,64,65,73) etc. can be prepared with cyclodextrins or their complexes. In this field rapid development can be prognosticated.

Table V. EXAMPLES FOR ELIMINATION OF REDUCTION OF UNDESIRABLE TASTES AND ODOURS IN FOODS AND FODDERS

	Reference
improvement of bad tastes:	
- caffee and tea	21
- protein hydrolyzates	22,25
- ginseng extract	23
- propylene glycol	24
deodorizing effect:	
- mutton and fish	26
- bone powder	27
- mouth deodorant	28
- yeast extract	30
- microbial cell mass (butyric or lactic acid bacteria)	31
- soybean lecithin	32
- body, refrigerator and lavatory deodorant	29,33
- aqueous caseinate solutions	34

Table VI. EXAMPLES FOR TECHNOLOGICAL IMPROVEMENTS BY UTILIZING CYCLODEXTRINS

	Reference
Prevention of:	
- fondants and candies from deliquescing by fatty acid-cyclodextrin complex	37
- precipitation of neohesperidine dihydrochalcone from soft drinks	38
- staling of noodles	41
Improvement of:	
- soft drinks (richness, mouth feel, flavour)	39
- egg-white beatability	40
- retention of volatile components	48,50
- bakery products with fatty acid complex	51
- ice creams	20
- water retention in emulsified food	57
Preparation of:	
- emulsified oils	42,43,49
- foaming baking powder (fatty acid or CO_2 complex)	44,75
- soy souce	45
- sausage casing	46
- phenylalanine free diet	47
Application of antifungal and microbicidal complexes:	
- iodine complex (for fish-paste, antiseptic sheets frozen marine prod.)	58,52,53,54
- ethanol complex for packaged foods	55
- benzoic acid, etc.	56

Table VII. EXAMPLES FOR APPLICATION OF CYCLODEXTRINS IN COSMETICS AND TOILETRY

	References
- applications in dentifrices	59,72,73
- in laundry preparations and detergents	67,68,69
- in emulsified cosmetic bases	63,64,65,73
- bath preparation	66
- solid perfumes, incenses and fragrant candles	60,61,62,71
- stabilization of royal yelly	70

The distribution of references cited in this paper according to their publication-year is as follows:

before 1975: 4 patents and papers
1975: 8
1976: 11
1977: 11
1978: 8
1979: 15
1980: 18

Until now Japan is the only country, where the application of cyclodextrin in food products is not limited, therefore it is no wonder, that the preponderant majority of the above cited patents and papers originates from Japan.

Literature Cited

1. Szejtli, J.: Cyclodextrins and their Inclusion Complexes. Akadémiai Kiadó, Budapest, 1981.
2. Rogers, W.I., Whaley, W.M.: US Patent 3, 061, 444 (1962)
3. Szejtli, J., Szente, L., Bánky-Előd, E.: Acta Chimica Acad.Sci. Hung., 101 27 (1979)
4. Szejtli, J., Szente, L., Kolta, R., Lindner, K., Zilahy, T., Kőszegi, B.: Hung. Patent, 174699 (1977)
5. Kernóczy, L., Tétényi, P., Mincsovics, E., Szejtli, J.: Quart.J. Crude Drug.Res., 16 153 (1978)
6. Pobozsny, K., Kernóczy, L., Tétényi, P., Héthelyi, I., Szejtli, J.: Planta Med., 42 255 (1981)

7. Lindner, K., Szente, L., Szejtli, J.: Acta Alimentaria, 10 175 (1981)

8. Szejtli, J., Bánky-Előd, E., Stadler, Á., Tétényi, P., Héthelyi, I Kernóczy, L.: Acta Chimica Acad.Sci.Hung., 99 447 (1979)

9. Yoneda, T., Matsukura, J.: Japan Patent 77, 18, 782 (1977)

10. Shimada, K., Tanaka, I., Nagahama, S.: Japan Kokai 75, 29, 724 (1975)

11. Suzuki, Y., Ikura, H.: Japan Kokai 75, 83, 454 (1975)

12. Osato, T., Takeuchi, S., Esumi, S., Higashikaze, C.: Japan Kokai 80, 38, 338 (1980)

13. Konishi, T., Komiya, J., Yoneda, T.: Japan Kokai 78, 18, 775 (1978

14. Miyake, T.: Japan Kokai, 78, 124, 657 (1978)

15. Szejtli, J., Szejtli, M., Szente, L.: Hung. Patent Appl. 02027 (1980)

16. Kawashima, F.: Japan Kokai, 79, 117, 536 (1979)

17. Suntory Ltd., Japan Maize Products Co., Ltd.: Japan Kokai 80, 114, 283 (1980)

18. Szejtli, J., Szente, L., Gál-Füzy, M., Szejtli, M., Kőszegi, B.: Hung. Patent Appl. 1287 (1980)

19. House Food Ind. KK.: Japan Kokai 51, 012, 970 (1976)

20. Lotte KK.: Japan Kokai 55, 034, 042 (1980)

21. Hamilton, R.M., Heady, R.E.: US Patent 3, 528, 819 (1970)

22. Suzuki, Y.: Japan Kokai 75, 69, 100 (1975)

23. Akiyama, Y.: Japan Kokai 79, 80, 463 (1979)

24. Asama Kasei KK.: Japan Kokai 80, 71, 456 (1980)

25. Helbig, N.B., Ho, L., Christy, G.E., Nakai, S.: J.Food Sci., 45, 331 (1980)

26. Nagano, P.: Japan Kokai 80, 77, 875 (1980)

27. Fukinbara, T.: Japan Kokai 80, 44, 305 (1980)

28. Yajima, M.: Ger.Offen 3, 008, 663 (1980)

29. Watanabe, T., Ishizone, H., Kawai, N.: Japan Kokai 78, 41, 440 (1978)

30. Toyo Jozo Co. Ltd.: Japan Kokai 80, 156, 564 (1980)

31. Miyairi, Y., Mori, S., Yoshimasu, T., Horikoshi, K.: Japan Kokai 79, 05, 092 (1979)

32. Sugiyama Ind.Chem.Inst.: Japan Kokai 80, 159, 760 (1980)

33. Toyo Ink MFG. KK.: Japan Kokai 52, 015, 809 (1977)
34. San-Ei Chem. Ind.: Japan Kokai 76, 006, 219 (1976)
35. Horiuchi, I.: Japan Kokai 53, 041, 440 (1978)
36. House Food Ind. KK.: Japan Kokai 51, 148, 052 (1976)
37. Nawata, Y., Yamamoto, K., Sano, M.: Japan Kokai 77, 12, 684 (1977)
38. Hashimoto, S., Katayama, A.: Japan Kokai 75, 35, 349 (1975)
39. Nakashima, T.: Japan Kokai 79, 145, 258 (1979)
40. Kuroda, N., Mogi, S.: Ger. Offen. 2, 406, 376 (1974)
41. Nakashima, T.: Japan Kokai 79, 145, 241 (1979)
42. Nawata, Y., Ofuji, T., Sano, M.: Japan Kokai 76, 126, 204 (1976)
43. Yanagise, M.: Japan Kokai 77, 96, 774 (1977)
44. Nakamura, S., Chujo, N., Takahashi, T.: Japan Kokai 79, 93, 686 (1979)
45. Higeta, S., Shogu Co. Ltd.: Japan Kokai 80, 71, 742 (1980)
46. Fujishiro, T., Nakamura, N., Matsuzawa, M.: Japan Kokai 79, 143, 458 (1979)
47. Szente, L., Szejtli, J., Rothe, M., Specht, M.: Hung. Patent Appl. CI-2001 (1979)
48. Hamano, M., Okayasu, M., Sugimoto, H.: Toketsu oyobi Kanso Kenkyukai Kaishi, 25 31 (1979)
49. Teijin KK.: Japan Kokai 52, 010 448 (1977)
50. Nishii KK.: Japan Kokai 81, 26, 147 (1981)
51. Sudo, M.: Japan Kokai 54, 113, 455 (1979)
52. Hirose, T.: Japan Kokai 76, 08, 625 (1976)
53. Hirose, T., Miwa, K.: Japan Kokai 76, 101, 123 and 76, 101, 124 (1976), (1976a)
54. Yoshitomi, T., Tanaka, Y.,Yoshida, R., Kokumai, J., Hirose, T.: Japan Kokai 76,140, 964 (1977) and Yoshitomi, T., Yamazaki,E., Tanaka, Y., Hirose, T.: Japan Kokai 77, 15, 809 (1977)
55. Asahi Denka Kogyo KK.: Japan Kokai 80, 87, 701 (1980)
56. Asahi Electrochem Ind. KK.: Japan Kokai 53, 113, 017 (1978)
57. Kanegafuchi Chem. Ind. KK.: Japan Kokai 52, 012, 955 (1977)
58. Toyo Ink MFG. KK.: Japan Kokai 51, 118, 859 (1976)
59. Hashimoto, S., Inoue, K., Watanabe, T.: Japan Kokai 77, 34, 941 (1977) and Hashimoto, S.: Japan Kokai 80, 05, 484 (1980)
60. Kokando Co. Ltd.: Japan Kokai 80, 78, 965 (1980)

61. Masuda, S., Ito, T., Matsuda, M., Kitano, H.: Japan Kokai 79, 80, 433 (1979)
62. Fukunaga, T.: Japan Kokai 78, 113, 040 (1978)
63. Kanebo Co. Ltd.: Japan Kokai 80, 64, 511 (1980)
64. Yokota, T., Kawahara, K.: Japan Kokai 79, 70, 434 (1979)
65. Yokota, H., Ito, K.: Japan Kokai 80, 11, 527 (1980)
66. Hirano, S., Tsumura, J., Imazaki, I., Ohuchi, M., Kito, H.: Japan Kokai 75, 64, 320; 75, 64, 418 (1975) and Hirano, S., Tsumura, I., Izeki, I., Kawamura, H., Ohhara,M.: Japan Kokai 75, 63, 126 (1975) and Hirano, S., Tsumura, J., Imazaki, I., Ohuchi, M., Yoneda, T.: Japan Kokai 77, 30, 822 (1977)
67. Honma, I.: Japan Kokai 76, 133, 960 (1976)
68. Nakagawa, Y., Honma, I.: Japan Kokai 76, 99, 195 (1976)
69. Koch, J.: Ger.Patent 3, 020, 269 (1981)
70. Akiyama, Y., Miyao, K.: Japan Kokai 77, 114, 060 (1977)
71. Kokando KK.: Japan Kokai 54, 080, 433 (1979)
72. Sun Star Dentifrice: Japan Kokai 52, 057, 338 (1977)
73. Sun Star Dentifrice: Japan Kokai 51, 121, 531 (1976)
74. MSC. YG.: Japan Kokai 53, 015, 467 (1978)
75. Schlenk, W.: US Patent 2, 830, 040 (1958)

I. Int. Symp. on Cyclodextrins
Budapest, 1981 6.2.

PRODUCTION AND APPLICATION OF β-CYCLODEXTRIN COMPLEXES IN THE FOOD INDUSTRY

GYÖRGY VENCZEL

CYCLOPACK Association for Research, Development and Production, Landler J. u. 26
H-1078 Budapest, Hungary

SUMMARY

The CYCLOPACK Research-Development-Production Corporation is engaged exclusively in the research, production and marketing of the food-industrial β-cyclodextrin aroma complex. The most important advantages of the CYCLOPACK products are the extreme stability, excellent, long-lasting storage, easy applicability equally in the largescale plants and in the households. Primary fields of application are expected in the meat-, canning-, sweets- and catering industries, as well as in the households and dietary catering of hospitals.

Verified informations are available today from several fields of interest regarding the advantages, potentialities of the β-cyclodextrin aroma complexes being used in the food industry.

Their extreme stability is verified by the instrument tests /see Figures 1-2/.

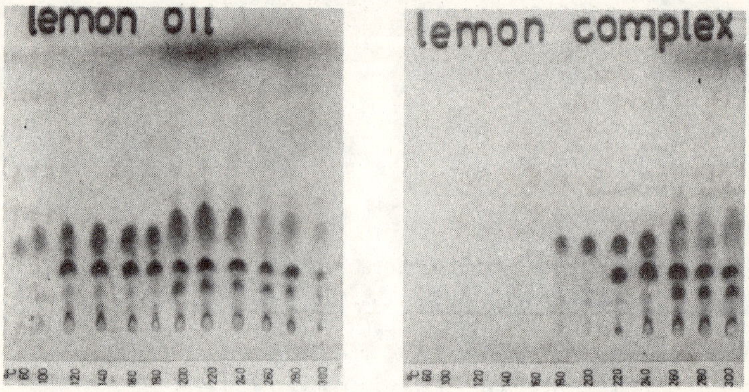

Figure 1:

Comparison with TAS technique of lemon oil and thermal stability of the β-cyclodextrin inclusion complex produced from it.

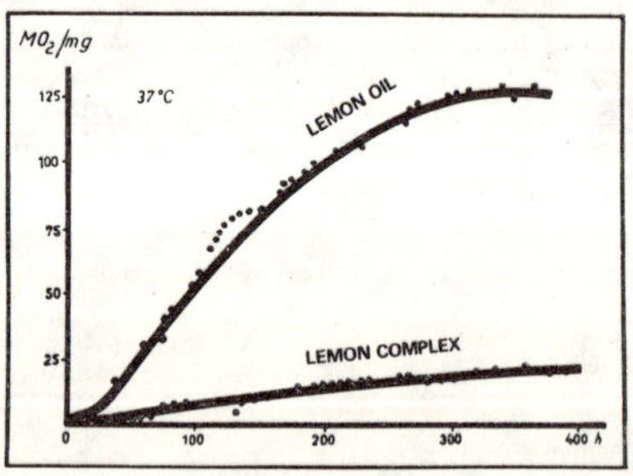

Figure 2:

Comparison with Warburg technique of lemon oil and resistance against oxidation of the β-cyclodextrin inclusion complex produced from it.

The Biochemical Research Laboratory of CHINOIN Pharmaceutical and Chemical Works achieved remarkable results in the field of research, laboratory production and testing of the β-cyclodextrin inclusion complexes. The general potentialities of the foodindustrial applicability existing at that time, urged COMPACK Commercial Packing Co. to establish initially a research-development co-operation with CHINOIN, then following the mutually successful phase, the two large enterprises decided on the foundation of a joint venture. The so-established

CYCLOPACK

Research-Development-Production Corporation has exclusive right for the research, development, and production, as well as for the marketing of the β-cyclodextrin inclusion complexes in the whole field of the food industry. The Corporation - with participation of the experts of the promoting companies and with outside institutes initiated the realization of the following objectives:

- starting the experimental production of the food industrial β-cyclodextrin aroma complexes (initially in a limited range at the end of 1981);
- further-development of the variety, meeting the foreign and domestic demands.

The initial range of utilization of the β-cyclodextrin aroma complexes is expected to be in the food industry /e.g.meat, canning, sweets industry/ catering industry, as well as in catering for the household and dietary food of hospitals. The initial product variety has been planned accordingly: onion, garlic, smoke, dill leaves, majoram, cumin, mustard, bergamot, jasmin, peppermint, vanilla, chamomile etc. β-cyclodextrin aroma complexes and the finished products with them by the promoting companies /such as aromatic teas, spiced salts etc./.

The advantages over the presently marketed powedered aromas /in which the aromas are dispergated in gum Arabicum/: safer storage /they can be stored without sticking, lumping, detrimental effect for several years, even without careful packaging/, and easy utilization /instant effect even in cold water/.

Proportion of the voletile oil content, i.e. the aroma substance of the complexes is about 10 %. According to the evidence of the comparative tests - depending on the given field of application - some of the β-cyclodextrine aroma complexes are equivalent to the natural materials in the following proportions:

Complexed volatile oil variety	Natural substance equivalent to 1 g β-cyclodextrin aroma complex/mg/
Onion	100 - 120
Garlic	30 - 40
Smoke	400 - 600

Outstanding results obtained from the experiments conducted in the food industry are the following :
Bergamot, jasmin and peppermint β-cyclodextrin complexes are well - and economically applicable for the production of aromatic teas. The granulated material made from the complexes with suitable process can be admitted into the filter tea by dry mixing. The required quantity is 1-4 % of the tea.

Favourable experiences have been acquired in several canning and meat industrial plants, mainly in the course of the technological production processes of meat products. The scope of applicability in this field will be determined by comparing the advantages and costs. Naturally they will have a decisive role against those aromas and volatile oils, which do not come up to the technological expectations.
A new range of application is in progress in the field

of flavouring the substance of chewing gums, chips and noodles.

The variable diet of patients suffering mainly from digestary disease is expected to be improved in the hospitals by use of the CYCLOPACK products. According to the research results so far the patients did not complain against spiced foods prepared with β-cyclodextrin complexes.

Bibliography

/1/ J.Szejtli, L.Szente, R.Kolta, K.Lindner, T.Zilahi, B.Kőszegi: Hungarian patent 174 699 /1977/

/2/ J.Szejtli: Die Stärke 30, 427 /1978/

/3/ Report on the research of the Commercial and Catering-industrial College /1980/

I.Int.Symp. on Cyclodextrins
Budapest, 1981

STABILISATION AND CONTROLLED RELEASE OF PERFUME IN DETERGENTS

J. KOCH

Unilever Forschungsgesellschaft GmbH, Behringstr. 154
D-2000 Hamburg 50, FRG

SUMMARY

The evaporation of perfumes from detergent powders is strongly inhibited by clathration with cyclodextrins (CD). The degree of evaporation from the dry clathrates depends more on the relative air humidity than on the temperature.

In aqueous solution the clathrated perfumes are displaced from the CD cavity by the surfactant molecules. The extent of this displacement varies largely with the chemical nature and/or steric factors of the competing molecules. The perfumes under investigation were released most easily by anionic surface active agents.

Thus CD provide good stabilisation of perfumes in detergent powders and fast and quantitative release upon addition of water.

INTRODUCTION

The suitability of the clathration with cyclodextrins for deodorisation and aroma or perfume stabilisation has been demonstrated and discussed in several publications (1-7). It has been shown that dry perfume clathrates are almost totally stable at ambient temperatures and do not release the included odours to any significant extent. In aqueous solution, however, the perfumes are liberated, although only with a fraction of their normal vapour pressure. This reduction of odour release may be suitable or acceptable in cases where a slow distribution over an extended period of time is desired (e.g. in some areas of toilet preparation and foods) or when dissolution in water is accompanied by an increase in temperature, which leads to an accelerated flavour release. But it is unfavourable when the stabilised perfume is incorporated in products from which it is supposed to be liberated quickly under the actual conditions of use, e.g. from soap or detergents during a low temperature washing process.

Earlier investigations carried out in a different context had shown us that clathrated perfumes may be displaced from the cyclodextrin cavity by other substances to an extent which varies with their chemical nature and sterical factors. There were indications that in case of mixtures of a surface active agent and a perfume competing for clathration by ß-cyclodextrin the equilibrium is far on the side of the detergent clathrates. Since a good protection of perfumes against evaporation or chemical degradation (e.g. by perborates) in the dry state (i.e. during storage) and a fast and almost complete release upon dissolution at all temperatures in the range from 30-90°C are indispensable attributes of a stabiliser for use in soaps and washing powders, it was the objective of the present investigation to study the influence of important components of washing powders, soaps, rinse conditioners and other detergent products on the release of perfumes present as clathrates. In addition some representatives of other relevant classes of surface active agents were included into these investigations.

METHODS AND RESULTS

Displacement of perfume from clathrates in aqueous surfactant solutions

In order to estimate the extent of perfume displacement in solution under simulated practical washing conditions we determined by GLC the perfume concentrations (total peak area of all components) in the headspace on top of aqueous solutions of free and clathrated perfumes in the presence and absence of surface active agents. The headspace bottles serving as reaction vessels were furnished with a magnetic stirrer bar and capped with a teflon-coated septum pierced by a capillary as pressure outlet in order to prevent an unrealistic pressure increase at elevated temperatures which does not occur in practice upon hand washing or in a washing machine. Temperatures were chosen according to the actual conditions of use, e.g. in case of a wash powder (DOBS) 30°C (for synthetics and woolens), 60°C (for coloured laundry) and 90°C (for white laundry). Equilibration times were approximately those needed, e.g. for machine-, hand- or bowl washing respectively. Perfume and detergent concentrations were similar to those present in actual detergent solutions. In a few cases, however, the influence of the detergent was followed over a wide (and unrealistic) concentration range.

Thus in a typical experiment simulating fabric bowl washing conditions the content of 4 headspace flasks consisting of

1) 1 µl perfume + 1 ml H_2O
2) 1 µl perfume + 1 ml H_2O + 20 mg ß-CD
3) 1 µl perfume + 1 ml H_2O + 3 mg DOBS
4) 1 µl perfume + 1 ml H_2O + 20 mg ß-CD + 3 mg DOBS

was stirred at 40°C for 5 min.

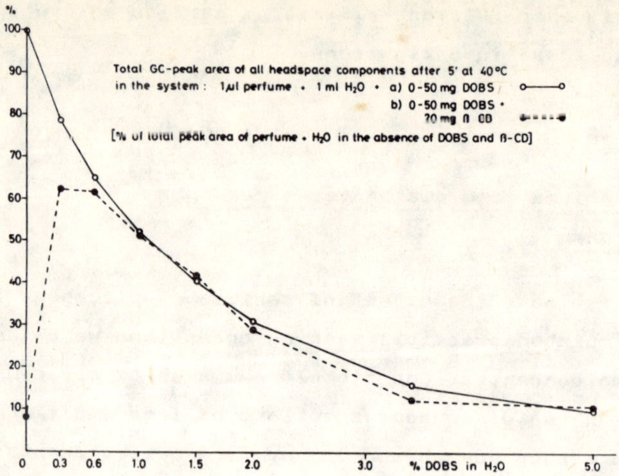

Fig. 1 Displacement of perfume from its ß-CD-clathrate by dodecylbenzene sulphonate (DOBS)

Fig. 1 shows that the residual partial vapour pressure of perfume in the presence of ß-CD (flask b) was approximately 9 % of that of the free perfume without CD. In the presence of 0.3 % DOBS the headspace concentration of the clathrated perfume increased to the 7-fold by displacement whereas that of the free perfume decreased to less than 80 % by solubilisation. Upon addition of 0.6 % DOBS both values were almost the same and steadily dropped on further increase of surfactant concentration remaining equal over the whole concentration range up to 15 % (in Fig. 1 only plotted up to 5 %). A similar curve was obtained at 40°C/5 min with lemon perfume as it is used in dishwashing powders. After 5 min at 25°C, however, the release of perfume was incomplete over the whole concentration range (Fig. 2) indicating that the displacement at room temperature is not quite fast enough to reach equilibrium within 5 min.

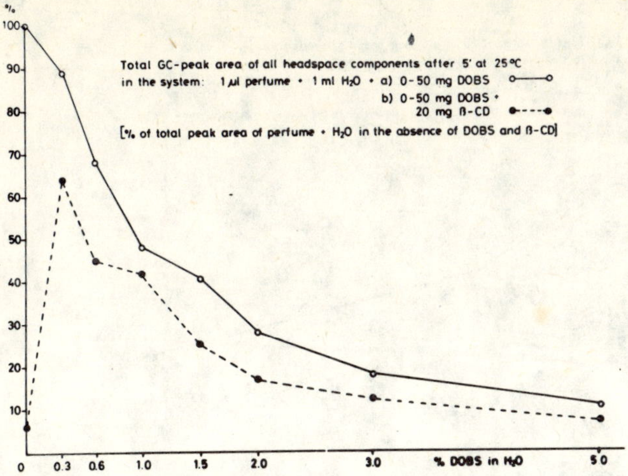

Fig. 2 Displacement of lemon perfume from its ß-CD-clathrate by dodecylbenzene sulphonate (DOBS)

For imitating fabric machine washing similar experiments with similar results were made at 30, 60 and 90°C/30 min.

Handwashing conditions were simulated by stirring the perfume/water/CD-mixtures with increasing quantities of soap for 3 min at 40°C. In this case 0.3 - 1 % soap was necessary for an almost complete perfume displacement (Fig. 3).

In contrary to the results obtained with anionic surfactants, no displacement was observed at r.t. with bulky quaternary ammonium compounds such as $(HT)_2N(CH_3)_2Cl$ (HT = hardened tallow alkyl-) (Fig. 4) used as rinse conditioners.

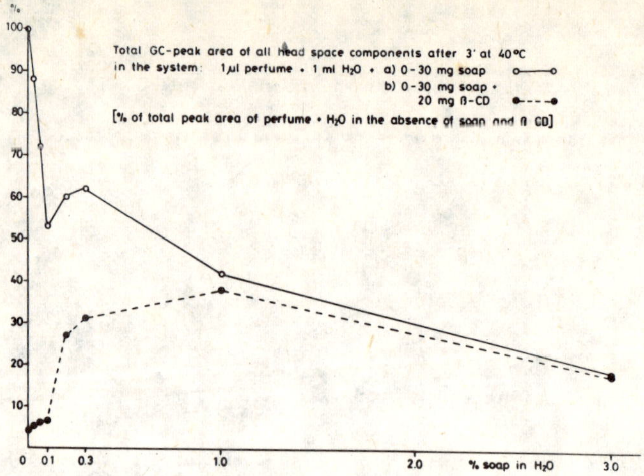

Fig. 3 Displacement of perfume from its ß-CD-clathrate by soap

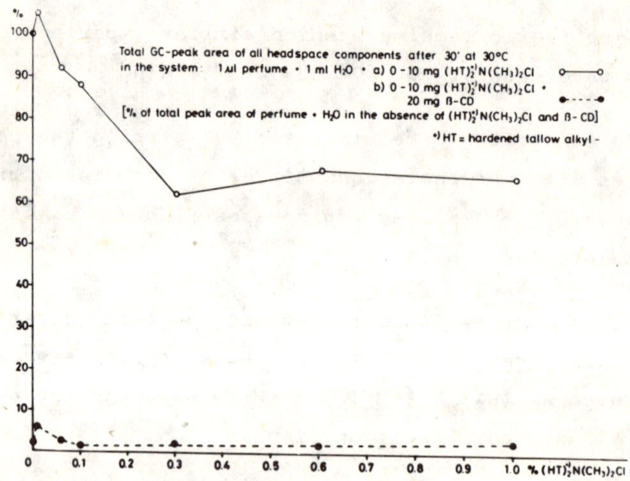

Fig. 4 Displacement of perfume from its ß-CD-clathrate by $(HT)_2 N(CH_3)_2Cl$

Table I

Displacement of perfume from its ß-CD clathrate by surface active substances

System a) 1 µl perfume + 1 ml H_2O + 10 mg surfactant

System b) 1 µl perfume + 1 ml H_2O + 10 mg surfactant + 20 mg ß-CD

Surface active agent	Perfume liberation*) (%) after 15 min at 30°C	60°C
Blank (no detergent)	4.5	29
Na-C_{12-18}-alkyl sulphonate	99	93
Na-dodecylbenzene sulphonate	103	103
Na-C_{12-18}-alkyl sulphate	90	95
80/20/5 soap base	90	102
Na-C_{12-14}-alkyl-3 EO-sulphate	65	112
Nonylphenol-10 EO	70	96
C_{12-18}-alkanol-5 EO	25	54
$C_{18:1}$-alkanol-8 EO	28	64
$C_{18:1}$-alkanol-20 EO	18	62
Polyoxyethylene sorbitan monostearate	33	49
Carboxylic acid polyglycolester	44	59
Coconut fatty acid ethanolamide-3 EO	30	87
$(HT)_2N(CH_3)_2$ Cl	4	60
$C_{16}H_{33}N(CH_3)_3$ Br	48	85
Na-dioctyl sulphosuccinate	35	72
Na-bis(tridecyl)-sulphosuccinate	36	65
Polyglycol 2000	5	34
Polyglycol 6000	3	38
Polypropyleneglycol	8	51
Na-polyphosphate	7	38
Na-perborate	9	37

*) perfume vapour pressure in headspace of system b) as % of a)

Table II

Stabilisation of perfume in dry washing powder by CDs under storage conditions

Residual perfume concentrations (% of original) in headspace over aqueous solutions of washing powder after storage

		Storage in open flasks			Storage in closed flasks		
		perfume			perfume		
Storage conditions	Storage time	sprayed on detergent	clathrated with ß-CD	clathrated with α-CD	sprayed on detergent	clathrated with ß-CD	clathrated with α-CD
38°C/ 75 % r.h.	1 d	13.2	92.5		92.4	100	
	3 d	0.5	27.3	56	61.6	56.8	
	7 d	0.3	3.4	52	33.1	50.4	100
	28 d			49	2.6	23.4	91
50°C/ 15 % r.h.	3 h	19.7	100				
	1 d	0.4	100		61.2	96.1	
	3 d	not detectable	86.4	99	22.1	63.1	
	7 d		72.5		2.1	42.6	
	28 d			96			100

Further results obtained with 1 % each of other surfactants and washing powder ingredients under standard conditions (15 min stirring at 30°C and 60°C respectively) are given in Table I. They show that in the presence of 1 % tenside an almost complete perfume displacement ($\geq$90 %) is observed in case of the anionic alkyl sulphonates and sulphates, alkylbenzene sulphonates and sodium soap. It is somewhat lower (57 - 86 %) for anionic and nonionic aromatic ethylene oxide (EO) adducts and - for practical purposes - insufficient (<50 %) for nonionic aliphatic EO-adducts, quaternary ammonium compounds and sulphosuccinates. At 60°C, however, an almost complete perfume release was achieved also by aromatic nonionic and anionic EO-adducts indicating that the incomplete release at 30°C is determined by the reaction rate rather than by the state of equilibrium. With polyglycols, polyphosphate and perborate no significant increase in perfume release compared to distilled water (blank) was observed.

Evaporation of clathrated perfume from detergent powders under rigorous storage conditions

In order to estimate the loss of free and clathrated perfume from detergent powders by evaporation during storage under critical climatic conditions a commercial detergent composition (10.5 % DOBS; 4.5 % C_{9-11}-alkanol-6 EO; 11.5 % Na-tallow/coconut soap 80:20; 9.0 % alkaline water-glass; 0.8 % carboxymethyl cellulose; 21.0 % Na-sulphate; 42.0 % Na-tripolyphosphate; 0.7 % sequestering agent, optical brightener, water) containing initially no perfume was mixed with 0.2 % of free perfume or perfume clathrated with α- and ß-CD respectively. 10-g quantities of these mixtures were stored in flasks which were either left open or closed tightly with washing powder carton (the ratio of carton surface/weight of detergent was approx. the same as for a 1-kg package) at elevated temperature (35°C and 50°C resp.) in a dry or humid atmosphere (15 % and 75 % r.h. resp.).

The results given in Table II show that the free perfume evaporates under all conditions much faster than the clathrated perfume. For the evaporation of the unstabilised perfume the temperature plays the dominant role, whereas in case of the ß-CD clathrate the influence of the relative humidity and the correlated water content of the clathrate are prevailing: the decrease of perfume concentration at 38°C/75 % r.h is much faster than at 50°C/15 % r.h.

The influence of humidity is illustrated even more clearly by the fact that in the dry atmosphere the loss of perfume from the open flask is significantly slower because the original water content of the CD-clathrate (approx. 10 % by weight at 50 % r.h.) is evaporating quickly, resulting in an inhibition of perfume liberation. The loss of perfume from the α-CD-clathrate is even slower: After 4 weeks at 38°C/75 % r.h. still half of the perfume was left in the open flask, whereas under all other conditions the losses were insignificant.

ACKNOWLEDGEMENT

The author wishes to thank Mr. R. Kröhn for the careful and diligent execution of the experimental work.

LITERATURE CITED

1. Bavley, A., Robb, E.W.: US-PS 3 047 431 (1962)
2. Suzuki, Y., Ikura, H.: J. Kokai-PS 83 454 (1975)
3. Hirano, S., Tsumara, J., Iziki, J., Kawamura, H. Ohhara, M.: J. Kokai-PS 63 126 (1975)
4. Yoneda, T., Matrukura, J.: J-PS 18 782 (1977)
5. Watanabe, T., Ishizone, H., Kawai, N.: J. Kokai-PS 41 440 (1978)
6. Szejtli, J., Szente, L., Kolta, R., Lindner, K. Zihaly, T., Köszegi, B.: DE-PS 28 26 724
7. Szejtli, J., Szente, L., Bánky-Elód, E.: Acta Chim. Acad. Sci. Hung. 101, 27 (1979)

I.Int.Symp. on Cyclodextrins 6.4.
Budapest, 1981

SEPARATION OF PHENYLALANINE AND TYROSINE FROM PROTEIN HYDROLYSATES

MANFRED SPECHT,[1] MANFRED ROTHE,[1] LAJOS SZENTE,[2]
JÓZSEF SZEJTLI[2]

[1] Central Institute of Nutrition Potsdam-Rehbrücke
Arthur-Scheunert-Allee 114/116
DDR-1505 Bergholz-Rehbrücke, GDR
[2] CHINOIN Pharmaceutical-Chemical Works, Tó u. 1-5
H-1045 Budapest, Hungary

Phenylalanine-free protein hydrolysates are needed for dietetic treatment of phenylketonuria. However, the phenylalanine-free protein hydrolysates commercially available deliver an unpalatable off-flavour. This is caused possibly by the sulfuric acid hydrolysis of proteins followed by calcium hydroxid neutralisation (1). This type of hydrolysis is necessary to get products without salt. Although several publications (2) e. g. discuss the purification of protein hydrolysates from off-flavour, the sensory quality of such products is yet insufficient and must be improved. Otherwise the patients show an aversion to the high daily intake of such products within the diet.

ß-Cyclodextrin-polymers (ß-CDP) (3, 4, 5, 6) as well as some types of dextran gels e. g. Sephadex G-10, G-15, G-25 (7) are well appropriated for the analytical separation of aliphatic and aromatic amino acids.

The use of ß-CDP for preparative separation is also possible (8). As a special advantage of this procedure sodium chloride does not interfere with the separation. Therefore the hydrochloric acid protein hydrolysates which are of pleasant taste may be applied as raw material. As can be seen from figure 1 the composition of the effluent can be calculated. In the fractions 12 - 18 33 % of the applied amino acids, 16 % of the phenylalanine and 10 % of the sodium chloride are eluated. Other ratios can be calculated, too. Table 1 shows the ideal amino acid composition before and after ß-CDP treatment. In this case the yield is about 25 %.

Literature Cited

1. Prendergast, K.: Food Trade Review 44 14, (1978)
2. Geipel, H., Rudnick, K.: DDR-WP 141 235
3. Zsadon, B., Szilasi, M., Tüdös, F., Fenyvesi, E., Szejtli, J.: Starch 31 11, (1979)
4. Szejtli, J., Fenyvesi, E., Zsadon, B.: Starch 30 127, (1978)
5. Zsadon, B., Szilasi, M., Otta, K., Tüdös, F., Fenyvesi, E., Szejtli, J.: Acta Chim. Acad. Sci. Hung. 100 265, (1979)
6. Wiedenhof, N.: Starch 21 163, (1969)
7. Gelotte, B.: J. Chromatography 3 330, (1960)
8. Specht, M., Rothe, M., Szente, L., Szejtli, J.: DDR-WP 147 615

Table 1

Relative amino acid composition of a hydrochloric acid casein hydrolysate before and after ß-CD-polymer treatment

Amino acid	Relative amino acid composition before chromatography [%]	after chromatography [%]
Lys	7.8	11.0
His	2.4	2.1
Arg	3.5	2.2
Asp	9.8	15.5
Thr	4.6	4.2
Ser	6.6	6.3
Glu	27.0	36.2
Pro	13.0	10.8
Gly	2.5	2.2
Ala	4.0	3.3
Val	4.4	3.2
Met	2.3	0.7
Ile	2.2	0.7
Leu	5.5	1.5
Tyr	1.3	0.1
Phe	3.1	0.1

figure 1 Preparative separation of phenylalanine from protein-hydrolysates on β - CDP - column (9,0×94,5 cm, Vt = 6000 ml)

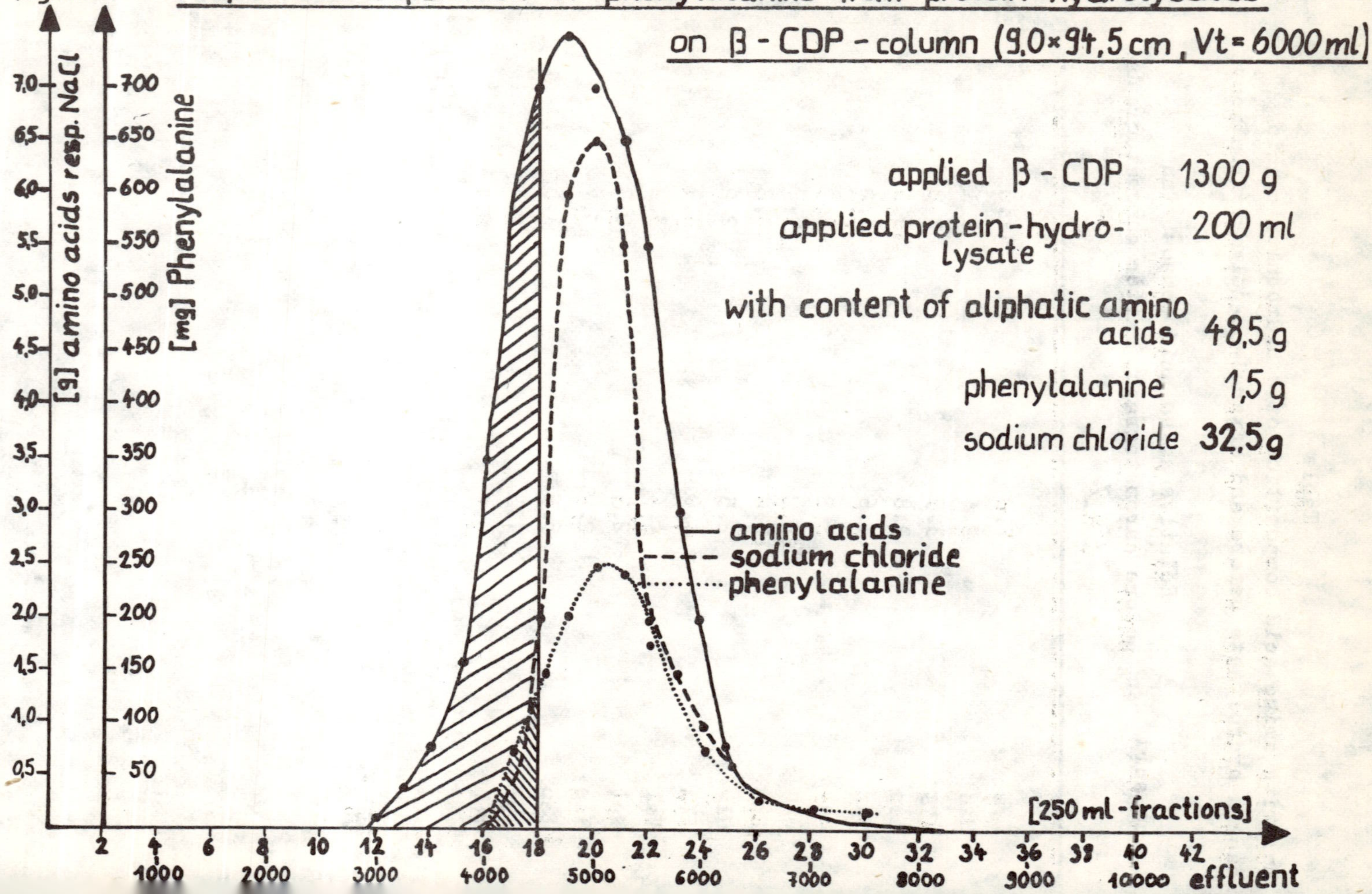

EFFECT OF CYCLODEXTRIN ON GROWTH PROCESSES AND PRODUCTIVITY OF PLANTS

P. TÉTÉNYI

Research Institute for Medicinal Plants
H-2011 Budakalász, Hungary

SUMMARY

Seeds of various cultivated plants have been treated by cyclodextrins before germination. The effect of treatment have been studied during whole life cycles of plants in growth chambers as well as in open air field experiments.

A 60-70% growth retarding effect was observed at 3-day-old seedlings with starchy seeds; which reverted afterwards: CD enhanced growth to 18-30% surpassing control at 4-week-old plants yet. An increased branching habit had been characteristic for treated cereal plants /barley, spring wheat, rye/, which resulted in higher crop yield /8-12% surplus in field trials/. Treatment of vegetable's seeds /radish, lettuce, parsley/ arrived to give not only a higher yield but also a raised content of dry material at harvest.

No unambiguous explanation can be given as yet for the observed effects, which can be due to competitive inhibition of amylases /30% retardation in the breakdown of starch/ followed by promotion of biochemical mechanism /respiration, protein synthesis, etc./. Taking into account the reverse effect produced by gibberellic acid an active involvement of cyclodextrin in the hormonal control of plant growth is supposed, which results in a lasting influence during the whole ontogeny.

The working out of processes consisting of seed treatment and plant spraying to increasing productivity has been tried in plant cultivation for ages. Trace elements, vitamins and growth regulators have been tried for this purpose. Recently, significantly positive effects were described after the use of Cycocel /1/, as well as of Ancymidol /2/. However, these procedures could not become wide-spread, because a resounding success in the practice may only be expected from a molecule that increases the productivity while being safe, harmless to health, cheap and available in bulk quantities.

The present communication reports on results obtained in the course of cyclodextrin treatment of plants with the above considerations in mind. The starting point was our suprising observation concerning the effect on plant growth of cyclodextrin /3,4,5/. This phenomenon was first found in a growth test where cyclodextrin was shown to inhibit the growth of oat coleoptyls. The degree of inhibition was 51% in a cyclodextrin solution of 1%, 17% in a solution of 0,1% and 5% in a solution of 0,01% which is corresponding to the growth inhibiting action of mannose, a growth inhibiting sugar known from the literature.

Similarly, a growth inhibition was observed on germinating grains /e.g. of spring barley, spring wheat, rye, Triticale/ which had been somehow pretreated with cyclodextrin e.g. by soaking, pilling or by fluidization/ /Figure 1./.The growth inhibition of seedling reached 60-80%, depending on the concentration and on the use of α- or β-cyclodextrin, respectively.

Fig.1.

THE INITIAL GROWTH INHIBITING ACTION OF CYCLODEXTRIN TREATMENT OF BARLEY SEEDLINGS /SUM OF THE DATA OF 30 PLANTS/

TREATMENT	3 DAYS		6 DAYS	
	ROOT mm	SHOOT mm	ROOT mm	SHOOT mm
CONTROL	5123	662	7714	2090
α-CYCLODEXTRIN	1056	96	2593	754
/Δ%/	/-79,4/	/-85,5/	/-66,4/	/-63,9/
β-CYCLODEXTRIN	1808	191	2277	556
/Δ%/	/-64,7/	/-71,2/	/-70,5/	/-73,4/
$SD_{5\%}$	900,6	124,3	814,4	223,9

Further interesting growth and development effects could be observed, when plants were cultivated from cyclodextrin-treated seeds in chambers /6/. It can be seen from Figure 2, that the vegetative growth of four weeks old wheat plants was not inhibited but even promoted by 18-30%, under the effect of treatment with cyclodextrin of the seeds /by soaking in a solution of 10^{-2} mole for 24 hours/.

Fig.2.

EFFECT OF CYCLODEXTRIN TREATMENT ON THE VEGETATIVE GROWTH OF 4 WEEKS OLD SPRING WHEAT /phytotron experiment at Martonvásár/

TREATMENT	ONE PLANT ROOT LENGTH/cm/	ROOT DRY MASS/g/	SHOOT LENGTH/cm/	SHOOT DRY MASS/g/
CONTROL	33,95	0,061	29,55	0,238
BCD	38,40	0,081	31,83	0,281
/ Δ % /	/+13,1/	/+32,8/	/+ 7,7/	/+18,1/
$SD_{5\%}$	1,63	0,019	2,26	0,049

The results of an experiment continued in a phytotron up to the harvest are shown in Figure 3: a significant promotion of 30-40% appeared in the branching and individual yield of plants at a significancy level of 0,1%, even in four repetitions.

Fig.3.

EFFECT OF CYCLODEXTRIN TREATMENT ON THE YIELD OF 4 MONTH OLD SPRING WHEAT
/phytotron experiment at Martonvásár/

TREATMENT	ONE PLANT EARS /pc/	GRAIN MASS /g/
CONTROL	2,8	4,15
BCD	3,7	5,87
/ Δ % /	/+32,1/	/+41,6/
$SD_{0,1\%}$	0,478	0,855

While approximating the practice with these experimental results, the possibilities of using cyclodextrin for seed treatments were studied. In addition to the soaking in solution carried out previously for experimental purposes, wet seed-dressing, being more useful for mechanical sowing, was tried on spring barley in a small-parcel field experiment with four repetitions. The effect on the vegetative growth of cyclodextrin was evaluated by measuring the overground parts from samples taken at one month before harvest. The other part of the plants were worked up concerning the crop at harvest time. It can be seen from the data of Figure 4, that our experiences obtained in the phytotron were supported by field conditions: both sprout formation and yield of grains were increased by cyclodextrin /6-19% and 8%, respectively/ as calculated either for the individual plant or for unit area. The seed-dressing seemed to be more useful for applying cyclodextrin than did soaking.

Starting from all these observations, a yield experiment was installed and planned for three years, with randomized block arrangement on small parcels of six experimental stations of our country. Our aim was to realize the effect of cyclodextrin treatment of the crop under conditions of agricultural practice. In the first part of the experiment, observations were made on the oat and spring barley. Two types of procedures of seed-treating were tried: wet seed-dressing and coating by sprayed fluidization. The amount of cyclodextrin applied on the seed surfaces during treatment reached about 2% of the seed weight in both cases. After sowing, the treated plants shared in the same usual agrotechnical procedure as the controls. The results obtained are shown in Figure 5. The yield increase appeared as a trend on the effect of cyclodextrin treatment as an average of the six stations: 1,2% with oat and 2,4% with spring barley, mainly by seed-dressing. It should be noted, that the effect of treatment as measured by crop differences was influenced in a different manner in the six regions of the country by the dry weather, that has been very unfavourable for cultivation of cereals this year. This is indicated by the results of the most balanced experiment at Kompolt, where the highest yield was obtained by cyclodextrin treatment: a significant increase in yield of 6,3% was reached with oat by seed-dressing, and 7,1% with spring barley by sprayed fluidization coating.

The cyclodextrin treatment was also tried by spraying in three

Fig.4.

SMALL - PARCEL FIELD BARLEX EXPERIMENT

/Budakalász/

CROP DATE / m²	1979 Ø	1979 BCD	1980 soaked Ø	1980 soaked BCD	1980 seed-dressed Ø	1980 seed-dressed BCD
At vegetation state:						
SHOOT /g/	-	-	670	680	530	610
/ Δ %/				/+1/		/+15/
At harvest time:						
SHOOT /g/	-	-	1090	1150	950	1130
/ Δ %/				/+6/		/+19/
EARS /pc/	263	334[+]	534	605	577	621
/ Δ %/				/+13/		/+ 8/
EARS MASS /g/	515	578	565	615	540	583
/ Δ %/		/+12/		/+ 9/		/+ 8/
GRAIN MASS /g/	453	508	497	539	479	516
/ Δ %/		/+12/		/+ 8/		/+8/
$^{+}SD_{5\%}$						

Fig.5.

APPLICATION OF CYCLODEXTRIN IN CEREAL'S CULTIVATION
/field experiment/

TREATMENT	YIELD			
	MAXIMUM /at Kompolt/		AVERAGE /six regions of the country/	
	kg/ha	%	kg/ha	%
Oat:				
CONTROL	5570	100	5120	100
BCD **/by fluidization/**	5600	100,5	5030	98,2
BCD **/by wet dressing/**	5920	106,3	5180	101,2
Spring barley:				
CONTROL	5810	100	4890	100
BCD **/by fluidization/**	6220	107,1	4970	101,6
BCD **/by wet dressing/**	5930	102,1	5010	102,4
$SD_{5\%}$	220		170	

different phases of plant development /branching, development of stem and ear formation/. The spraying was carried out by 1% cyclodextrin solutions on separate half parcels and the treatments were combined. The results are summarized in Figure 6: the plant-height and whole amount of overground were mostly increased by the combined treatment at all the three stages. The number of ears was most favourably influenced by spraying during stem and ear development, while grain crop was highest by spraying during branching and ear development, i.e. at a different stage.

Thus, the yield increasing action of cyclodextrin treatment seems to be proved in the cultivation of cereals, whereas further careful experiments are needed to determine the way of application, i.e. seed treatment by soaking or by wet dressing or by spraying fluidization, or the cyclodextrin should be applied by spraying on the plants at the critical stages.

During the study of effect of the cyclodextrin treatment on plant growth and development, other plants and ways of use were also tried. On the basis of our results obtained on germination and seedling development in Petri-dishes, different vegetables /celery, radish, lettuce, parsley, carrot, paprika/ were treated with cyclodextrin to obtain different growth and development effects. The cyclodextrin was added to the pericarp during pilling, or it was applied to the plant by spraying at a defined stage of crop maturation. Several characteristic cyclodextrin effects on vegetables are shown in Figure 7. The individual **mass** gain on effect of cyclodextrin is obvious: the yield increased by 18-30% at lettuce, 50-60% at celery and by 7-16% at paprika. Some other favourable effects of cyclodextrin have also been observed e.g. the improvement of germination /10-16% on lettuce and paprika/ as well as the intensification of the colour /25% on paprika/.

In order to explain all these empirical phenomena the fate of cyclodextrin should be explored after e.g. seed treatment. In the course of translocation studies performed with ^{14}C-labelled β-cyclodextrin, it was stated, that the amount of cyclodextrin taken up during 24 hours soaking is 80 µg/seed which can be detected /by autoradiography/ in the highest concentration in the seedling and mainly in the radicle. 85% of the total activity found in the seeds

Fig.6.

EFFECT OF CYCLODEXTRIN SPRAY ON SPRING BARLEY

/Budakalász/

TIME OF BCD TREATMENT			HEIGHT		SHOOT		EARS		YIELD	
BRANCHING	SHOOTING	EARING	cm	%	kg/ha	%	pc/m²	%	kg/ha	%
CONTROL			48,9	100	2600	100	174	100	220,0	100
+			45,9	94	2625	101	137	79	171,6	78
	+		50,0	102	2875	120	157	90	286,0	130
		+	47,5	97	3325	128	195	113	305,6	139
+	+		44,4	91	2675	103	127	73	242,6	110
	+	+	47,2	96	2775	107	225	129	390,6	177
+		+	48,2	98	3775	145	205	118	453,0	206
+	+	+	51,2	105	3960	152	202	116	400,7	182

Fig.7.

EFFECT OF CYCLODEXTRIN TREATMENTS /PILLING, SPRAYING/ ON THE VEGETABLE CULTIVATION

VEGETABLES	ONE PLANT YIELD				GERMINATION %		PIGMENT ‰	
	FRESH MASS/g/ Ø	FRESH MASS/g/ BCD	DRY MASS /g/ Ø	DRY MASS /g/ BCD	Ø	BCD	Ø	BCD
LETTUCE	195,1	229,3	12,1	16,3	87	97	-	-
SWEET PEPPER	75,6	87,8	-	-	66	73	2,44	3,05
RED PEPPER	20,8	22,3	-	-	77	89	3,16	3,91
CELERY								
shoot	18,4	33,2	3,4	5,9	-	-	-	-
root	8,4	11,1	1,9	2,1	-	-	-	-

are removed by respiration, during the first five days of germination. Thus, it can be concluded that cyclodextrin exerts its deep influence on the living plant organism directly after application at the beginning of the growth process. This can only be interpreted in such a way, that cyclodextrin influences a partial key-process at the beginning of growth, or it has an effect on the growth in cooperation with some regulating system. Thus, in the following I may only discuss the germination connected tightly with the phenomenon of growth and the joint action with growth hormones controling growth. May I not mention our several observations obtained on interdependence of cyclodextrin with the plant /e.g. respiratory stimulation influence on biosynthesis - discussed previously, before my communication/.

The enzymatic regulation of the mobilization of nutritive materials plays an important role in the germination of seeds. It is known from the literature, that the action of β-amylase is competitively inhibited by cyclodextrins /7/. On comparison of various enzyme sources /8/. It was also stated that the β-amylase of batata is more sensitive to the inhibition by cyclodextrin than that of Bacillus polymixae.

On this basis the idea was obvious, that cyclodextrins could be used for regulating the germination of starch-storing seeds of cereals by means of their inhibiting action on amylase catalyzing starch hydrolysis. The delay of starch decomposition of barley seeds treated with cyclodextrin have been proved. It appears that the action of cyclodextrin inhibiting growth can be related to hindered amylase activity at the beginning of germination of starch seeds.

It could be stated during our further study on the mode of action in plants of cyclodextrins, that the cyclodextrin inhibition of growth was antagonized by gibberellic acid. When isolated radish cotyledons had been pretreated with cyclodextrin, then the mass gain in the Loomis' growth test approximated the normal value after incubation with gibberellic acid under light. Thus, it can be concluded, that cyclodextrin exerts its action in a direct cooperation with hormones participating of the plant growth system.

No literature antecedents were found concerning the growth promoting and crop increasing effect of cyclodextrin. From the literature, retardants can only be mentioned, e.g. CCC and Ancymidol, which have analogous /growth promoting and crop enhancing/ actions and simultaneously, they are known inhibitors of the gibberellic acid biosynthesis.

It can be stated however, on the basis of our experimental results that cyclodextrins are suitable for application in agriculture because they regulate the growth and development processes of plants in such a way, that this is expressed by a better productivity.

Literature Cited

1. El-Sharkawy, M.A., Yousef, M.E., Nagi, M.R.: Libyan J.Agric.7 1,/1978/
2. Brückner, U., Hőfner, W.: Z.Acker.Pflbau 149 251, /1980/
3. Szejtli, J., Tétenyi, M., Tétényi, P.: Hung.Patent, 176764,//1978/
4. Tétényi, M., Szejtli, J.: Proc. 20th Hung.Annu.Meet.Biochem., Siófok 1980; p. 231
5. Szejtli, J., Tétényi, M.: Proc. 8th Meet.Cereal, Bergholz-Rehbrücke DDR. 1980; p. 276.
6. Szejtli, J., Tétényi, M., Rajki, E.: Acta Agr.Hung. in press /1982/
7. Thoma, A.J., Koshland, D.E.: J. Am.Chem.Soc. 82 3329, /1960/
8. Marshall, J.J.: Mol. Cell.Biochem. 7 127, /1975/

MANIFESTATION OF GROWTH CONTROL EFFECT OF CYCLODEXTRIN (CD) IN BIOSYNTHESIS OF DNA, RNA AND PROTEINS IN GERMINATING BARLEY

J. HOLLAND,[1] L. KŐRÖSI,[1] M. TÉTÉNYI,[2] J. SZEJTLI[2]

[1] "Frédéric Joliot-Curie" National Research Institute for Radiobiology and Radiohygiene, Pentz K. u. 5
H-1221 Budapest, Hungary
[2] CHINOIN Pharmaceutical-Chemical Works, Tó u. 1-5
H-1045 Budapest, Hungary

SUMMARY

The effect of beta-cyclodextrin on DNA, RNA and protein-synthesis of germinating barley was investigated. The results of macromolecule biosynthetic examinations performed in the early period of plant development, proved that first inhibition then intensity elevation of processes was induced by CD treatment. In this respect the critical point is on days 7 - 10. The parallelism appearing in the tendency of precursor uptake and incorporation change suggests the mutual participation of transport and synthesis processes in the components of the effect.

As is presented in details in the other part of this volume P.Tétényi and co-workers proved that cyclodextrin induces contrary growing effect in some plants during the various stages of development /1/. **Experiments to be reported on** aimed at studying whether beta-cyclodextrin causes any estimable change in the biosynthesis of macromoleculco of the germinating barley in the early stage of development of the plant.

Methods

In case of the experimental sample, the swelling of the seeds took place in 10^{-2}M beta-cyclodextrin containing solution for 24 hr. By the termination of the swelling period, seeds were washed in water, planted in phytobox and let them grow further on, for 5 days in dark, thereafter ensuring the necessary light-periods.

DNA, RNA and protein synthesis was examined with the incorporation of precursors labelled with radioactive isotope as follows:

Compound	label	specific radioactivity	radioactive concentration
Thymidine	methyl-^{14}C	2,2 GBq/mM /59 mCi/mM /	55,5 kBq/ml /1,5 μCi/ml/
Uridine	2-^{14}C	2,0 GBq/mM /54 mCi/mM /	18,5 kBq/ml /0,5 μCi/ml/
Protein hydrolysate /from algae/	U-^{14}C	37 MBq/mg /1 mCi/mg /	37 kBq/ml /1 μCi/ml /

In the selected time points of germinating and development, respectively, plants were picked from the earth and incubated in aqueous medium for a fixed period, in the presence of the relevant radioactive precursor at room temperature. The optimal incubation period was settled in pre-examinations within it an equilibrium can be reached between the synthesis and the decomposition of the de novo formed macromolecules.

In one incubation period each incorporation activity of 8 - 12 plants was examined together. Plants were washed carefully, then root and stem separated from each other and uniting them one by one, after threefold freezing and melting, they were mechanically homogenized in distilled water. Macromolecules were precipitated with cold trichloracetic acid. The proteins were separated according to Siekewitz /2/ the nucleic acids by the method of Schmidt and Thannhäuser /3/ then both kinds of biopolimers were isolated on glass fibre filter according to our procedure. The radioactivity containing precipitate - taken with the filter - was placed in a scintillation cocktail and counted in a liquid-scintillation spectrometer. Radioactivity was related to the DNA, RNA or protein content, respectively, determined in an aliquot part of the total homogenate /4, 5/.

Results and Discussion

R N A s y n t h e s i s /Fig.1/

The intensity of RNA synthesis was investigated on days 3rd and 5th in time kinetics ranging from three to 48 hr considering shorter and longer syntheses and life-time RNA-species. Divergence was observed at the plants of three days-old, namely CD-induced inhibition, which, especially at the root resulted the development of completely another sort of synthesis-kinetics.

D N A s y n t h e s i s /Fig.2/

Both DNA and protein-synthesis investigations were completed with the determination of the labelling of appropriate precursor-pool - nucleoside and amino acid, respectively indicated as a term "uptake". Accordingly, the decision making becomes possible, whether any change appearing in the intensity of biosynthesis, can be deducted, fully or partly, to the change of cell-wall permeability, or not.

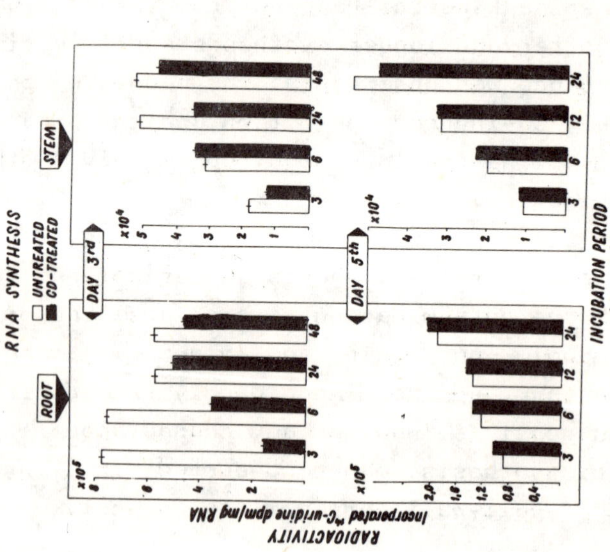

Figure 1

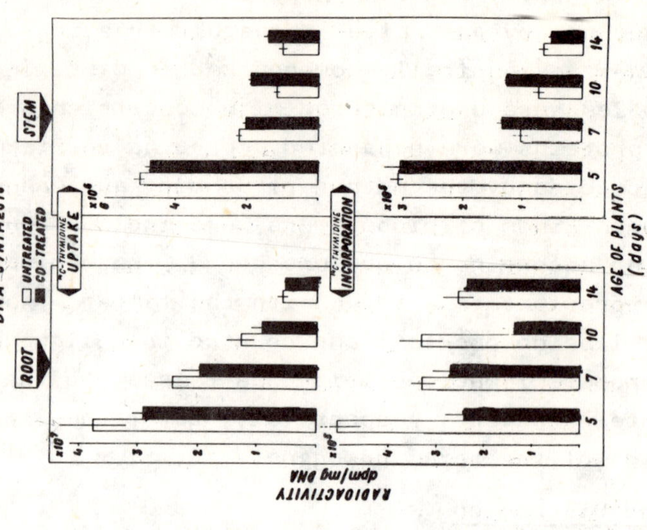

Figure 2

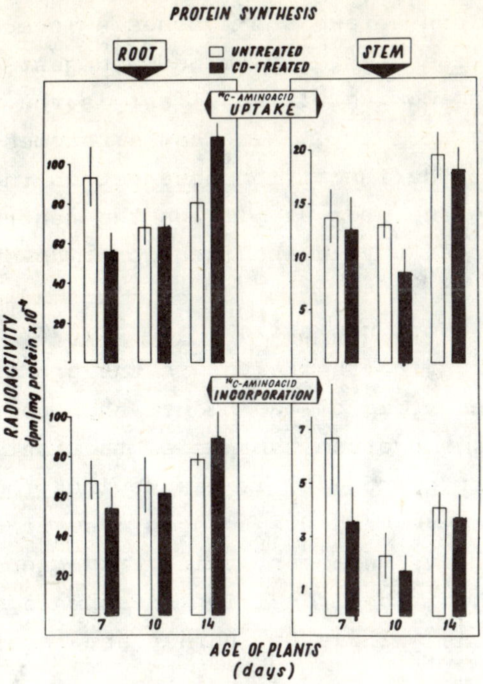
PROTEIN SYNTHESIS
ROOT
UNTREATED
CD-TREATED
STEM
14C-AMINOACID
UPTAKE
14C-AMINOACID
INCORPORATION
RADIOACTIVITY
dpm/mg protein x 10-4
AGE OF PLANTS
(days)

Figure 3

The ^{14}C-thymidin uptake and synthesis is equally the most intensive in case of the examined youngest /5-day-old/ plant-group and the CD inhibiting effect is the most prominent in the root activity. This change becomes normal till the day 14th. The inhibition observed in the root can not be observed in the stem at all, on the contrary, the 7 - 10 days equivocal uptake and incorporation-elevation appear.

P r o t e i n - s y n t h e s i s /Fig.3/

Changes manifesting themselves in the growth, mass development and the yield of the plant, show correlation with the intensity of the protein synthesis, the most. In examining this biosynthetic process, it was obvious that there is an activity minimum on day 10th both in case of non-treated and CD-treated barley. Moreover, this time coincides with the CD-induced inhibition-transition to enhancement in the root of the plant as well as the uptake and incorporation.

Literature Cited

1. Tétényi,P.: Effect of cyclodextrin on growth processes and productivity of plants. Symposium on Cyclodextrin /in this Volume/ 1981
2. Siekewitz,P.J.; J.Biol.Chem. 195 549,/1952/
3. Schmidt,G.,Thannhäuser,S.J.:J.Biol.Chem. 161 83,/1945/
4. Ceriotti,G.:J.Biol.Chem. 214 59,/1955/
5. Lowry,O.H.,Rosebrough,N.J.,Farr,A.L.,Randall,N.J.: J.Biol.Chem. 193 265,/1951/

BIOLOGICAL EFFECTS OF THE CYCLODEXTRIN COMPLEXES OF 2-CHLOROMETHYLPHOSPHONIC ACID

M. TÉTÉNYI

CHINOIN Pharmaceutical-Chemical Works, Tó u. 1-5
H-1045 Budapest, Hungary

SUMMARY

A crystalline inclusion complex containing 29% of 2-chloroethylphosphonic acid /CEPA/ in α-cyclodextrin similarly to free CEPA releases ethylene in plant as well.

Following the ethylene production by manometric method /Warburg technique/ a retarded ethylene release was observed with the inclusion complex, as compared with the non-complexed CEPA. Complex-treated plant leaves showed a 60% active substance penetration in tissues by the 6. hour, compared with leaves treated with free CEPA showing 90%. Assuring "prolonged ethylene effect" in plant tissues, this retarded absorption and the retarded ethylene-release caused increased physiological ethylene-reaction: defoliation occured earlier and in a greater percentage in the abscission-test of the complex than in that of the free form of active substance.

Agronomical use of 2-chloroethylphosphonic acid α-cyclodextrin complex may result a decrease of effective dose /saving material, lowering the pollution of the environment/. This retarded ethylene-effect assures the "physiological ripeness" at those vegetables /like sweet pepper, red pepper/ where the usual acceleration for "market ripening" causes deterioration of quality.

The 2-chloroethylphosphonic acid /CEPA, Ethrel, Ethephon, etc/ is a plant growth regulator /1/, widely utilised in the agriculture /e.g. to control flowering, ripening, abscission, etc/. This compound is stable in acidic solutions, however rapidly decomposes over p H 3,5 releasing ethylene. The plant-hormon effects are attributed wholly to the released ethylene. The 2-chloroethylphosphonic acid applied on the plants, diffuses into the cells, where the environmental pH leds to a rapid ethylene release resulting in rapid /sometimes too rapid!/, hardly controllable biological effects. Therefore a controlled ethylene release, i.e. a retarded absorption and decomposition of 2-chloroethylphosphonic acid seemed to be purposeful.

A possible approach to this task is the preparation of the cyclodextrin inclusion complex of 2-chloroethylphosphonic acid /CEPA-CD/. This new complex /2, 3/ is a watersoluble crystalline powder, containing 29 w/w%, 10-15 w/w%, or 20-30 w/w% CEPA in α-, β-, or γ-cyclodextrin complex, respectively.

Following the ethylene production by manometric method /Warburg technique.

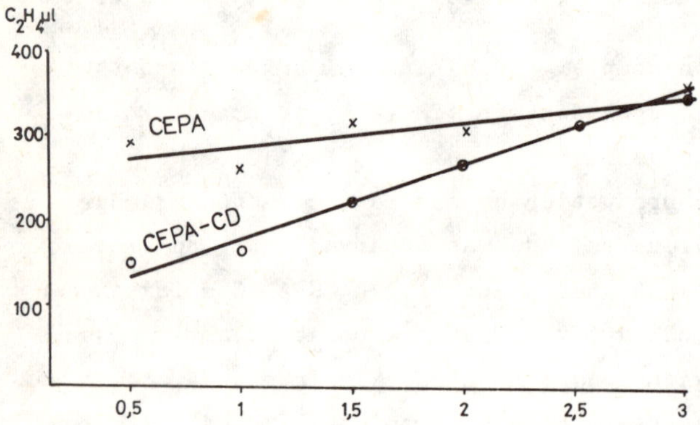

Figure 1.

A retarded ethylene release was observed with the inclusion complex, as compared with the non-complexed CEPA

Applying this complex on plant leaves, the decomposition of the active ingredient, i.e. the ethylene release is decelerated as compared to the non-complexed ethylene-precursor.

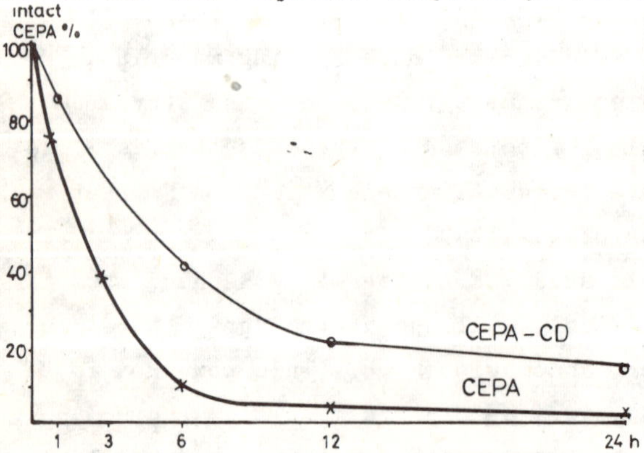

Figure 2.

It can be seen from Figure 2.,that complex-treated plant leaves showed a 60% active substance penetration in tissues by the 6.hour, while leaves treated with free CEPA showed 90%.

This retardation of the ethylene release results in a protracted ethylene-hormon effect, which is manifested in an enhanced biological effect.

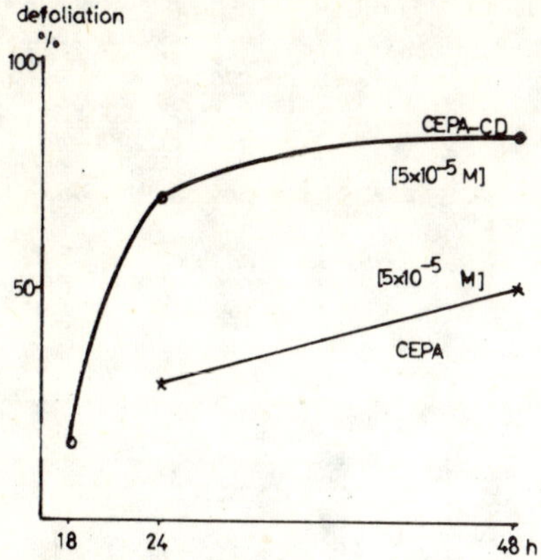

Figure 3.

At identical doses of complexed and non-complexed CEPA, the complexed substance is more potent in defoliation: it is illustrated in Figure 3. by a shorter lag-time, and the higher effect.

The CEPA-CD can be applied also in cultivation of vegetables, the results are summarized in Figure 3. and Figure 4.

Fig.4.

ACCUMULATION OF ORGANIC SUBSTANCE IN PARSLEY

TREATMENT	LEAVES/PLANT	
	FRESH MASS /g/	DRY MASS /g/
COATED SEEDS WITHOUT CEPA-CD	24,0	4,79
COATED SEEDS WITH CEPA-CD	28,51	5,81

Fig.5.

CONTROL THE RIPENING OF SWEET PEPPER AND RED PEPPER

Treatment	SWEET PEPPER FOR SEED PRODUCTION PIECE % ripe	halfr.	markr.	AV.WEIGHT g	YIELD to/ha	GERM. %	RED PEPPER PIECE % ripe	halfr.	green	AV.WEIGHT g	PIGM. ‰	GERM. %
WATER	0	50	50	75,6	38,2	66	0	53	47	20,8	-	77
BCD	8	54	38	87,8	40,7	73	3	57	40	22,3	-	81
CEPA	4	50	46	84,4	42,0	71	43	35	22	22,0	2,25	70
CEPA-CD	0	75	25	90,0	42,7	83	23	50	27	23,3	3,67	82

markr. = marketripe

The results of treatment on plant with CEPA-CD: improvement of "biological ripeness" /better germination of the seeds, enhanced pigment content/ of the "market ripened" fruits, uniform ripening, higher yield, improved quality of horticultural products, less CEPA results in satisfactory effects and less pollution of environment.

Therefore, the use of α-cyclodextrin inclusion complexes of 2-chloroethylphosphonic acid retarded release of ethylene, prolonged effect of ethylene and finally increased biological effectivity.

Literature Cited

1. Abeles, F.B.: Ethylene in plant biology, Academic Press, New York-London 1973; p.73.
2. Szejtli, J., Budai, Zs., Tétényi, M., Pap, G.: Hung.Patent Appl. CI-2000, /1979/
3. Budai, Zs., Szejtli, J.: Acta Chim. Hung. 107 231. /1981/

I.Int.Symp. on Cyclodextrins
Budapest, 1981

RECOVERY OF SOLVENT VAPOURS FROM GASEOUS PHASE BY CYCLODEXTRIN SOLUTIONS

ZS. BUDAI, J. SZEJTLI

CHINOIN Pharmaceutical-Chemical Works, Tó u. 1-5
H-1045 Budapest, Hungary

SUMMARY

The majority of organic solvents and other volatile compounds can form crystalline inclusion complexes with cyclodextrins. Such compounds can be absorbed also from gaseous phase by aqueous cyclodextrin solutions.

By heating the aqueous suspension of crystalline complexes to the appropriate temperature the complexes dissociate, the solvents are recovered by destillation, the cyclodextrin solution can be recycled.

This technical possibility seems to be utilizable in environment protection for diminishing the air pollution, recovering expensive volatile compounds from highly diluted gaseous phase, etc.

A method has been elaborated for the study of the formation of crystalline β-cyclodextrin inclusion complexes (1). The turbidity of the aqueous solution containing of β-cyclodextrin and the guest molecule was recorded continuously from 60 °C downwards.

No crystal formation was observed in 1 and 2 % β-cyclodextrin solutions under the conditions mentioned down to 2 °C (Fig.1.). The crystallization of the 3 % solutions starts at 13.0 °C and stops at 10 °C. In more concentrated (4-7 %) solutions, crystallization starts at higher temperatures and the maximum turbidity is also achieved within a 2 °C interval.

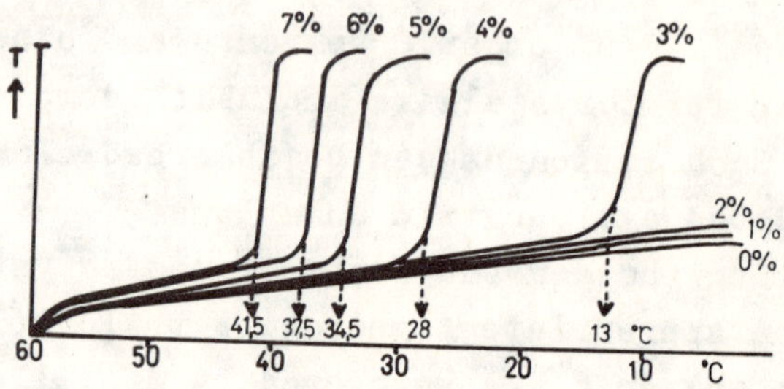

Fig.1.
Turbidity curves of 1-7 % aqueous β-cyclodextrin solutions in function of temperature from 60 °C downwards

Cyclodextrin inclusion complexes are less soluble in water than cyclodextrin itself. The formation of inclusion complexes with liquid guest molecule is proved unambiguously if crystallization occurs at a higher temperature than with cyclodextrin. Therefore, the appearance of turbidity caused by crystal formation before the characteristic temperature of β-cyclodextrin crystallization (i.e. at higher temperatures) is considered as evidence for inclusion complex formation.

Some of the solvents studied (acetylacetone, methanol, ethanol, n-propanol, etc.) results in no crystal formation over the temperature of crystallization of β-cyclodextrin. The majority of organic solvents and other complex-forming volatile compounds results in formation of crystalline complexes already at relatively high temperatures, e.g. benzene, toluene, n-hexane, carbontetrachloride, etc. The Fig.2. shows the crystallization curves of β-cyclodextrin complexes of some solvents. The composition of solvent complexes is near to the 1:1 molar ratio.

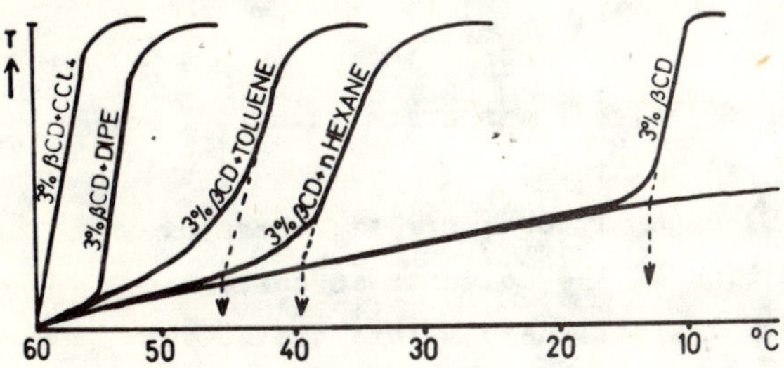

Fig.2.
Turbidity curves of 3 % β-cyclodextrin solution in presence of various solvents (solvent cc. about 0,5 w %)

Such compounds are readily absorbed and complexed also from gaseous phase by aqueous cyclodextrin solutions. By heating the aqueous suspensions of such micro-crystalline complexes to the appropriate temperature they dissociate, the solvents can be recovered by distillation, the cyclodextrin solution can be recycled (2,3). The schematic illustration of the constructed device is seen on Fig.3.

The capacity of solvent-recovery is limited by the solubility of cyclodextrin (at 20-30 °C 1,5-4 % aqueous solution can be prepared).

Using water soluble β-cyclodextrin polymer, solution can be prepared up to 40 g/100 ml (equivalent to about 25 % β-cyclodextrin), also the polymer-complex is characterized by excellent solubility at all temperatures (4). Most convenient is the application of β-cyclodextrin by admixing tensides (e.g. 0,5-1,5 % cetyl-trimethyl-ammonium bromide) the solubility strongly increases, up to 15 %, the milky complex slurry does not block up the tubing. In consequence the surface active compound, about one fifth volume of solvents remain in emulsified phase without any foaming.

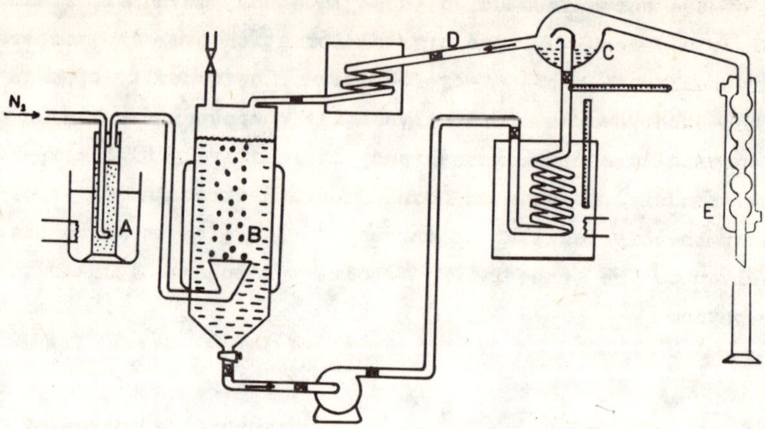

Fig.3. About 200 ml/hour of solvent was evaporated (A), and passed through about 5 liter of recirculated aqueous β-CD solution (B) by a nitrogen stream, where an overwhelming fraction of the solvent was converted immediately into a crystalline complex: the solvent and the CD solution was separated in a heat-exchanger (C): the CD solution was recycled (D) and the solvent recovered (E). Under the given conditions 1-2 liter/8 hours of appropriately selected solvents could be recovered with a yield of 60-80 %.

This method - studied on laboratory scale - may open new ways to recover, separate and recycle several solvents, aromas and other volatile compounds.

Several examples for solvents, forming inclusion complexes with β-cyclodextrin:

diisopropylether (DIPE)	chloroform
carbontetrachloride	toluene
trichloroethylene	benzene
1,2-dichloroethane	n-hexane
dichloromethane	tert-buthanol
cyclohexane	chlorobenzene
3,4-dichlorotoluene	buthylisocyanate
chloroformic-acid-ethyl-ester,	etc.

This technical possibility seems to be utilizable in environment protection for diminishing the air pollution, reducing the possibility of explosion, recovering expensive volatile compounds from highly diluted gaseous phase.

Literature cited

1. Szejtli, J., Budai, Zs.: Acta Chimica Acad.Sci.Hung. 99 (4) 433-446 (1979)
2. Budai, Zs., Szejtli, J.: Magyar Kémikusok Lapja XXXVI. (4-5) 248-254 (1981)
3. Szejtli, J., Budai, Zs., Hegedüs, G., Kovács, A., Frisch, M.: Hung.Patent Appl. 18023 (1979)
4. Fenyvesi, É., Szilasi, M., Zsadon, B., Szejtli, J., Tüdős, F.: Water-soluble cyclodextrin polymer as inclusion complex forming agent
   I.International Cyclodextrin Symposium 30.09-2.10. 1981, Budapest

I.Int.Symp. on Cyclodextrins
Budapest, 1981

QUALITY CONTROL OF THE β-CYCLODEXTRIN INCLUSION COMPLEXES OF VOLATILE OILS AND MATERIALS

T. ZILAHY, E. KOTSIS

CHINOIN Pharmaceutical-Chemical Works, Tó u. 1-5
H-1045 Budapest, Hungary

SUMMARY

Industrial scale production of β-cyclodextrin inclusion complexes of various flavours and fragrances is planned. Considering that these products will be utilized by food industry and catering, strict health-authority regulations have to be considered. Quality requirements and appropriate analytical methods have been elaborated for quality control of these products.

The poster demonstrates those important quality parameters of β-cyclodextrin complexes of ten volatile oils that might affect their use, processing and flavouring.

AUTHOR INDEX

SUBJECT INDEX